The Satellite Communication Applications Handbook

For a complete listing of the *Artech House Telecommunications Library,*
turn to the back of this book.

The Satellite Communication Applications Handbook

Bruce R. Elbert

Artech House
Boston • London

Library of Congress Cataloging-in-Publication Data
Elbert, Bruce R.
    The satellite communication applications handbook / Bruce R. Elbert.
        p. cm.
    Includes bibliographical references and index.
    ISBN 0-89006-781-3 (alk. paper)
    1. Artificial satellites in telecommunication.   I. Title.
TK5104.E43 1997
384.5'1—dc20   96-28225
                                                                    CIP

British Library Cataloguing in Publication Data
Elbert, Bruce R.
    Satellite communication systems
    1. Artificial satellites in telecommunication
    I. Title
    621.3'825

    ISBN 0-89006-781-3

Cover design by Kara Munroe-Brown

International Standard Book Number: 0-89006-781-3
Library of Congress Catalog Card Number: 96-28225

10 9 8 7 6 5 4 3 2

To Cathy, my one and only

Contents

Preface

The satellite communication industry has experienced tremendous growth over the past decade, surpassing the expectations of all who have contributed to its success. This book is offered as a comprehensive review of the applications that have driven this growth. It discusses the technical and business aspects of the systems and services that operators and users exploit for strategic advantage or just plain fun.

The book is organized into four parts that deal with the most fundamental areas of concern to application developers and users: the technical and business fundamentals, the application of simplex (broadcast) links to multiple users, duplex links that deliver two-way interactive services, and regulatory and business affairs that drive investment and financial performance. The eleven chapters of the book fall nicely into these general categories.

Anyone entering this exciting field at this time has a lot of options to consider and many avenues to follow. Fortunately, there is a great deal of useful information and experience available to anyone who wishes to do the research and explore its many dimensions. The origin of this book comes from the author's journey over the past thirty years, at Hughes, COMSAT, Western Union, and, of course, the U.S. Army Signal Corps (where one really learns how to communicate). The idea and structure of the book evolved from a series of satellite communication seminars that the author presented in Paris, London, and Munich, through auspices of the Technology Training Corp. The ideas and materials were refined over a five-year period while teaching a satellite communication engineering course at UCLA Extension.

Teachers and other presenters may contact the author for additional help in using this book as a text for a technical or business course on satellite communication. This, after all, is its origin and fundamental use.

The author wishes to express his appreciation to his technical reviewer, Ray Sperber, for doing such an excellent job of providing additional ideas and asking the difficult questions. Also, Paul Resch provided some of the nice figures in Chapter 4 as well as the ideas referenced therein. Several of the photos and figures were derived from materials provided by DIRECTV, Inc., Hughes Space and Communications, and Hughes Network Systems.

Part I
Systems Considerations

Evolution of Satellite Technology and Applications

Communication satellites, whether in *geostationary Earth orbit* (GEO) or *low Earth orbit* (LEO), provide an effective platform to relay radio signals between points on the ground. The users who employ these signals enjoy a broad spectrum of telecommunication services on the ground, the sea, and in the air. In recent years, such systems have become practical to point where a typical household can have its own satellite dish. Whether we reach the stage of broad acceptance depends on the competition with the more established broadcasting media, including over-the-air and cable TV. In any case, GEO and LEO satellites will continue to offer unique benefits for other services such as mobile communications and multimedia, but it is up to us to exploit them.

To this end, this book shows how satellite technology can meet a variety of human needs, the ultimate measure of the effectiveness. Our first work, *Introduction to Satellite Communication* [1], established the foundation for the technology and its applications. These have progressed significantly since the late 1980s; however, the basic principles remain the same. Satellite communication applications (which we will refer to simply as *satellite applications*) extend throughout human activity—both occupational and recreational. Many large companies have built their foundations on satellite services such as cable TV, data communications, information distribution, maritime communications, and remote monitoring. For others, satellites have become a hidden asset by providing a reliable communications infrastructure. In the public sector, satellite applications are extremely effective in emergency situations where terrestrial lines and portable radios are not available or ineffective for a variety of reasons.

So, we can conclude that there are two basic purposes to creating and operating satellite applications, namely, to make money from selling systems and services and to meet a vital communications need. The composition of satellite communication markets has changed over the years. Rates of return are, on average, comparable to other segments of the telecommunications industry. In the 1990s, video transmission established itself as the hottest application, with data communications gaining an important second place position. Voice services are no longer the principle application in industrialized countries but retain their value in rural environments and the international telecommunications field. Special-purpose voice applications like mobile telephone and emergency communications continue to expand. The very fact that high-capacity fiber optic systems are extending literally everywhere in the world makes satellite applications that much more important as a supplementary and backup medium. Satellites are enjoying rapid adoption in regions where fixed installations are impractical. For example, ships at sea no longer employ the Morse code because of the success of the Inmarsat system.

For satellite operators to succeed in a competitive marketplace, they must attract a significant quantity of users. The fixed ground antennas that become aligned with a given satellite create synergy and establish a "real estate value" for the orbit position. Ultimately, one can create a "hot bird" that attracts a very large user community of antennas and viewers. Galaxy I, a very successful cable TV hot bird, established the first shopping center in the sky, with "anchor" tenants like HBO and ESPN and "boutiques" like A&E and The Discovery Channel.

From this experience, those of us who have offered satellite services to large user communities know that the three most important words in satellite service marketing are LOCATION, LOCATION, LOCATION! What this means is that satellite "tenants" all want to be in the best "neighborhood." Consider as an example the U.S. geostationary orbital service arc shown in Figure 1.1. Highlighted are the most popular cable hot birds—Galaxy I-R, Satcom C-1, Galaxy 5, and Satcom C-3. These 4 satellites collect revenue comparable to the remaining 25 satellites! Users of hot birds pay a premium for access to the ground infrastructure of cable TV receiving antennas much like tenants in a premium shopping mall pay to be in an outstanding location and in proximity to the most attractive department stores in the city. In the case of cable TV, access is everything because the ground antenna is, in turn, connected to households where cable services are "consumed" and paid for. For a new satellite operator to get into such an established market often requires them to subsidize users by paying some of the switching costs out of the expected revenues.

Satellite operators, who invest in the satellites and make capacity available to their customers, generally prefer that users own their own Earth stations. This is because installing antennas and associated indoor electronics is costly for satellite service providers. Once working, this investment must be maintained and upgraded to meet evolving needs. On the other hand, why would users want to make such a commitment? There are two good reasons for this trend toward ownership of the ground segment by the user: (1) The owner/user has complete control of the network resources, and (2) the cost and complexity of ownership and operation have been greatly reduced because of advances in

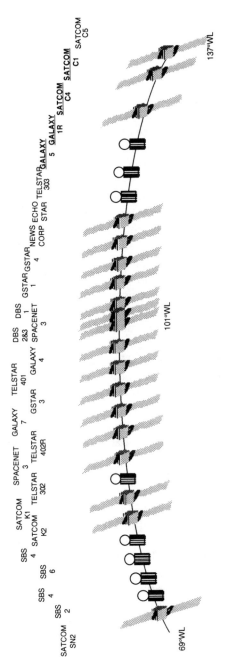

Figure 1.1 An illustration of the U.S. GEO orbital arc as of 1996, highlighting the most popular cable TV satellites.

microcircuitry and computer control. A typical small Earth station is no more complex than a cellular telephone or VCR. Larger Earth stations such as TV uplinks and international telephone gateways are certainly not a consumer item, but the advantages of control typically motivate leading TV networks and communication companies to take these on.

Happily for the new user, satellite communications can reduce entry barriers for many information industry applications. As a first step, a well-constructed business plan based on the use of existing satellites could be attractive to venture capitalists. (More on finance in Chapter 11.) Satellite capacity may be taken as a service, eliminating a large financial commitment of owning and operating repeater stations in space.

The history of commercial satellite communications includes some fascinating startup services that took advantage of the relatively low cost of entry. The following three examples, taken from the United States in the mid-1980s, illustrate the range of possibilities. The Discovery Channel made the substantial commitment to a Galaxy I transponder but thereby gained access to the most lucrative cable TV market in North America. Another startup, Equatorial Communications, pioneered *very small aperture terminal* (VSAT) networks to deliver financial data to investors. Their first receive-only product was a roaring success, and the company became the darling of venture capitalists. Unfortunately, they broke their sword trying to move into the much more complicated two-way data communication market. Their technology failed to gain acceptance, and the company disappeared through a series of mergers. Argo Communications entered the newly created long-distance telephone market in 1984 using a wideband satellite network. They were up and running in less than six months, which is something that would have been impossible with any kind of terrestrial network. Their demise was due to the success of U.S. Sprint, which had a head start with its advanced fiber optic network. Argo Communications may have been a technical success, but they failed in the market dominated by much more substantial players who could exploit the technical and economic benefits of long-haul fiber.

Several U.S. corporations attempted to introduce direct-to-home (DTH) satellite broadcasting at a time when cable TV was still establishing itself. The first entrants experienced great difficulties with limited capacity of existing low- and medium-power Ku-band satellites, hampering the capacity of the networks and affordability of the home receiving equipment. Europe and Japan had problems of their own finding the "handle" on viable DTH systems, choosing first to launch high power Ku-band satellites with only a few operating channels. It was not until BSkyB and NHK were able to bring attractive programming to the public exclusively on their respective satellites that consumers moved in the millions of numbers.

In the United States, the only viable form of DTH to emerge in the 1980s was through the backyard C-band satellite dish that could pull in existing cable TV programming from hot birds like Galaxy I and Satcom 3R. This clearly demonstrated the principle that people would vote with their money for a wide range of attractive programming, gaining access to services that were either not available or priced out of reach. Early adopters of the dishes purchased these somewhat expensive systems because the signals were not scrambled at the time. Later, HBO and other cable networks scrambled their programming using the Videocipher 2 system, resulting in a halt to expansion of backyard dishes. This market settled back into the doldrums for several years.

In 1994, Hughes Electronics introduced its DIRECTV service through a high-power satellite (labeled DBS-1 in Figure 1.1). With over 150 channels of digitally compressed programming, DIRECTV programming and another service provider, USSB, demonstrated that DTH could be both a consumer product and a viable alternative to cable. An older competing service, PrimeStar, was first introduced by TCI and other cable operators as a means to serve users who were beyond the reach of their cable systems. This subsequently provided the platform for a new and improved PrimeStar, rebuilt along lines very similar to DIRECTV. By converting from an analog to a digital system and expanding programming, the PrimeStar service has seen much greater demand than when it was first introduced in 1992.

Satellite communication applications can establish a solid business for companies that know how to work the details to satisfy customer needs. Another example is IDB Communications Group. From a very modest start in the audio distribution business, IDB branched out into every conceivable area of the telecommunications industry. Whether the requirement is to cover a media event in Antarctica or to connect Moscow to the rest of the world with voice and video links, IDB proved its worth. They capitalized on the versatility of satellite communications in the pursuit of a growing company and a demanding base of investors. Sadly, IDB ran into financial difficulties in 1994 and had to merge with LDDS, the fourth largest long-distance carrier in the United States. IDB's early success points up the facts that needs can be met and money is to be made in the field of satellite communications.

1.1 SATELLITE NETWORK FUNDAMENTALS

Every satellite application achieves its effectiveness by building on the strengths of the satellite link. A satellite is capable of performing as a microwave repeater for Earth stations that are located within its coverage area, determined by the altitude of the satellite and the design of its antenna system. The arrangement of three basic orbit configurations is shown in Figure 1.2. A GEO satellite can cover nearly one-third of the Earth's surface, with the exception of the polar

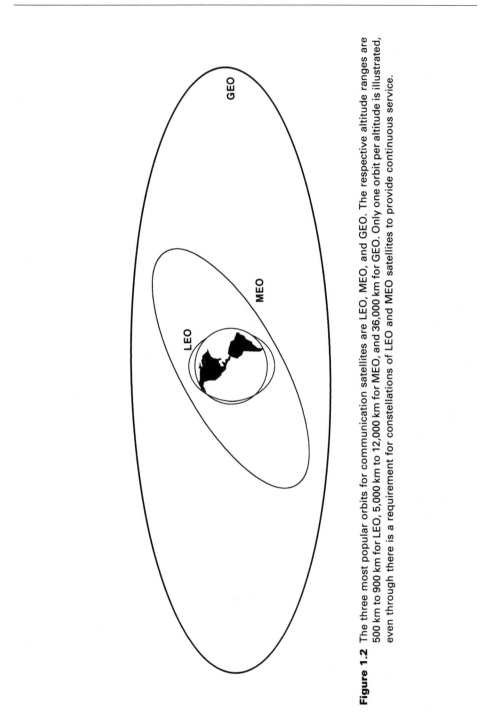

Figure 1.2 The three most popular orbits for communication satellites are LEO, MEO, and GEO. The respective altitude ranges are 500 km to 900 km for LEO, 5,000 km to 12,000 km for MEO, and 36,000 km for GEO. Only one orbit per altitude is illustrated, even through there is a requirement for constellations of LEO and MEO satellites to provide continuous service.

regions. This includes more than 99% of the world's population and essentially all of its economic activity.

The LEO and *medium Earth orbit* (MEO) approaches require more satellites to achieve this level of coverage. Due to the fact that the satellites move in relation to the surface of the Earth, a full complement of satellites (called a constellation) must be operating to provide continuous, unbroken service. The tradeoff here is that the GEO satellites, being more distant, incur a longer path length to Earth stations, while the LEO systems promise short paths not unlike those of terrestrial systems. The path length introduces a propagation delay since radio signals travel at the speed of light. Depending on the nature of the service, the increased delay of MEO and GEO orbits may impose some degradation on quality or throughput. The extent to which this materially affects the acceptability of the service depends on many factors, such as the degree of interactivity, the delay of other components of the end-to-end system, and the protocols used to coordinate information transfer and error recovery.

At the time of this writing, only one non-GEO system, Orbcom, has entered commercial service with its first two satellites. The applications of Orbcom include paging and low-speed burst data. Systems capable of providing telephone service, such as Iridium and I-CO communications, will achieve operation around the year 2000. On the other hand, GEO satellite systems continue to be launched with operating lives that will carry users through the year 2010 at the least.

These satellites have microwave repeaters that operate over an assigned segment of the 1- to 30-GHz frequency range. As microwaves, the signals transmitted between the satellite and Earth stations propagate along line-of-sight paths and experience free-space loss that increases as the square of the distance. The spectrum allocations are given in the following approximate ranges:

- L band: 1.5–1.65 GHz;
- S band: 2.4–2.8 GHz;
- C band: 3.4–7.0 GHz;
- X band: 7.9–9.0 GHz;
- Ku band: 10.7–15.0 GHz;
- Ka band: 18.0–31.0 GHz.

Actual assignments to satellites and Earth stations are further restricted to permit different services and a large community of users to share this valuable resource.

Applications are delivered through a network architecture that falls into one of three categories: point-to-point (mesh), point-to-multipoint (broadcast), and multipoint interactive (VSAT). Mesh type networks mirror the telephone network. They allow Earth stations to communicate directly with each other on a one-to-one basis. To make this possible, each Earth station in the network must have sufficient transmit and receive performance to exchange information with its least effective partner. Generally, all such Earth stations have similar antennas and transmitter systems, so their network is completely balanced. Links between pairs of stations can be operated on a full-time basis for the transfer of broadband information like TV or multiplexed voice and data. Alternatively, links can be established only when needed to transfer information, either by user scheduling (reservation system) or on demand (demand assignment system).

A broadcast of information by the satellite is more efficient than terrestrial arrangements using copper wires, fiber optic cables, or multiple wireless stations. By taking advantage of the broadcast capability of a GEO satellite, the point-to-multipoint network supports the distribution of information from a source (the hub Earth station) to a potentially very large number of users of that information (the remote Earth stations). Any application that uses this basic feature will tend to provide value to the business.

Many applications employ two-way links, which may or may not use the broadcast feature. The application of the VSAT to interactive data communication applications has proven very successful in many lines of business. As will be covered in Chapter 7, a hub and spoke network using VSATs competes very effectively with any terrestrial network topology that is designed to accomplish the same result. This is because the satellite provides the common point of connection for the network, eliminating the requirement for a separate physical link between the hub and each remote point or a less reliable string of links called a multidrop line. Other interactive applications can employ point-to-point links to mimic the telephone network, although this tends to be favored for rural and mobile services. The incoming generation of Ka-band satellite networks that involve very low-cost VSATs is expected to reduce barriers to mass market satellite networks.

The degree to which satellite communications is superior to terrestrial alternatives depends on many interrelated factors. Experience has shown that the following features tend to give satellite communication an advantage in appropriate applications:

- Wide area coverage of a country, region, or continent;
- Wide bandwidth available throughout;
- Independent of terrestrial infrastructure;
- Rapid installation of ground network;
- Low cost per added site;
- Uniform service characteristics;
- Total service from a single provider;
- Mobile/wireless communication, independent of location.

While satellite communications will probably never overtake terrestrial telecommunications on a major scale, these strengths can produce very effective niches in the marketplace. Once the satellite operator has placed the satellite into service, a network can easily be installed and managed by a single organization. This is possible on a national or regional basis (including global in the case of non-GEO systems). The frequency allocations at C, Ku, and Ka bands offer effective bandwidths of 1GHz or more per satellite, facilitating a range of broadband services that are not constrained by local infrastructure considerations. The satellite delivers the same consistent set of services at costs that are potentially much lower than those of wireline systems. For the long term, the ability to serve mobile stations and to provide communications instantly are features that will gain strength in a changing world.

Originally, Earth stations were large, expensive, and located in rural areas so as not to interfere with terrestrial microwave systems that operate in the same frequency bands. These monsters had to use wideband terrestrial links to reach the closest city. Current emphasis is on customer premise Earth stations— simple, reliable, low cost. An example of modern small VSAT is illustrated in Figure 1.3. Home receiving systems for DTH service are also low in cost and quite inconspicuous. Expectations are the coming generation of Ka-band VSATs will reduce cost and increase use in smaller businesses and ultimately the home.

The other side of the coin is that as terminals have shrunk in size, satellites have grown in power and sophistication. There are three general classes of satellites used in commercial service, each designed for a particular mission and capital budget. Smaller satellites, capable of launch by the Delta II rocket or dual-launched on the Ariane 4 or 5, provide a basic number of transponders usually in a single-frequency band. Satellite operators in the United States, Canada, Indonesia, and China have established themselves in business through this class of satellite. The Measat satellite, illustrated in Figure 1.4, is an example of this class of vehicle. The introduction of mobile service in the LEO will involve satellites of this class as well. Moving up to the middle range of spacecraft, we find designs capable of operating in two frequency bands simultaneously. Galaxy VII, shown in Figure 1.5, provides 24 C-band and 24 Ku-band transponders to the U.S. market. A dual payload of this type increases capacity and decreases the cost per transponder. First-generation DTH and MSS applications are also provided by these spacecraft.

Finally, the largest satellites serve specialized markets where the highest possible power or payload weight are required. The generation of GEO mobile satellites that are capable of serving handheld phones fall into this classification. An example of one of these satellites, APMT, is shown with its 12m antenna deployed in Figure 1.6. Also, the trend to use the smallest possible DTH home receiving antenna and to cover the largest service area combine to demand the

Figure 1.3 An example of a small customer premise VSAT used in data communication (photograph courtesy of Hughes Network Systems).

largest possible spacecraft. The total payload power of such satellites reaches 8 kW, which is roughly eight times that of Measat. Designs are on the drawing boards at the time of this writing for satellites that can support payload powers of up to 16 kW.

While most of the money in satellite communications is derived from the broadcast feature, there are service possibilities where the remote Earth stations must transmit information back to the hub Earth station. Examples of such return link applications include:

- Control signals to change the content of the information being broadcast (to achieve narrow casting on a broadcast link);
- Requests for specific information or browsing of documents (to support Internet or intranet services);

Figure 1.4 Measat, first launched in 1996, is a small satellite of the Hughes HS-376 class. With 1100W of payload power, it provides both low-power C-band and medium-power Ku-band transponders to Malaysia and surrounding countries in the Asia-Pacific region (photograph courtesy of Hughes Space and Communications Company).

Figure 1.5 Galaxy VII, launched in 1993, is a medium-power Hughes HS-601 satellite that provides 24 low-power C-band and 24 medium-power Ku-band transponders to the U.S. market. It is used by cable TV and over-the-air broadcasting networks to distribute programming to affiliates (photograph courtesy of Hughes Space and Communications Company).

Figure 1.6 APMT is a regional GEO mobile communications satellite of the largest class produced by Hughes. It serves the Asia-Pacific region and employs a 12m reflector to
serve mobile and fixed users, many of which will use handheld user terminals
(photograph courtesy of Hughes Space and Communications Company).

- Responsive information that the hub needs to update the record for a particular customer;
- Point-to-point information that the remote wishes to be routed to another remote user (like e-mail).

Adding the return link to the network tends to increase the cost of the remote Earth station by a significant amount since both a transmitter and controller are required. However, there are many applications that demand an interactive or two-way communication feature.

The relative amount of information (bandwidth) on the forward and return links can be quantified for the specific application, as suggested in Figure 1.7. Most of the bandwidth on GEO satellites is consumed in the forward direction, as indicated by the area in the lower right for TV broadcast or distribution. There are also uses for transmitting video in both directions, which is indicated in the upper right-hand corner. Cutting the bandwidth back on the forward link but not on the return link supports an application where bulk data is transferred from a remote to a centralized host computer. Reducing the

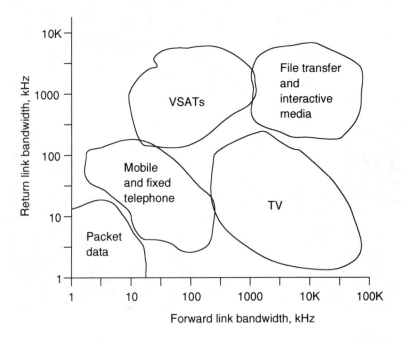

Figure 1.7 The approximate relationship of bandwidth usage between the forward link (hub transmit) and return link (remote transmit) in satellite applications.

bandwidth equally in both directions to expand the user quantity offers low data rate switched service that is directly applicable to the fixed and mobile telephone markets.

These general principles lead to a certain set of applications that serve telecommunication users. In the next section, we review the most popular applications in preparation for the detailed evaluations in the remaining chapters.

1.2 SATELLITE APPLICATION TYPES

Applications in satellite communications have evolved over the years to adapt to competitive markets. Evolutionary development, described in [1], is a natural facet of the technology because satellite communications is extremely versatile. This is important to its extension to new applications in the coming century as well.

1.2.1 Television and Video Services

Commercial TV is the largest segment of the entertainment industry; it also represents the most financially rewarding user group to satellite operators. The four fundamental ways that the satellite transfers TV signals to the ultimate consumer are:

- Point-to-multipoint distribution of TV network programming from the studio to the local broadcast station;
- Point-to-point transmission of specific programming from an event location to the studio (alternatively, from one studio to another studio);
- Point-to-multipoint distribution of cable TV programming from the studio to the local cable TV system;
- Point-to-multipoint distribution of TV network and/or cable TV programming from the studio directly to the subscriber (i.e., DTH).

It may have taken ten or more years for the leading networks in the United States and Europe to adopt satellites for distribution of their signals, but since 1985, it has been the main stay. Prior to 1985, pioneering efforts in Indonesia and India allowed these countries to introduce nationwide TV distribution via satellite even before the United States made the conversion from terrestrial microwave. European TV providers pooled their resources through the *European Broadcasting Union* (EBU) and the EUTELSAT regional satellite system. Very quickly, the leading nations of Asia and Latin America adopted satellite TV delivery, rapidly expanding this popular medium to global levels.

1.2.1.1 Over-the-Air TV Broadcasting

The first of the four fundamental techniques is now standard for TV broadcasting in the VHF and UHF bands, which use local TV transmitters to cover a city or area. The satellite is used to carry the network signal from a central studio to multiple receive Earth stations, each connected to a local TV transmitter. This has been called TV distribution or TV rebroadcast. When equipped with transmit uplink equipment, the remote Earth station can also transmit a signal back to the central studio to allow the station to originate programming for the entire network. U.S. TV networks like CBS and Fox employ these reverse point-to-point links for on-location news reports. The remote TV uplink provides a transmission point for local sporting and entertainment events in the same city. This is popular in the United States, for example, to allow baseball and football fans to see their home team play an away-from-home game in a remote city. More recently, TV networks employ fiber-optic transmission, but the satellite continues to be the flexible routing system.

Revenue for local broadcast operations is available from two potential sources: advertisers and public taxes. Since its beginnings in the United States, TV provided an excellent medium to influence consumer purchase behavior. In exchange for watching commercials for soap, airlines, and automobiles, the consumer is entertained for nothing. This has produced a large industry in the United States as stations address local advertisers and the networks promote nationwide advertising.

An alternative approach was taken in many European countries and Japan, where government-operated networks were the first to appear. In this case, the consumer is taxed on each TV set in operation. These revenues are then used to operate the network and to produce the programming. The BBC in the United Kingdom and NHK in Japan are powerhouses in terms of their programming efforts and broadcast resources. However, with the rapid introduction of truly commercial networks, cable TV, and DTH, these tax-supported networks are experiencing funding difficulties.

Public TV in the United States developed after commercial TV was well established. Originally called Educational TV, this service existed in a fragmented way until a nonprofit organization called the *Public Broadcasting Service* (PBS) began serving the nation by satellite in 1978. The individual stations are supported by the local communities through various types of donations. Some are attached to universities; others depend on donations from individuals and corporations. PBS itself acquires programming from the member stations and from outside sources like the BBC. Programs are distributed to the members using satellite transponders purchased by the U.S. government. It must therefore compete with other government agencies for Congressional support. PBS programming is arguably of better quality than some of the popular shows on the commercial networks. Even though PBS addresses a relatively narrow segment, its markets are now under attack by even more targeted cable TV networks like A&E, The Discovery Channel, The Learning Channel, and a more recent entry, The History Channel. All of these competitors built their businesses on satellite delivery to cable systems and DTH subscribers.

The local airwaves provide a reasonably good medium to distribute programming and the added benefit of allowing the local broadcaster to introduce local programs and advertising. Satellite transmission, on the other hand, is not limited by local terrain and thus can be received outside the range of terrestrial transmitters, extending across a nation or region.

1.2.1.2 Cable TV

Begun as a way to improve local reception in rural areas, cable TV has established itself as the dominant force in many developed countries. This was facilitated by organizations that used satellite transmission to distribute unique programming formats to cable subscribers. The cable TV network was pioneered by HBO in the 1970s. Other early adopters of satellite delivery include Turner Broadcasting, Warner Communications, and Viacom. By 1980, 40% of urban U.S. homes were using cable to receive the local TV stations (because the cable provided a more reliable signal); at the same time, the first nationwide cable networks were included as a sweetener and additional revenue source. During the 1980s, cable TV became an eight billion dollar industry and the prototype for this medium in Europe, Latin America, and the developed parts of Asia.

Cable TV networks, discussed in Chapter 4, offer programming as a subscriber service to be paid for on a monthly basis or as an almost free service like commercial TV broadcasting. HBO and the Disney Channel are examples of premium (pay) services, while The Discovery Channel and TBS are examples of commercial channels that receive most of their revenue from advertisers. The leading premium channels in North America and Europe are immensely successful in financial terms, but the business has yet to be firmly established in economies with low income levels.

Cable TV became the first to offer a wide range of programming options that are under the direct control of the service provider. The local cable system operator controls access and can therefore collect subscription fees and service charges from subscribers. If the fees are not paid, the service is terminated. Wireless cable, a contradiction in terms by nevertheless a viable alternative to wired cable, is gaining popularity in Latin America and Asia. The principle here is that the cable channels are broadcast locally in the microwave band to small dishes located at subscribers. Just as in the case of DTH, wireless cable depends on some form of conditional access control that allows the operator to electronically disconnect a nonpaying user. Theft of signals, called piracy, is a common threat to the economic viability of wired and wireless cable.

1.2.1.3 Direct to Home

The last step in the evolution of the satellite TV network is DTH. After a number of ill-fated ventures during the early 1980s, DTH appears to be ready to take off. BSkyB in the United Kingdom, NHK in Japan, DIRECTV in the United States, and STAR TV in Asia are now established businesses, with other broadcasters following suite. Through its wide-area broadcast capability, a GEO satellite is uniquely situated to deliver the same signal throughout a country or region at an attractive cost per user. The particular economics of this delivery depend on the following factors.

- The size of the receiving antennas: Smaller antennas are easier to install and maintain and are cheaper to purchase in the first place. They are also less noticeable (something that is desirable in some cultures but not necessarily in all others).
- The design of the equipment, which is simple to install and operate (this author's *Digital Satellite System* (DSS) installation, needed to receive DIRECTV, took only 2 hours—that is, 105 minutes—to run the cables and 15 minutes to install and point the dish).
- Several users can share the same antenna. This is sensible if the antenna is relatively expensive, say in excess of $1,000; otherwise, each user can afford his or her own. A separate receiver is needed for each independent TV watcher (i.e., for private channel surfing).
- The number of transponders that can be accessed through each antenna: The trend is now toward locating more than one satellite in the same orbit position in order to increase the available number of transponders. The more channels that are available at the same slot, the more programming choices that the user will experience.
- The number of TV channels that can be carried by each transponder: Capacity is multiplied through digital compression techniques discussed in Chapter 5. This multiplies the quantity of simultaneous programs by five or more.

The ideal satellite video network delivers its programming to the smallest practical antenna on the ground, has a large number of channels available (greater than 100), and permits some means for users to interact with the source of programming. A simple connection to the public-switched telephone network (PSTN) allows services to be ordered directly by the subscriber. Thus, the system emulates a direct point-to-point connection between user and programming supplier, which is something that broadband fiber optic networks promise but have yet to deliver on a widescale basis.

1.2.2 Trends in Satellite Voice Communications

Voice communications are fundamentally based on the interaction between two people. It was recognized very early in the development of satellite networks that the propagation delay imposed by the GEO tends to degrade the quality of interactive voice communications, at least for some percentage of the population. However, voice communications represent a significant satellite application due to the other advantages of the medium. For example, many developing countries and lightly inhabited regions of developed countries continue to use satellite links as an integral part of the voice network infrastructure. Furthermore, an area where satellite links are essential for voice communications is the mobile field. These developments are treated in detail in Chapters 8 and 9.

The PSTN within and between countries is primarily based on the requirements of voice communications, representing something in the range of 60% to 70% of all interactive traffic. The remainder consists of facsimile (fax) transmissions, low- and medium-speed data (both for private networks and access to public network services such as the Internet), and various systems for monitoring and controlling remote facilities. The principal benefit of the PSTN is that it is truly universal. If you can do your business within the limits of 3,000 Hz of bandwidth and can tolerate the time needed to establish a connection through its dialup facility, the PSTN is your best bet.

Propagation delay has become an issue when competitively priced digital fiber optic networks are introduced. Prior to 1985 in the United States, AT&T, MCI, and others were using a significant amount of analog telephone channels both on terrestrial and satellite links. An aggressive competitor in the form of U.S. Sprint invested in an all-digital network that employed fiber optic transmission. Sprint expanded their network without microwave or satellite links and introduced an all-digital service at a time when competition in long distance was heading up. Their advertising claimed that calls over their network were so quiet "you can hear a pin drop." This strategy was so successful that both MCI and AT&T quickly shifted their calls to fiber, resulting in rapid turndown of both satellite voice channels and analog microwave systems.

The economics of satellite voice communications are substantially different from that of the terrestrial PSTN, even given the use of digital technology with both approaches. With low-cost VSAT technology and high-powered satellites at Ku and Ka bands, satellite voice is the cheapest and quickest way to reach remote areas where terrestrial facilities are not available. It will be more attractive to install a VSAT than to extend a fiber optic cable over a distance greater than a few hundred meters. A critical variable in this case is the cost of the VSAT, which is expected to drop from the $10K level to perhaps as low as $1.5K before 2000. Fiber, however, is not the only terrestrial technology that can address the voice communication needs of subscribers. There is now a new class of terrestrial wireless system called fixed cellular. Low-cost cordless phones or simple radio terminals are placed in homes or offices, providing access to the PSTN through a central base station. The base stations are concentrating points for traffic and can be connected to the PSTN by fiber or even satellite links. The cost of the base station and network control is kept low by not incorporating the automatic hand-off feature of cellular mobile radio. Instead, user terminals of different types make the connection through the closest base station, which remains in the same operating mode throughout the call.

New classes of public network services are expected in coming years, under the general category of broadband communications. The underlying technology is *asynchronous transfer mode* (ATM), a flexible high-speed packet-switched architecture that integrates all forms of communications [2]. ATM services can be delivered through fiber optic bandwidths and advanced digital switching systems and include the following, among others:

- High-speed data on demand (384 Kbps to 155 Mbps, and greater);
- Multichannel voice;
- Video services on demand;
- High-resolution fax (300 dots per inch and better);
- Integrated voice/data/video for enhanced Internet services.

Due to the high cost of upgrading the terrestrial telephone plant to provide ATM services on a widespread basis, many of these services will not appear in many places for some time. However, they represent the capability of the coming generation of public networks being implemented around the globe.

Satellite communications currently addresses these services as well and, in fact, has been the only means available to provide them on a consistent basis. In private network applications, satellite communications has a big advantage for delivering broadband services to a large quantity of sites. Fiber optic networks are attractive for intra- and intercity public networks and can offer broadband point-to-point transmission that is low in cost per user. Yet this is easier to do with a satellite because it provides a common traffic concentration point in the sky. The bandwidth is used more effectively (a principle of traffic engineering), and therefore the network can carry more telephone conversations and generate more revenue.

Satellite networks are very expandable because all points are independent and link local terrain does not influence performance. Consider the example of the largest German bank, Deutsche Bank, which needed to offer banking services in the new states of the former East Germany. The telecom infrastructure in East Germany in 1990, while the best in the Soviet Block, was very backward by western European standards. Deutsche Bank installed medium-sized Earth stations at new bank locations and was then able to offer banking services that were identical to those of their existing branches in the west. Another excellent example is Wal-Mart Department Stores, reviewed later in this chapter.

1.2.3 Data Communications

Satellite networks are able to meet a wide variety of data communication needs of businesses and noncommercial users such as government agencies. The wide-area coverage feature combined with the ability to deliver relatively wide bandwidths with a consistent level of service make satellite links attractive in the developing world as well as in the geographically larger developed countries and regions.

The data that is contained within the satellite transmission can take many forms over a wide range of digital capacities. The standard 36-MHz transponder, familiar to users of C band worldwide, can transfer 60 Mbps, which is suitable for supercomputer applications and multimedia. Most applications do not need this type of bandwidth; therefore, this maximum is often divided up among multiple users who employ a multiple access system of some type. In fact, the multiple access techniques used on the satellite mirror the approaches used in *local area networks* (LANs) and *wide area networks* (WANs) over terrestrial links. As in any multiple-access scheme, the usable capacity decreases as the number of independent users increases. Satellite data networks employing VSATs offer an alternative to terrestrial networks composed of fiber optics and microwave radio. There is even a synergy between VSATs and the various forms of terrestrial networks, as both can multiply the effectiveness of their counterpart.

Some of the must successful users of VSATs are familiar names in North American consumer markets. Wal-Mart, the largest U.S. department store chain, was an early adopter of the technology and has pushed its competitors to use VSATs in much the same manner that they pioneered. With their Earth station hub at the Arkansas headquarters, Wal-Mart centralizes its credit authorization and inventory management functions. Chevron Oil likewise was first among the gasoline retailers to install VSATs at all of their company-owned filling stations to speed customer service and gain a better systemwide understanding of purchasing trends.

While voice networks are very standardized, data networks cover an almost infinite range of needs, requirements, and implementations. In business, *information technology* (IT) functions are often an element of competitive strategy

[3]. In other words, a company that can use information and communications more effectively than its competitors could enjoy a stronger position in the ultimate market for its products or services. A data communication system is really only one element of an architecture that is intended to perform business automation functions.

A given IT application using client/server computing networks or broadband multimedia will demand efficient transfer of data among various users. Satellite communication introduces a relatively large propagation delay, but this is only one factor in the overall response time. There are many contributors to response time and throughput, measured in usable bits per second; for example, the input data rate (also in bits per second), the propagation delay through the link, the processing and queuing delay in data communication equipment, and any contention for a terrestrial data line or computer processing element. This is shown in Figure 1.8. Each of these contributors must be considered carefully when comparing terrestrial and satellite links.

Propagation delay from a satellite link could reduce the throughput if there is significant interaction between the two ends of the link. The worst case condition occurs where the end devices need to exchange control information to establish a connection or confirm receipt. Modern data communication standards, like *transmission control protocol/Internet protocol* (TCP/IP) and *systems network architecture/synchronous data link control* (SNA/SDLC), guard against the loss of throughput by only requesting retransmission of blocks of data that have errors detected in them by the receiver. To optimize throughput, there is a need to test and tune the circuit for the delay, type of information and error rate. Even still, suppliers of VSAT network hardware and software include protocol processing (called *spoofing*) to compensate for the added delay of a GEO satellite connection.

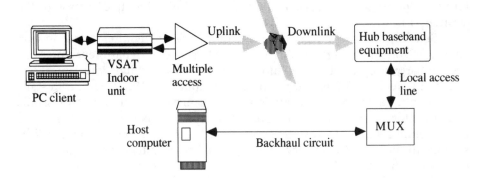

Figure 1.8 Contributors to the throughput and end-to-end delay of a data communication link involving a VSAT and a hub.

End-to-end data transfer delay on a GEO satellite link can be made to be lower than on a terrestrial telephone circuit. Consider a telephone circuit with an input rate of 4.8 Kbps (we ignore the effect of data compression such as that provided by the V.42 standard); this will take slightly more than 0.4 seconds (400 ms) to pass a block of 2,000 bits. This is calculated by dividing the number of bits by the data rate in bits per second (2000/4800). To this must be added the propagation time: 260 ms for a GEO satellite link and less than 50 ms for a terrestrial link of 4,000 km. A comparison of the total delay in this case is 660 ms for the satellite link and only 450 ms for the terrestrial link. Now, suppose we were to expand the satellite link to an input rate of 256 Kbps. The amount of time to transfer the 2,000 bits is 2,000/256,000, which is approximately 8 ms. The total delay of 268 ms of the satellite link is substantially less than the 450 ms for the terrestrial connection.

One can have a requirement that favors the terrestrial link. Suppose the PSTN is all fiber and has excellent transmission properties (which is the case in North America, Japan, and much of western Europe). Now, it is possible to input data at 19.2 Kbps. The total terrestrial delay is only about 150 ms, which is significantly shorter than for the satellite link. However, we must actually obtain 19.2 Kbps throughout the PSTN, something that is not particularly consistent. If you establish an international connection to many countries in the world, you would find that 19.2 Kbps is not feasible on an end-to-end basis. The satellite systems of the world, however, can produce links of sufficient quality to permit data transfer at 256 Kbps or higher.

The main deterrent to wider usage of digital satellite links is the cost of the terminal equipment, which today is much higher than that of a personal computer (PC). True low-cost VSATs were yet to appear at the time of this writing. Once the US$2,000 barrier is broken, many residential and small business users will be able to afford to purchase their own systems. This is not unlike the situation with backyard dishes, which during the 1980s were selling above $2,000. Through innovations like DIRECTV's DSS, which sells for around $500, the home dish has become as attainable as the VCR. This also has benefits for data applications. If the bulk of the data transfer is outbound only (so that the user installation is receive-only from the satellite), the DSS type of installation easily can deliver over 20 Mbps of downlink capacity. A key example is the Direct PC service described in Chapter 7.

1.2.4 Mobile and Personal Communications

The world has experienced an explosion of the demand for mobile telephone and data communications. The basis for this is the technology of cellular radio systems, which allow handheld mobile phones to connect to the public network as if they had access by wire [4]. Availability of cellular radio at a reasonable cost has allowed a much larger subscriber base to develop than was even possible during earlier generations of mobile phone technology. Now, we are prepared for the advance of cellular-like systems that connect users directly to orbiting satellites, thus eliminating the terrestrial base station (a much smaller quantity of gateway Earth stations are still needed to connect mobile calls to PSTN subscribers).

Satellite-based mobile radio was first introduced to the commercial market in the late 1970s in the form of *maritime mobile satellite service* (MMSS) by COMSAT. Within a ten-year period, MMSS overtook the radio telegraph as the primary means of allowing ships to communicate. Ships ceased carrying conventional radio stations in the 1990s, which have been completely replaced by MMSS terminals. Today, the Inmarsat System serves commercial ships and has been extended to commercial airlines as well.

Like INTELSAT, Inmarsat serves the needs of its owners and their respective clients. Within the last five years, Inmarsat introduced land-based communications in a service called *land mobile satellite service* (LMSS). The first LMSS terminals are portable, so they can be set up quickly from a stationary location.

In 1994, Optus Communications of Australia began providing the Mobilesat LMSS service to its domestic market. Mobilesat follows the model of the second generation of LMSS phones, allowing subscribers to communicate from moving vehicles. Optus is the pioneer in this type of system, having directed the development of the ground infrastructure and mobile terminals. Additional information on Mobilesat is provided in Chapter 9. The Optus B series of satellites includes the necessary L-band payload for Mobilesat and also carries Ku-band transponders to connect to gateways in the PSTN. The *American Mobile Satellite Corp.* (AMSC) and Telesat Mobile, Inc., extended this concept to the North American market in 1996. In this case, a satellite is dedicated to providing mobile telephone services.

A new generation of LMSS is expected before the year 2000, capable of supporting handheld phones similar to those used in cellular. In fact, the promoters of this advanced class of satellite network show models of modified cellular phones to promote their concept. The two basic approaches being taken are either the use of GEO, with satellites that serve specific countries or regions, and the non-GEO approach, where an orbiting constellation of satellites covers the entire globe (see Figure 9.4). The advantage of the GEO approach is its lower initial cost and possibly lower risk, while the advantage of the LEO or MEO approach is the lower propagation delay and ability to connect users to each other anywhere in the world.

While the satellite industry has been working to compete with conventional cellular telephone, the telephone and mobile radio business has been working to produce a more friendly and less expensive cellular system called *personal communications services* (PCS). Although not originally intended as a mobile system, PCS attacks the high cost and poorer voice quality of most cellular networks. The ultimate objective of PCS is to give each subscriber a single phone number for wireline and wireless telephone service. In exchange for these benefits, PCS might not provide mobile communications but instead behaves like a public version of a cordless phone. At the time of this writing, early versions of PCS were introduced in the United Kingdom and were promised in the United States. However, their properties are more like standard cellular and do not offer a universal number.

While PCS will replace some wired telephone usage in the coming years, it cannot provide the extensive coverage that LMSS offers. Perhaps there will be a way to merge the two after they have established their positions in the market.

Many of these wireless networks have value to business as a vehicle for extensive data communications and paging services. Mobile data services have been around in the United States for over a decade, first introduced for private purposes. Federal Express equips their drivers with radio-based data communications terminals, and today a majority of competing delivery companies have also adopted the technology. An explosion may occur before the year 2000 due to the creation of several satellite mobile data networks in the United States and other countries. There is now the prospect of providing a public data network service to all users who wish to access information databases and other facilities while on the move. This is something that the cellular network has had difficulty demonstrating due to technical issues of maintaining a solid circuit connection during the entire data call.

All together, the MSS is poised to become a substantial contributor to the future of the satellite industry. Spacecraft manufacturers are developing technical designs, which even a few years ago looked like science fiction (see Figure 1.6). Once again, today's science fiction becomes tomorrow's science fact.

References

[1] Elbert, B. R., *Introduction to Satellite Communication*, Norwood, MA: Artech House, 1987.

[2] Elbert, B. R., and B. Martyna, *Client/Server Computing—Architecture, Applications, and Distributed Systems Management*, Norwood, MA: Artech House, 1994.

[3] Elbert, B. R., *Networking Strategies for Information Technology*, Norwood, MA: Artech House, 1992.

[4] Macario, R. C. V. (on behalf of the Institution of Electrical Engineers), *Personal and Mobile Radio Systems*, London: Peter Peregrims, 1991.

Satellite Links and Access Methods

Satellite links employ microwave frequencies in the 1-GHz to 30-GHz range. The engineering process is no different from the practice developed during and immediately after World War II, when the application of this medium was accelerated for radar and communications. While the principles remain the same, many innovations in digital processing and array antennas are allowing more options for new applications. In this chapter, we briefly review the basics of the satellite link and relate it as much as possible to the needs of the application. Multiple-access systems, including FDMA, TDMA, and CDMA, are also discussed and their strengths and weaknesses identified. Once this review is completed, we consider the popular frequency bands used in commercial satellite communication (i.e., L, S, C, X, Ku, and Ka) and how they are best applied in practice.

The information that follows introduces the engineering side of developing satellite communication applications. Readers who have a technical background should have little difficulty following along with the format and content of the link budget. This is approached as a review and is not a substitute for a more detailed engineering evaluation of the specific losses and impairments that would be experienced in a particular design (in particular, see [1] and [2]). Nontechnical readers may wish to peruse the text concerning microwave link engineering and instead focus on the final sections that consider multiple-access methods and frequency band selection.

2.1 DESIGN OF THE SATELLITE LINK

The satellite link is probably the most basic in microwave communications since a line-of-sight path typically exists between the Earth and space. This means that an imaginary line extending between the transmitting Earth station and the satellite antenna passes only through the atmosphere and not ground obstacles. Such a link is governed by free space propagation with only limited variation with respect to time. Attenuation is dominated by the inverse square law, which states that the power received is inversely proportional to the square of the distance. The same law applies to the amount of light that reaches our eyeballs from a distant point source such as an automobile headlight or star. There are, however, a number of additional effects that produce a significant amount of degradation and time variation. These include rain, terrain effects, and some less-obvious impairments produced by unstable conditions of the air and ionosphere.

It is the job of the communication engineer to identify all of the significant contributions to performance and make sure that they are properly taken into account. This guarantees that the application will go into operation as planned, meeting its objectives for quality and reliability.

2.1.1 Basics of Microwave Link Engineering

The first step in designing the microwave link is to identify the overall requirements and the critical components that determine performance. For this purpose, we use the basic arrangement of the link shown in Figure 2.1. This example shows a large hub type of Earth station in the uplink and a small VSAT in the downlink. Transmission in the other direction is possible as well. Each element contributes a gain or loss to the link, resulting in the overall performance in terms of *carrier-to-noise* ratio (C/N) and, ultimately, information quality.

2.1.2 Definition of the Power Balance and Path Loss

The link between the satellite and Earth station is governed by the basic microwave radio link equation

$$P_r = \frac{P_t G_t G_r c^2}{(4\pi)^2 D^2 F^2} \tag{2.1}$$

where P_r is the power received by the receiving antenna; P_t is the power applied to the transmitting antenna; G_t is the gain of the transmitting antenna, as a true ratio; G_r is the gain of the receiving antenna, as a true ratio; c is the speed of light (i.e., 300×10^6 m/s); D is the path length in meters; and F is the frequency in hertz.

Almost all link calculations are performed after converting from products and ratios to logarithms. This uses the popular unit of decibels, which is discussed in detail in a following section. The same formula, when converted into decibels, has the form of a power balance:

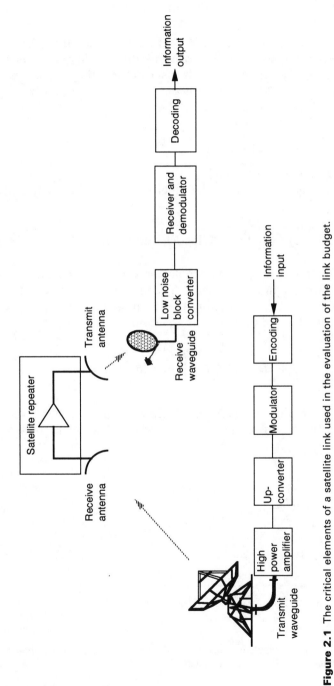

Figure 2.1 The critical elements of a satellite link used in the evaluation of the link budget.

$$P_r = P_t + G_t + G_r - 10 \log(F^2D^2) + 147.6 \qquad (2.2)$$

The last two terms together represent the free-space path loss (A_0) between the Earth station and the satellite. If we assume that the frequency is 1 GHz and that the distance is simply the altitude of a GEO satellite (e.g., 35778 km), then the path loss equals 183.5 dB (expressed as a positive number to reflect the fact that it is a loss); that is,

$$P_r = P_t + G_t + G_r - 183.5 \qquad (2.3)$$

for $F = 1$ GHz and $D = 35788$ km.

We can correct the path loss for other frequencies and path lengths using the formula

$$A_0 = 183.5 + 20 \log(F) + 20 \log(D/35788) \qquad (2.4)$$

where A_0 is the free-space path loss in decibels, F is the frequency in gigahertz, and D is the path length in kilometers. The term on the right can be expressed in terms of the elevation angle from the Earth station toward the satellite, as shown in Figure 2.2 and given in the equation

$$D = R_0 \left[\sqrt{\left(\frac{R}{R_0} \right)^2 - \cos^2\theta} - \sin^2\theta \right] \qquad (2.5)$$

where R_0 is the radius of the Earth at the equator (e.g., 6378 km), R is the distance from the center of the Earth to the satellite (e.g., 42166 km), and θ is the elevation of the Earth station antenna in the direction of the satellite.

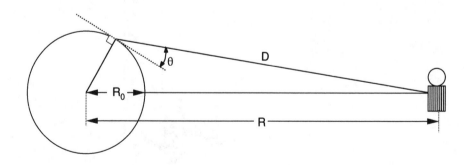

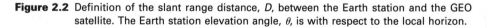

Figure 2.2 Definition of the slant range distance, *D*, between the Earth station and the GEO satellite. The Earth station elevation angle, θ, is with respect to the local horizon.

Substituting for D in (2.4), we obtain the correction term in decibels to account for the actual path length. This is referred to as the slant range adjustment and is plotted in Figure 2.3 as a function of the elevation angle, θ.

2.1.3 Additional Link Attenuation Factors

The link power balance relationship in (2.2) considers only the free-space loss and ignores the effects of the different layers of the Earth's atmosphere. The following listing identifies the dominant effects that introduce additional path loss, which can vary with time. Some are due to the air and water content of the troposphere, while others result from the ionosphere.

- *Absorption by air and water vapor (noncondensed):* this is nearly constant for higher elevation angles, adding only a few tenths of decibels to the path loss. It generally can be ignored at frequencies below 15 GHz.
- *Refractive bending and multipath at low elevation angles:* Earth stations that must point near the horizon to view the satellite are subject to wider variations in received or transmitted signal and therefore require more link margin.
- *Rain attenuation:* this important factor increases with frequency and rain rate. Additional fade margin is required for Ku- and Ka-band links, based on the statistics of local rainfall. This will require careful study for services that demand high availability, as suggested in Figures 2.4 and 2.5.
- *Faraday rotation of linear polarization:* most pronounced at L and S bands, with significant impact at C band during the peak of sun-spot activity. Not a significant factor at Ku and Ka bands.
- *Ionosphere scintillation:* most pronounced in the equatorial regions of the world (particularly along the geomagnetic equator). Like Faraday rotation, this source of fading decreases with increasing frequency, making it a factor for L-, S-, and C-band links.

At frequencies above C band (i.e., above 7 GHz), rain introduces a substantial amount of loss that must be taken into account in the link design. Regions with intense thunderstorm activity, particularly in the tropics, offer some challenge at C band as well. As discussed in the next section, proper engineering practice includes the provision of several decibels of margin above the minimum required. Margin represents extra power in the link that gives a cushion against fades and other errors. More recently in mobile communications, it has become a practice to include margin for short-term terrain blockage, as when a vehicle travels under trees and past a tall building. If the mobile link is stable and not experiencing a fade, then the margin shows up as better signal quality.

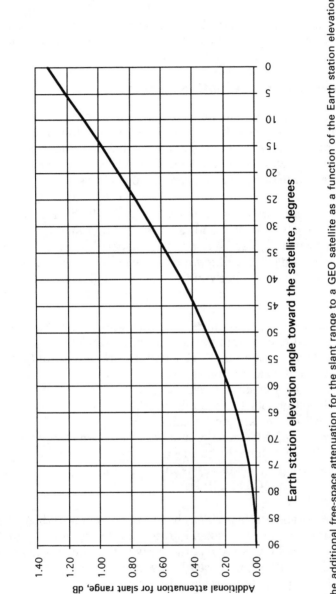

Figure 2.3 The additional free-space attenuation for the slant range to a GEO satellite as a function of the Earth station elevation angle.

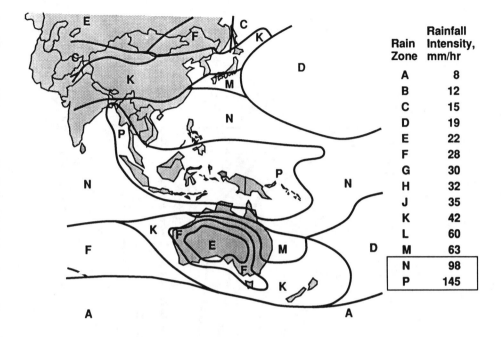

Rain Zone	Rainfall Intensity, mm/hr
A	8
B	12
C	15
D	19
E	22
F	28
G	30
H	32
J	35
K	42
L	60
M	63
N	98
P	145

Figure 2.4 An example of the rain climatic zones in the Asia-Pacific region. The rain rate for each zone corresponds to a probability of 0.03% (according to the CCIR). Region P is typical of the highest rainfall in the world.

2.1.4 Rain Attenuation

Rainfall is not a predictable phenomenon from year to year. On average, the statistics follow some patterns that have aided in the design of links in rainy regions, like those indicated in Figure 2.4. This information, based on a rain attenuation model developed by the CCIR (now ITU-BR), provides an indication of how the typical rainfall of a region must be considered in link design. Properly engineered satellite links at the higher frequencies may be as dependable as one operating in the popular C band. In Figure 2.5, rain attenuation is plotted for the worst rain regions (N and P) at C and Ku bands. Readers should keep in mind that this information applies for a general case in Asia Pacific, based on the studies of the CCIR. A particular design should involve a thorough study of local rainfall statistics for a system operating a Ku and Ka band, where rain attenuation could be substantially different from a general case.

The satellite application engineer can select a desired value of availability such as 99.9% (along the x-axis of Figure 2.5), which determines the required amount of link margin to counter the most extreme rain attenuation condition

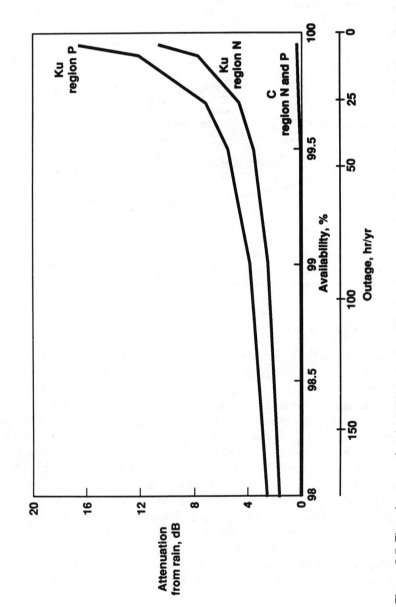

Figure 2.5 The rain attenuation, in decibels, for C and Ku bands, as related to the rain climatic zones in Figure 2.4.

(along the y-axis). The link should experience outage for 0.1% of the time during the rainiest month. This equates to 72 minutes during that month or a maximum of 8.8 hours for the year. It is clear from the example in Figure 2.5 that more rain margin must be provided at Ku band than at C band for the same link availability. The link budget provides the vehicle for finding the best combination of power and gain to achieve this result.

2.1.5 Meaning and Use of the Decibel

Satellite application engineers are more comfortable working with sums and differences than with multiplication and division (which is how the link actually relates to the real world). The decibel (dB) has been settled upon as the most convenient metric because, once converted into decibels, complex factors can be added and subtracted on paper with a calculator or even in your head and because any number that is either a ratio of two powers or a power expressed in watts can be converted to decibels.

The precise formulation for this is

$$\text{Power ratio in decibels} = 10 \cdot \log_{10}(P_2/P_1) \tag{2.6}$$

where P_2 is the output of the device or link and P_1 is its input.

Table 2.1 provides this conversion for integer values of the ratio of P_2 to P_1. The first entry for the number 1.0 represents the ratio of two equal power values. The power does not change for a power ratio of 1, and the corresponding decibel value is 0. If the power increases from this point by 12%, the power ratio is 1.12, which corresponds to a 0.5-dB increase. All decibel values are relative to the starting point; so if we increase the power again by 12% we must add another 0.5 dB, giving a total increase of 1 dB. According to Table 2.1, a 1-dB increase corresponds to a change of 26% or, equivalently, multiplication of the original value by 1.26. This is also equal to 1.12 multiplied by 1.12.

The table can be used to represent changes that go in the opposite direction, that is, where the ratio of the first power to the second is less than one. For example, if the output is 1 and the input is 10, then the ratio is 0.10. Using the formula, the decibel value is −10 dB. When we invert the relationship between the first and second power values, we simply put a minus sign in front of the decibel value. In Table 2.1, dividing by 2 (instead of multiplying by 2) produces as −3-dB change (instead of a +3-dB change).

An experienced satellite application engineer maintains this table in his or her head and uses basic arithmetic to quickly analyze the main factors in a particular link. We call this *dB artistry*, the rules for which are as follow.

1. $10 \cdot \log_{10}(x) = X$ expressed in decibels (note the use of *common* logarithms).
2. Power ratios are expressed in decibels.
3. Voltages must first be squared to convert to power; that is, $10 \cdot \log_{10}(V_2/V_1)^2$ or, equivalently, $20 \cdot \log_{10}(V_2/V_1)$.
4. Increases are positive; decreases are negative.
5. $10 \cdot \log_{10}(1/x) = -10 \cdot \log_{10}(x)$.
6. By definition, 0 expressed in decibels is $-\infty$.
7. For each multiplication by 10, add 10 dB; for each division by 10, subtract 10 dB.
8. Power relative to 1W is expressed as dBW; that is, 1W is identically equal to 0 dBW.
9. Power relative to 1 mw (0.001W) is expressed in dBm; that is, 1 mw is identically equal to 0 dBm; also, 0 dBW = 30 dBm.

A proficient dB artist can do a rough link analysis in his or her head by remembering these rules along with a few values from Table 2.1. For example, if the power at the transmitter is doubled, then it has increased by 3 dB. If the power is cut in half, then it decreases by the same 3 dB (rule 5). The net result

Table 2.1
Conversion Between Power Ratios, Percentage
Changes, and Decibels

Power Ratio	% Change	dB
1.0	0	0
1.12	12	0.5
1.26	26	1.0
1.41	41	1.5
1.59	58	2.0
2.00	100	3.0
2.51	151	4.0
3.16	216	5.0
3.98	298	6.0
5.01	401	7.0
6.31	531	8.0
7.94	694	9.0
8.91	791	9.5
10.0	900	10.0
20.0	—	13.0
100.0	—	20.0
1000.0	—	30.0

is no change at all; that is , +3, −3, nets 0. A 10-dB increase simply means that the power has increased by a factor of ten (rule 7). If we multiply by 100, then we have introduced 20 dB, which is two increases of 10 dB (rule 7 again).

The following is a practical example that demonstrates the power of dB artistry. Say that a satellite radiates a power of 100W toward us on the Earth. If we want the effective power to be 200W instead, then we are looking for a 3-dB increase. This is found by taking the ratio 200/100 = 2, which is converted to 3 dB in Table 2.1. If we need the effect of 2000W, then according to rule 7, we must add another 10 dB, giving a total increase from 100W of 13 dB.

In summary, the essential value of decibels is that they help to express what is important in the microwave link. You can easily adjust powers and gains up and down using decibel values obtained from a scientific calculator; alternatively, you can memorize the first five to ten values of Table 2.1 and do the calculations in your head. It has been the experience of this author that it is adequate to work in tenths of decibels, as this represents the smallest increment that can be measured in practice. The overall link, to be discussed in the next section, tends to be forgiving as long as you make a good effort to identify and quantify the more significant elements.

2.2 LINK BUDGETS

Satellite application engineers need to assess and allocate performance for each source of gain and loss. The link budget is the most effective means since it displays the components of the power balance equation, expressed in decibels. Despite over 25 years of commercial experience, there still is no standard format. What we will do instead is to evaluate a specific example using a typical link budget in the process. This will provide the reader with a typical format and some guidelines for a practical approach. We divide the overall evaluation into three distinct budgets: the downlink budget, the uplink budget, and the overall link budget. These are often shown in a single table, but dividing them should make the development of the process clearer for readers. The real engineering comes into play with the development of each entry of the table. This particular example is for a digital video link at 40 Mbps, which is capable of transmitting four to six TV channels.

2.2.1 Downlink Budget

Table 2.2 presents the downlink budget in a manner that identifies the characteristics of the satellite transmitter and antenna, the path, the receiving antenna, and the expected performance of the Earth station receiver. The latter contains the elements that select the desired radio signal (i.e., the carrier) and demodu-

Table 2.2
Link Budget Analysis for the Downlink (3.95 GHz, C band)*

Item	Link Parameter	Value	Units	Computation
1	Transmit power (10W)	10.0	dBW	
2	Transmit waveguide losses	1.5	dB	
3	Transmit antenna gain	27.0	dBi	
4	Satellite EIRP (toward LA)	35.5	dBW	1 − 2 +3
5	Free-space loss	196.0	dB	(2.4)
6	Atmospheric absorption (clear air)	0.1	dB	
7	Receive antenna gain (3.2m)	40.2	dBi	
8	Receive waveguide loss	0.5	dB	
9	Received carrier power	−121.7	dBW	4 − 5 − 6 + 7 − 8
10	System noise temperature (140K)	21.5	dBK	
11	Earth station G/T	18.2	dB/K	7 − 8 − 10
12	Boltzmann's constant	−228.6	dBW/Hz/K	
13	Bandwidth (25 MHz)	74.0	dB Hz	
14	Noise power	−133.1	dBW	10 + 12 + 13
15	Carrier-to-noise ratio	11.4	dB	9 − 14

*Earth station location in Los Angeles, CA.

lates the useful information (i.e., the baseband). Once converted back to baseband, the transmission can be applied to other processes, such as demultiplexing, decryption, and digital-to-analog conversion (D/A conversion).

Figure 2.6 provides the horizontal downlink coverage of Galaxy VI, a typical C-band satellite that serves the United States. Each contour shows a constant level of saturated *effective isotropic radiated power* (EIRP) (the value at saturation of the transponder power amplifier). Assuming the receiving Earth station is in Los Angeles, it is possible to interpolate between the contours and estimate a value of 35.5 dBW. A wideband carrier for TV transmission could consume all of this power. Alternatively, the power could be shared using one of the multiple-access methods described in Section 2.3. We review the details of this particular link budget for those readers who wish to gain an understanding of how this type of analysis is performed. This material may be skipped by the nontechnical reader.

The following parameters relate to the significant elements in the link (Figure 2.1) and the power balance equation. Most are typically under the control of the satellite engineer.

- Transmit power (P_t);
- Antenna gain (G_t) and beamwidth (Θ_3 dB);

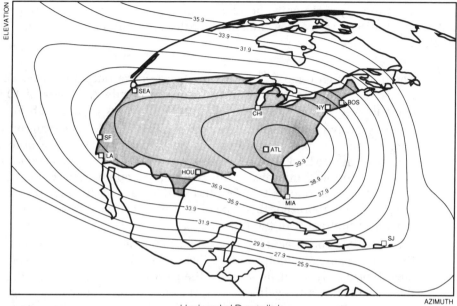

Horizontal Downlink

Note: Actual values may vary by ± 1 dB.

Figure 2.6 The downlink coverage footprint of the Galaxy VI satellite, located at 74 deg west longitude. The contours are indicated with the saturated EIRP in decibels referred to 1W (0 dbW) (graphic courtesy of Hughes Communications, Inc.).

- Feeder waveguide losses (L_t);
- EIRP in the direction of the Earth station (Figure 2.6);
- Receiver noise temperature (T_0); also noise figure (F) or noise power spectral density (N_0);
- Earth station figure of merit: $G/T = G_r - 10 \cdot \log_{10}(T_0)$. This combines two factors in the receiving Earth station (or satellite), providing a standard specification at the system level. The same can be said of EIRP.

Each of the link parameters relates to a specific piece of hardware or some property of the microwave path between space and ground. The best way to develop the link budget is to prepare it with a spreadsheet program like Lotus 1–2–3 or Microsoft Excel. This permits the designer to include the various formulas directly in the budget, thus avoiding the problem of external calculation or the potential for arithmetic error (which still exists if the formulas are wrong or one adds losses instead of subtracting them).

The following comments and clarifications relate to each item in Table 2.2, with additional references made to the rules of dB artistry in the previous section.

1. The transponder onboard the satellite has a power output of 10W. This is a factor of ten greater than 1W. According to rule 7, a factor of 10 is equivalent to 10 dB; therefore, we represent this power as 10 dBW.

2. The microwave transmission line between the satellite power amplifier output and the spacecraft antenna absorbs about 40% of the output, converting it into heat. This total loss of 1.5 dB (Table 2.1, line 4) includes some absorption in microwave filters used to combine carriers and other microwave components that are part of the waveguide assembly. It is the practice to represent this type of loss as a positive number and then to simply subtract it in the link budget.

3. The satellite is engineered to cover a particular area of the Earth, called the coverage area or footprint (see Figure 2.6). The size of the coverage area is primarily what determines the gain of the antenna—the larger the area, the lower the gain. A value of 27 dBi results for an area roughly the size of the United State or Indonesia. The letter "i" after "dB" on line 3 stands for *isotropic* and indicates that the gain is given relative to that of an antenna that uniformly radiates in all directions (i.e., around a sphere). It is a fiction used as a standard for comparison. For example, a gain of 27 dBi means that the power radiated in the direction of interest is 500 times that from an isotropic antenna that is fed with the same input power.

4. This EIRP specifies the maximum radiated power per transponder in the direction of a specific location on the Earth. When we look at the downlink footprint of a satellite, we see contours of constant EIRP. Figure 2.6 is actually a contour map of antenna gain, measured at an antenna test range. The contours are labeled in dBW of EIRP by simply adding the input power to the antenna gain.

5. Free-space loss is the primary loss in the satellite link. For frequencies between 1 GHz and 30 GHz for a GEO satellite, it ranges between 183 dB and 213 dB. Included is an additional loss to account for the real path length (e.g., the slant range), provided in Figure 2.3. To use this figure, we need to know the elevation angle from the Earth station toward the satellite. This is determined from the latitude and longitude of the Earth station and the longitude of the satellite. For Los Angeles and Galaxy VI, the elevation angle is approximately 30 deg. The shortest possible path corresponds to an elevation angle of 90 deg, where the Earth station is located on the equator directly below the satellite (i.e.,

at the same longitude). The worst case value of slant range loss adjustment is 1.6 dB for a 0-deg elevation angle toward the satellite.

6. At C band, the elements of clear air absorb a small amount of microwave energy as the wave passes through the lower atmosphere. This absorption loss increases as the elevation angle to the satellite decreases; that is, the more air there is to go through, the greater the loss. This loss also increases significantly with moisture content, particularly rain, although this is treated separately in the link budget. The value of 0.1 dB in the table is typical for an elevation angle greater than 30 deg under the condition of normal humidity. Recognizing that this loss is practically insignificant, many analysts simply leave it out of the budget.

7. In this example, the receiving antenna has a diameter of 3.2m (10 ft). Either of the following formulas provides a good estimate:

$$G = 10 \cdot \log(10 \cdot \eta \cdot f^2 D^2)$$

where f is in gigahertz, D is in feet, and η is the antenna efficiency $(\eta \le 1)$ or

$$G = 10 \cdot \log(108 \cdot \eta \cdot f^2 D^2)$$

where f is in gigahertz and D is in meters. A typical value for M of 0.6 (60%) is used in this example. Again, we state the gain in terms of dBi to indicate that we are comparing this antenna to the isotropic model.

8. Waveguide or cable loss between the antenna feed and LNC reduces the received signal and increases link noise by the same proportion. We have included 0.5 dB of loss for this effect. The cable that connects the low noise converter (LNC) to the receiver does not directly impact the performance because it is after the relatively high gain (usually greater than 50 dB) provided by the LNC.

9. Received carrier power is calculated directly by the power balance method. This computed value of −121.7 dBW includes all of the gains and losses in the link. It is an absolute measure in terms of power; however, we cannot tell at this point if the signal strength is sufficient for good reception. This will have to wait until we consider the uplink and the threshold performance of the demodulator.

10. The noise that exists in all receiving systems is the main cause of degradation. The system noise temperature includes contributions from the LNC, antenna, and transmission line. The LNC is rated in terms of its noise temperature, typically in the range of 60° to 100° Kelvin at C band. (The Kelvin scale starts at absolute zero, which is where noise is nonexistent.)

The antenna itself collects background noise from space and the local terrain, typically adding about 40°K at C band. We have assumed a combined system noise temperature of 140°K, allowing 60°K for the LNC, 45°K for the antenna, and 35°K for the feeder line. The noise from the feeder line is calculated from $(L - 1) \cdot 290$, where $L = 10^{(a/10)}$, a is in decibels, and L is less than or equal to 1.

11. Earth station G/T is the difference in decibels between the net antenna gain and the system noise temperature converted to decibels, that is,

$$G/T = G - a - 10 \cdot \log_{10}(T_{sys})$$

where a is the waveguide loss between the antenna and the LNC and T_{sys} is the receiving system noise temperature.

12.–14. The noise power that reaches the receiver is equal to the product kTB, where k is Boltzmann's constant, T is the equivalent noise temperature, and B is the noise bandwidth of the carrier in hertz. In the link budget, we use the decibel equivalents of these factors and hence can use addition instead of multiplication. The noise bandwidth for the digital carrier in this carrier is assumed to be 25 MHz for a QPSK signal, which corresponds to a data rate of approximately 40 Mbps. Our resulting noise power is −133.1 dBW within this bandwidth.

15. The difference in decibels between the received carrier power and the noise power is the carrier-to-noise ratio. The value in the table of 11.4 dB is considerable, demonstrating a solid link. As mentioned in 9, we cannot determine at this point if 11.4 is adequate for the overall link.

The downlink budget is now complete and may be put aside for the moment as we proceed to the uplink.

2.2.2 Uplink Budget

We next perform nearly the same calculation for the uplink, providing an estimate of the carrier-to-noise ratio as measured at the output of the spacecraft antenna system. Table 2.3 presents the link budget from the transmitting Earth station to the satellite, including a few new terms.

Each of the items in Table 2.3 is reviewed in this section. Since this link budget is very similar to that of the downlink, we occasionally refer to the previous explanations for items 1 through 15.

16. The Earth station *high-power amplifier* (HPA) provides sufficient power to operate the satellite transponder at saturation. Here, 850W is assumed.

Table 2.3
Link Budget Analysis for the Uplink (6.175 GHz, C band)

Item	Link Parameter	Value	Units	Computation
16	Transmit power (850W)	29.3	dBW	
17	Transmit waveguide losses	2.0	dB	
18	Transmit antenna gain (7m)	50.6	dBi	
19	Uplink EIRP from Boston	77.9	dBW	16 − 17 + 18
20	Spreading loss	162.2	dB(m²)	
21	Atmospheric attenuation	0.1	dB	
22	Flux density at the spacecraft	−84.4	dBW/m2	19 − 20 − 21
23	Free-space loss	200.0	dB	
24	Receive antenna gain	26.3	dBi	
25	Receive waveguide loss	0.5	dB	
26	System noise temperature (450K)	26.5	dB(K)	
27	Spacecraft G/T	−0.7	dB/K	24 − 25 − 26
28	Received C/T	−122.9	dBW/K	19 − 23 − 21 + 27
29	Boltzmann's constant	−228.6	dBW/Hz/K	
30	Bandwidth (25 MHz)	74.0	dB Hz	
31	Carrier-to-noise ratio	31.7	dB	28 − 29 − 30

17. An allocation of 2 dB is made to account for the loss between the HPA and the Earth station antenna feed.

18. A 7m Earth station antenna diameter, provides 50.6 dBi of C-band gain, according to formula given in item 7.

19. Uplink EIRP must be sufficient to saturate the satellite transponder. We determine its value from the saturation flux density requirement of the satellite in item 22.

20. The spreading loss allows us to convert from Earth station EIRP to the corresponding value of flux density at the face of the satellite receive antenna. It is calculated as $10 \cdot \log10(4\pi D^2)$, where D is the slant range. The units are in dB(m²).

21. Atmospheric loss at 6 GHz is essentially equal to that at 4 GHz.

22. It is customary in commercial satellite communications to specify the uplink driving signal to the transponder in terms of the flux density. If the particular value is that which causes the transponder to transmit the maximum EIRP in the downlink, then we refer to this as the *saturation flux density* (SFD). Once you know the SFD for the satellite, you can compute the required EIRP for the Earth station through the reverse of the calculation. Figure 2.7 provides the SFD for Galaxy VI in the form

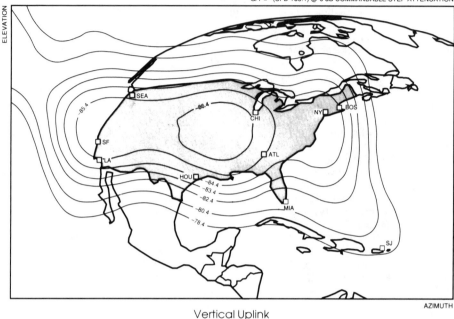

Vertical Uplink

Figure 2.7 The uplink coverage footprint of the Galaxy VI satellite, located at 74 deg west longitude. The contours are indicated with the SFD in the direction of the Earth (graphic courtesy of Hughes Communications, Inc.).

of a coverage footprint. This is based on the antenna gain and takes account of the repeater gain between the antenna and the transponder power amplifier.

23. At this point, we revert to a direct calculation of carrier power received at the satellite. We use the free-space loss in lieu of the spreading loss for certain types of uplink calculations. Free-space loss is calculated according to the method in item 5 and (2.4). Alternatively, we could have used the spreading loss and added a term that is the area of an isotropic antenna at this frequency, i.e., $10 \cdot \log(\lambda^2/4\pi)$.

24. The spacecraft antenna is designed to cover a particular geographic area, here assumed to be the United States. The typical design provides a minimum value of 27 dBi.

25. An allocation of 0.5 dB is made for the loss between the spacecraft antenna and the receiver front end.

26. The typical C-band satellite has a system noise temperature of 450°K (equivalently, 26.5 dBK), which includes 270°K for the antenna temperature (microwave brightness from the Earth), 50°K for the waveguide line, and 130°K for the receiver itself.

27. The third key satellite performance parameter is the G/T, or receiving system figure of merit. As stated in the explanation of item 10, G/T is the difference between the net antenna (including waveguide loss) and the system noise temperature expressed in decibels. The G/T and SFD differ by a fixed constant since they vary by the gain of the spacecraft antenna alone. From the specific example in Figure 2.7, $G/T = -(SFD + 85.1)$ dB/K.

28. The value of C/T received by the satellite is calculated from the power balance as

$$C/T = EIRP - A_0 - A_{at} + G/T$$

where A_0 is the free-space loss and A_{at} is the atmospheric loss.

29.–31. These values are considered in the same manner as items 12 through 15. The value of uplink C/N presented in item 31 (i.e., 31.7 dB) is substantially higher than the downlink value in item 15 (i.e., 11.4 dB). Under this condition, the downlink will dominate the overall link performance, as discussed for the overall link budget.

2.2.3 Overall Link

The last step in link budgeting is to combine the two link performances and compare the result against a minimum requirement. Table 2.4 presents a detailed evaluation of the overall link under the conditions of line-of-sight propagation in clear sky. We have included an allocation for interference coming from sources such as a cross-polarized transponder and adjacent satellites. This type of entry is necessary because all operating satellite networks are exposed to one or more sources of interference. The bottom line represents the margin that is available to counter rain attenuation and any other losses that were not included in the link budgets. We have included itemized remarks as for the previous examples.

32. The uplink C/N (line 15) is converted to a true value (not decibels) using the transformation, $x = 10^{(X/10)}$.

33. This is simply the inverse, which provides the uplink noise in a normalized form.

34.–35. See comments for items 32 and 33.

Table 2.4
Combining the Uplink and the Downlink to Estimate Overall Link Performance

Item	Link Parameter	Value	Units	Computation
32	Uplink C/N (31.7 dB)	1479.1	Ratio	31
33	N_u/C	0.000676	Ratio	
34	Downlink C/N (11.4 dB)	13.8	Ratio	15
35	N_d/C	0.0724	Ratio	
36	Total thermal noise (N_{th}/C)	0.0731	Ratio	33 + 35
37	Total thermal C/N_{th}	13.7	Ratio	
38	Total thermal C/N_{th}	11.4	dB	
39	Interference C/I (18.0 dB)	63.1	Ratio	Assumption
40	I/C	0.015848	Ratio	
41	Total noise $(N_{th} + I)/C$	0.0889	Ratio	36 + 40
42	Total $C/(N_{th} + I)$	11.2	Ratio	
43	Total $C/(N_{th} + I)$	10.5	dB	
44	Required C/N	8.0	dB	Equipment
45	Margin	2.5	dB	43 − 44

36. This is an important step in the overall evaluation. The normalized uplink and downlink noise terms (items 33 and 35) are added together. We see that the downlink noise is much larger than the uplink noise (because the downlink C/N is 20.3-dB lower). In fact, the uplink only contributes about 1% of the total.

37. The inverse of item 36 is the total C/N_{th}, in normalized units (not decibels). The subscript, th, indicates that the noise comes from thermal sources in the Earth station and satellite receivers.

38. This is the combined C/N_{th} in decibels.

39. An estimated C/I value of 18 dB is shown on this line. In reality, we would perform several interference calculations based on the likely sources of interference into the system. Cross-polarized interference will come from transmissions to and from the satellite on the same frequency but in the opposite polarization. Adjacent satellite interference would be determined through the frequency coordination process, discussed in Chapter 10. At C band there might also be terrestrial interference since this is a shared band.

40. See comments for items 32 and 33.

41.–43. With all of the noise and interference sources included, we now have the combined C/N for the uplink and downlink. This is the value that we expect to measure at the input to the receiving Earth station demodulator.

44. The required value of *C/N* is determined by the performance of the receiver digital demodulator. This characteristic is determined from the demodulator design and can be verified in the laboratory. Digital receivers usually include forward error correction and therefore can operate at lower values of *C/N* than analog demodulators. The actual operating *C/N* might be set by a maximum allowable error rate for the type of service provided. This cannot be specified in general as it depends heavily on the type of coding and the tolerance of the end-user device to errors and other signal degredations. Some specific guidelines for digital video are provided in Chapter 5.

45. The link margin is simply the difference between the total *C/N* and the required value. It provides a cushion against variations in the link that the budget does not include directly. Obviously, if we included every possible degrading factor, we would not need any margin. Things that can pull the link down from the value on item 45 include rain attenuation (the biggest single factor), atmospheric fading due to ducting, ionospheric scintillation, and antenna misalignment.

2.2.4 Additional Sources of Noise and Interference

Since most satellites will have neighbors in orbit that operate on the same frequencies, we need to include the contribution of orbital interference. Chapter 10 contains a more detailed discussion of this subject. In a typical link, orbital interference could add 20% to the total thermal noise, although the precise amount is determined by the number of satellites, their spacing and the types of signals in operation on both the desired and interfering satellites. Another source of noise is intermodulation distortion produced in the transponder and the Earth station HPA. The contribution here could be as much as 100% of total thermal noise (i.e., it could be equal). C-band links also include the interference caused by terrestrial microwave stations that are within range of the receiving Earth station. The obvious advantage of using a band not shared with such microwave stations is that this particular interference source is not present.

We can extend the budget shown in Table 2.4 to include the additional sources of link noise. In general, we can apply the following formula for total link *C/N* (thermal noise, distortion, and interference):

$$\frac{C}{N_{\text{total}}} = \left\{ \left[\frac{N_d}{C}\right]^{-1} + \left[\frac{N_u}{C}\right]^{-1} + \left[\frac{N_{im}}{C}\right]^{-1} + \left[\frac{N_{oi}}{C}\right]^{-1} + \left[\frac{N_{ti}}{C}\right]^{-1} \right\}^{-1}$$

This rather imposing formula simply states that you can combine all of the contributions together by first converting each *C/N* to a true number, invert each to show its relative noise contribution, add up the noise contributions in this form, and then invert the sum. The result is the total C/N as a true number. As a last step, convert the total C/N to a decibel value and compare it to the requirement, as in items 43 to 45.

Link budget analysis is probably the most important engineering discipline in designing a satellite application. We recommend that you take the time to get a basic understanding so that you may at least be able to ask the right questions.

2.3 MULTIPLE ACCESS SYSTEMS

Applications employ multiple-access systems to allow two or more Earth stations to simultaneously share the resources of the same transponder or frequency channel. These include the three familiar methods: *frequency division multiple access* (FDMA), *time division multiple access* (TDMA), and *code division multiple access* (CDMA). In [3] we suggested that there was another multiple-access system, namely, *space division multiple access* (SDMA). In practice, SDMA is not really a multiple-access method but rather a technique to reuse frequency spectrum through satellite antenna polarization and beam-to-beam discrimination. Because every satellite provides some form of frequency reuse, SDMA is an inherent feature in all applications. We are now including CDMA as a distinct multiple-access system because of an acceleration in its use. TDMA and FDMA require a degree of coordination among users, while CDMA permits users to transmit independently of each other.

Multiple access is always required in networks that involve two-way communications among multiple Earth stations. The selection of the particular method depends heavily on the specific communication requirements, the types of Earth stations employed, and the experience base of the provider of the technology. All three methods are now used for digital communications because this is the basis of a majority of satellite networks. The digital form of a signal is easier to transmit and is less susceptible to the degrading effects of the noise and interference. Once in digital form, the information can be compressed to remove unnecessary bits and forward error correction (FEC) is usually provided to reduce the required carrier power even further. The only significant analog application at this time is the transmission of cable TV and broadcast TV. These networks are undergoing a slow conversion to digital as well, which may in fact be complete around the year 2000.

2.3.1 Frequency Division Multiple Access

Nearly every terrestrial or satellite radio communications system employs some form of FDMA to divide up the available spectrum. The areas where it has the strongest hold are in *single channel per carrier* (SCPC) voice telephone systems, VSAT data networks, and some video networking schemes. Any of these networks can operate along side other networks within the same transponder. Users need only acquire the amount of bandwidth and power that they require to provide the needed connectivity and throughput. Also, equipment operation is simplified since no coordination is needed other than assuring that each Earth station remains on its assigned frequency and that power levels are properly regulated.

Stated another way, FDMA is always combined with TDMA or CDMA. This reduces the required transmit power from the Earth station, thus saving on electronics and transmission costs. Networks can also share the same transponder, which is another cost reduction strategy. The satellite operator divides up the power and bandwidth of the transponder and sells off the capacity in attractively priced segments. Users pay for only the amount that they need. If the requirements increase, additional FDMA channels can be purchased. The *intermodulation distortion* (IMD) that FDMA produces within a transponder must be quantified and controlled; otherwise, service quality and capacity will degrade rapidly as users attempt to compensate by increasing uplink power further. The big advantage, however, is that each Earth station has its own independent frequency on which to operate. A bandwidth segment can be assigned to a particular network of users, who subdivide the spectrum further based on individual needs. A discussion of the business aspects of FDMA services is provided in Chapter 11.

2.3.2 Time Division Multiple Access and ALOHA

TDMA is a truly digital technology, requiring that all information be converted into bit streams or data packets before transmission to the satellite. Contrary to most other communication technologies, TDMA started out as a high-speed system for large Earth stations. As the cost and size of digital electronics came down, it became practical to build a TDMA Earth station into a compact package. While the first experiments with TDMA were conducted prior to 1970, it was not until the late 1980s that TDMA came into its own as a practical approach for low-cost VSAT networks. For the cost of the RF electronics and of using the satellite to be reasonable, the speed of transmission had to be reduced substantially. TDMA stations therefore cover a wide range of speeds, from the high-capacity 60-Mbps systems to the current VSAT standard around 128 Kbps.

TDMA signals are restricted to assigned time slots and therefore must be transmitted in bursts. The Earth station equipment takes one or more continuous streams of data, stores them in a buffer memory, and then transfers the output toward the satellite in a burst at a higher speed. At the receiving Earth station, bursts from one or more Earth stations are received in sequence by the downlink

Earth station, selected for recovery if addressed for this station, and then spread back out in time in an output buffer. It is vital that all bursts be synchronized to prevent overlap at the satellite. Individual time slots are preassigned to particular stations by a master control. Since much of the traffic requires consistent or constant timing (e.g., voice and TV), the time slots repeat at a constant rate.

There is an adaptation of TDMA, called ALOHA, that uses burst transmission but eliminates the synchronization from a master control. ALOHA is a powerful technique for low-cost data networks that need minimum response time. Throughput is less than 20% if the bursts come from stations that are completely uncoordinated because there is the potential for overlap (called a collision). After a station determines that its burst has collided with another station's burst, it performs a retransmission. This introduces time delay. The 20% limit is based on there being a high percentage of collisions and resulting retransmissions, resulting in delay that is unacceptable to the application. If the stations follow a common timing standard (but are not controlled), the throughput can be effectively doubled. In Chapter 7, we review the performance of ALOHA and compare it to TDMA.

TDMA can come very close to 100% throughput, the only reduction due to the requirement for some guardtime between bursts. The corresponding time delay is proportional to the number of stations sharing the same channel. This is because each station must wait its turn to use the shared channel. ALOHA, on the other hand, allows stations to transmit immediately upon need. Time delay is minimum, except when you consider the effect of collisions and the resulting retransmission times.

The standard digital modulation used is *phase shift keying* (PSK), with the most popular form being *quadra-phase PSK* (abbreviated QPSK). The advantage of QPSK is that it doubles the number of bits per second that are carried within a given amount of bandwidth. QPSK modems are now integrated into cellular phones, Inmarsat terminals, and VSATs; the receive portion is also a standard feature of every DSS receiver. Modulator and demodulator design are important to link operation and performance. In FDMA, the modem operates more or less continuously and threshold performance can be optimized. Modems for TDMA must operate in the burst mode, meaning that the demodulator must acquire and re-acquire the signal rapidly to capture data from different Earth stations operating on the same frequency.

The most important parameter in digital transmission is the *bit error rate* (BER). Common requirements in satellite communications are digitized telephone (voice) at 10^{-4}, medium-speed data at 10^{-7}, and digital video at 10^{-6}. The BER can be reduced (e.g., going from 10^{-4} to 10^{-7}) by using *forward error correction* (FEC) within the modem, a feature that takes advantage of custom VLSI chips. These codes automatically correct errors in the received data, yielding a significant improvement in the error rate that is delivered to the end user. The tradeoff with FEC is an increase in data rate (to include the extra FEC bits) in exchange for a decrease in the error rate by at least three orders of magnitude. The precise improvement depends on the percentage of extra bits and the FEC coding and decoding algorithms. Performance has been further improved by performing two encodings of the data through a process called concatenation (see the discussion of the DVB standard in Chapter 5).

Data communication systems use protocols to control transfer of information. These are arranged in a hierarchy of layers that define how communicating nodes and end computing devices control data transfer and verify that no data are corrupted [4]. The standard layers, according to the Open Systems Interconnection model, are: (1) the physical layer, (2) the link layer, (3) the network layer, (4) the transport layer, (5) the session layer, (6) the presentation layer, and (7) the application layer. Each layer provides services, defined by the protocol, to the layer immediately below it. For satellite applications, the physical layer refers to the actual Earth station equipment that multiplexes and modulates/demodulates the information, while the link layer defines the protocol structure of every block of data and how requests for retransmission are processed. Detection of errors at the link and network layers is afforded by parity check bits and *cyclic redundancy check* (CRC) computations. For data, error is effectively eliminated by the end-to-end devices that employ automatic retransmission (character or block oriented). Block-oriented protocols like the *high-level datalink control* (HDLC), which use the "look back N" scheme, are the most effective for satellite links, which have significant propagation delay. Fortunately, these tolerant link layer protocols have become standard in all data communication applications. The transport layer like TCP of the Internet protocol provides a connection between the end user devices. Higher layers are outside of the network and are associated with the software applications that require communication services.

TDMA is a good fit for all forms of digital communications and should be considered as one option during the design of a satellite application. The complexity of maintaining synchronization and control has been overcome through miniaturization of the electronics and by way of improvements in network management systems. With the rapid introduction of TDMA in terrestrial radio networks like the GSM standard, we will see greater economies of scale and corresponding price reductions in satellite TDMA equipment.

2.3.3 Code Division Multiple Access

CDMA, also called spread spectrum communication, differs from FDMA and TDMA because it allows users to literally transmit on top of each other. This feature is allowing CDMA to gain importance in commercial satellite communication. It was originally developed for use in military satellite communication where its inherent antijam and security features are highly desirable. More recently, CDMA has been advocated as an advanced communication technology that increases capacity and reduces equipment costs relative to established FDMA and TDMA systems. Some of these claims are founded; however, it has not been possible to prove that CDMA is inherently superior as this depends on the specific requirements.

A simplified block diagram of a basic spread spectrum link is provided in Figure 2.8 The basic principle of operation is that an input data stream of R_b bps at A is mixed with a pseudorandom scrambling sequence at B with a rate n times R_b. The value of n is generally in the range of 100 to 1000, which has the effect of multiplying the bandwidth of the output at C by the same factor. A modulator converts the baseband version of the signal to an RF carrier at D using standard PSK or QPSK. The output is the spread spectrum signal that can be subjected to link noise and interference, indicated at E. Often, the bandwidth of an interfering carrier is much less than that of the spread spectrum signal.

At F, the combination of spread spectrum signal, noise, and interference is applied to a standard PSK demodulator to recover the baseband spread spectrum signal. A mixer is used to multiply the baseband by the same pseudorandom sequence that was used in the transmitter. However, the difference here is that the signal baseband at G has been modulated by the original data. This difference is what allows the receiver to recover the original data at I. For this to occur, the two pseudorandom sequences must be perfectly aligned in time (a process called correlation). Once synchronized, the bit timing circuitry can recover the original pulses and square shaping is restored. The output data stream at J contains the original data plus occasional errors (inverted bits) that result from noise and interference that remains after the despreading process. However, there is a significant reduction in the effect of interference-induced errors because the despreading process does just the opposite to any interference carrier; namely, it spreads it out over a bandwidth related to n times R_b.

As an introduction to nontechnical readers, consider the following summary of the features of spread spectrum technology.

- *Simplified multiple access:* no requirement for coordination among users.
- *Selective addressing capability:* chip code provides authentication.
- *CDMA:* stations are separated by their unique code sequence.
- *Low-power spectral density:* bandwidth is spread by the code over a bandwidth, which is n times that of the original data; this reduces the power spectral density in inverse proportion.
- *Ability to secure from eavesdroppers:* direct sequence modulation by the pseudorandom code makes detection difficult.

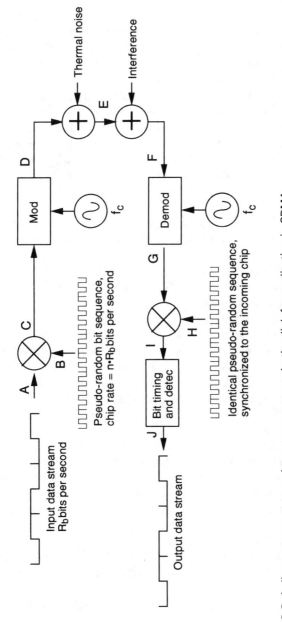

Figure 2.8 A direct-sequence spread spectrum communication link for application in CDMA.

- *Interference rejection:* the spread-spectrum receiver suppresses narrow-band interference.

Selective addressing means that each transmission is automatically identi-fied as to its source by the specific code that spreads the signal in the first place. Further addressability comes from a unique word that is attached to a data header, identifying the source or recipient. Through this and the remaining features, spread spectrum permits many stations to operate on the same fre-quency channel because only one of the signals will be detected by any given receiver. However, the undesired spread spectrum signals will appear as noise that will still degrade the total *C/N* performance. The consequence of this particular feature is that it is extremely difficult to accurately determine the maximum number of signals to simultaneously share the same channel. This makes it difficult to quantify one of the most touted benefits of CDMA, namely, high multiple-access capacity.

In reality, CDMA has special features that make it advantageous under certain conditions. For example, in a multibeam satellite system, CDMA permits the same frequencies to be used in adjacent beams. Unlike FDMA or TDMA, however, the CDMA signals from these beams will add to the total noise budget. If these conditions of appropriateness are not satisfied, then CDMA may not excel in capability and, in fact, can bring with it a penalty relative to FDMA or TDMA. The difficult part is determining the real situation well ahead of an expensive implementation project. This is because the loading of CDMA transmissions on top of each other may follow an unpredictable pattern. Individ-ual carriers must be controlled in power since the carriers by their nature look to the system like noise. A given carrier that is, say, 6 dB above the proper level will introduce four times the expected interference (the original plus the effect of three more). On average, the network could experience an excessive loading of self-interference.

The detailed design of the spread-spectrum receiver is critical to proper CDMA operation because it incorporates all of the features of a good PSK modem plus the ability to acquire the spread signal. A significant part of the challenge is that the signal power density, measured in watts per Hertz, is already less than the noise power density in the receiver. Stated another way, the RF *C/N* is actually less than one (i.e., negative in terms of decibels) because the informa-tion is below the noise level.

A typical CDMA receiver must carry out the following functions in order to acquire the signal, maintain synchronization, and reliably recover the data.

- Synchronization with the incoming code through the technique of correla-tion detection;
- Despreading of the carrier;

- Tracking the spreading signal to maintain synchronization;
- Demodulation of the basic data stream;
- Timing and bit detection;
- Forward error correction to reduce the effective error rate;
- Cyclic redundancy checking of data blocks.

The first three functions are needed to extract the signal from the clutter of noise and other signals. The processes of demodulation, bit timing and detection, and FEC are standard for a digital receiver, regardless of multiple-access method. The cyclic redundancy check provides an additional (higher) level of assurance that the data is good.

The bottom line in multiple access is that there is no single system that provides a universal answer. FDMA, TDMA, and CDMA will each continue to have a place in building the applications of the future. They can all be applied to digital communications and satellite links. When a specific application is contemplated, our recommendation is to perform the comparison to make the most intelligent selection.

2.4 FREQUENCY BAND TRADEOFFS—L, S, C, X, KU, AND KA

Satellite communication is a form of radio or wireless communication and therefore must compete with other existing and potential uses of the radio spectrum. During the initial ten years of development of these applications, there appeared to be more or less ample bandwidth, limited only by what was physically or economically justified. In the coming years, the allocation of spectrum has become a domestic and international battlefield as service providers fight among themselves, joined by their respective governments when the battle extends across borders. So, we must consider all of the factors when selecting a band for a particular application.

The most attractive portion of the radio spectrum for satellite communication lies between 1 GHz and 30 GHz. The relationship of frequency, bandwidth, and application are shown in Figure 2.9. The scale along the x-axis is logarithmic in order to show all of the satellite bands; however, observe that the bandwidth available for applications increases in real terms as one moves toward the right (i.e., frequencies above 3 GHz). Also, the precise amount of spectrum that is available for services in a given region or country is usually less than Figure 2.9 indicates. Please refer to the latest edition of the ITU Radio Regulations or relevant domestic allocations in the country of interest. Chapter 10 provides a review of the regulatory process.

The use of letters probably dates back to World War II as a form of shorthand and simple code for developers of early microwave hardware. Today, the

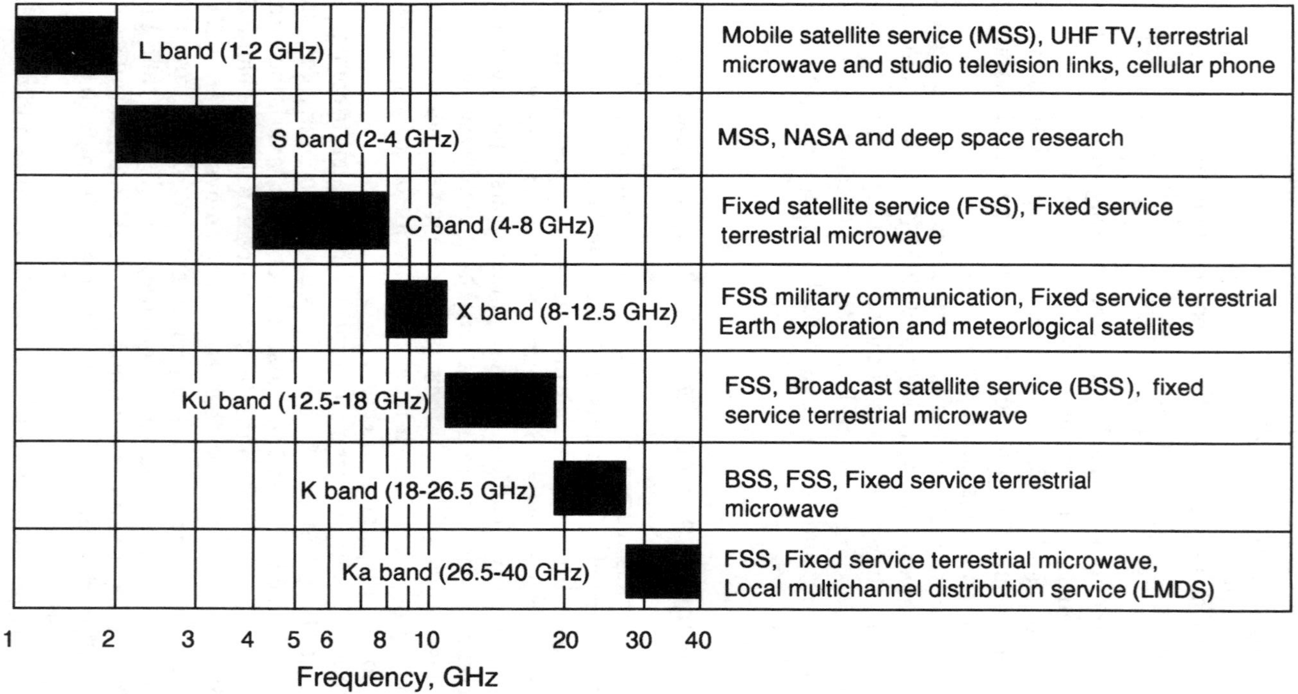

Figure 2.9 The general arrangement of the frequency spectrum that is applied to satellite communications and other radiocommunications services. Indicated are the short-hand letter designations along with an explanation of the typical applications. Note that the frequency ranges indicated are the general ranges and do not correspond exactly to the ITU frequency allocations and allotments.

letter designations continue to be the popular "buzzwords" that identify band segments that have commercial application in satellite communications. The international regulatory process, maintained by the ITU, does not consider these letters but rather uses band allocations and service descriptors.

The general properties of these bands are reviewed in [3]. Suffice it to say, the lower the band in frequency, the better the propagation characteristics. This is countered by the second general principle, which is that the higher the band, the more bandwidth that is available. The *mobile satellite service* (MSS) is allocated to the L and S bands, where propagation is most forgiving. Yet, the bandwidth available between 1 GHz and 2.5 GHz, where MSS applications are authorized, must be shared not only among MSS applications, but with all kinds of radio broadcast and point-to-point services as well. The competition is keen for this spectrum due to its excellent space and terrestrial propagation characteristics. The rapid rollout of wireless services like cellular radiotelephone and PCS will probably conflict with advancing GEO and non-GEO MSS systems.

On the other hand, wideband services like DTH and Broadband ISDN can be accommodated at frequencies above 3 GHz, where there is more than ten times the bandwidth available. Add to this the benefit of using directional ground antennas that effectively multiply the unusable number of orbit positions. Some wideband services have begun their migration from the well-established world of C band to Ku band and Ka band. In the following paragraphs we provide some additional comments about the relative merits of these bands. These should be considered as starting points for evaluating the proper frequency band and are not substitutes for a detailed evaluation of the relative cost and complexity of different approaches. Higher satellite EIRP used at Ku band allows the use of relatively small Earth station antennas. On the other hand, C band should maintain its strength for video distribution to cable systems and TV stations, particularly because of the extensive investment in uplink and downlink antennas and electronic equipment.

2.4.1 L Band

Frequencies between 1 GHz and 2 GHz are usually referred to as L band, a segment not applied to commercial satellite communication until the late 1970s. Within this 1 GHz of total spectrum, only about 30 MHz of uplink and downlink, each, was initially allocated to satellite mobile applications by the ITU. First to apply L band was COMSAT with their Marisat satellites. Constructed primarily to solve a vital need for UHF communications by the U.S. Navy, Marisat also carried an L-band transponder for early adoption by the commercial maritime industry. COMSAT took a gamble that MSS would be accepted by commercial vessels, which at that time relied on high-frequency radio and the Morse Code.

Over the ensuing years, Marisat and its successors from Inmarsat proved that satellite communications, in general, and MSS, in particular, are reliable and effective. By 1993, the last commercial HF station was closed down in favor of satellite links.

As is familiar to readers, early MSS Earth stations required dish antennas that had to be pointed toward the satellite. The equipment also was quite large, complex, and expensive. Real demand for this spectrum began to appear as portable, land-based terminals were developed and supported by the network. Moving from rack-mounted to suitcase-sized to attaché case and finally hand-held terminals, the MSS has grown remarkably.

The most convenient L-band ground antennas are small and do not require pointing toward the satellite. We are all familiar with the very simple cellular whip antennas used on cars and handheld mobile phones. L-band antennas are not quite so simple because of the requirement to provide some antenna gain in the direction of the satellite. Additional complexity results from a dependence on circular polarization to allow the mobile antenna to be aligned along any axis (and to allow for the effect of Faraday rotation). First-generation L-band rod or mast antennas are approximately 1m in length and 2 cm in diameter. This is to accommodate the long wire coil (a bifilar helix) that is contained within. The antenna for the handheld phone is more like a fat fountain pen. We anticipate improvements in L-band antennas in coming years as more effort is applied to reach an expanding MSS market.

While there is effectively no rain attenuation at L band, the ionosphere does introduce a source of significant link degradation. This is in the form of rapid fading called ionospheric scintillation, which is the result of the RF signal being split into two parts—the direct path and a refracted (or bent) path. At the receiving station, the two signals combine with random phase. Then, the signals may cancel, producing a deep fade. Ionospheric scintillation is most pronounced in equatorial regions and around the equinoxes (March and September). Also, the ionosphere will rotate linear polarization, which is another reason why circular polarization is preferred at L band. Both scintillation and Faraday rotation decrease in importance as frequency increases and are nearly negligible at Ku band and higher.

From an overall standpoint, L band represents a regulatory challenge but not a technical one. There are more users and uses for this spectrum than there is spectrum to use. Over time, technology will improve spectrum efficiency. Techniques like digital speech compression and CDMA can improve the utilization of this very attractive piece of spectrum. Additional spectrum will be opened up between the period 1996 to 2005 as terrestrial users are moved out of the spectrum that was reallocated to MSS by WARC-92. However, there will be situations in many countries where the band is supposed to be available to MSS but remains restricted if and until incumbents move out or are shut down.

2.4.2 S Band

S band was adopted early on for space communications by NASA and other governmental space research activities around the world. It has an inherently low background noise level and suffers less from ionospheric effects than L band. DTH systems at S band were operated in past years as experiments by NASA and as an operational service by the Indian Space Research Organization. More recently, the ITU allocated a segment of S band for MSS and future land mobile applications. This holds the greatest prospect for expanded commercial use on a global basis.

As a higher frequency band than L band, it will suffer from somewhat greater atmospheric loss and less ability to adapt to local terrain. LEO and MEO satellites are probably a good match to S band since the path loss is inherently less than for GEO satellites. One can always compensate with greater power on the satellite, a technique used very effectively at Ku band.

2.4.3 C Band

Once viewed as obsolete, C band remains the most heavily developed and used piece of the satellite spectrum. During recent World Radiocommunication Conferences, discussed in Chapter 10, the ITU increased the available uplink and downlink bandwidth from the original allocation of 500 MHz to almost 800 MHz. This spectrum is effectively multiplied by a factor of two with dual polarization and again by 180, assuming 2-deg spacing between satellites. Further reuse by a factor of between two and five takes advantage of the geographic separation of land coverage areas. The total usable bandwidth is therefore in the range of 568 GHz to 1.44 THz, which compares favorably with land-based fiber optic systems. The added benefit of this bandwidth is that it can be delivered across an entire country at once.

Even though this represents a lot of capacity, there are situations in certain regions where additional satellites are not easily accommodated. In North America, there are more than 35 C-band satellites in operation across a 70-deg orbital arc (Figure 1.1). This is the environment that led the U.S. *Federal Communications Commission* (FCC) to adopt the radical policy of 2-deg spacing. The GEO orbit segments in Western Europe and East Asia are becoming just as crowded as more countries apply and consider applying satellite technology. European governments had the foresight to mandate the use of Ku band for domestic satellite communications, delaying somewhat the day of reckoning. Asian countries favor C band because of reduced rain attenuation as compared to Ku and Ka bands, making this a vital issue in that region.

C band is a good compromise between radio propagation characteristics and available bandwidth. Service characteristics are excellent because of the modest amount of fading from rain and ionospheric scintillation. The one drawback is the somewhat large size of Earth station antenna that must be employed. The 2-deg spacing environment demands antenna diameters greater than 1m, and in fact 3m is more the norm. This size is also driven by the relatively low power of the satellite, itself the result of sharing with terrestrial microwave. High-power video carriers must generally be uplinked through antennas of between 7m and 13m; this assures an adequate signal and minimizes radiation into adjacent satellites and terrestrial receivers.

The prospects for continued use of C band are good and could improve again as digital compression is adopted for video transmission and terrestrial microwave radios are shut down. Video applications like DTH and video teleconferencing can make good use of Ku and Ka bands, where interactive VSAT networks are feasible. The resulting space at C band might even be exploited for mobile and personal communications applications.

2.4.4 X Band

Government and military users of satellite communication established their fixed applications at X band. This is more by practice than international rule, as the ITU frequency allocations only indicate that the 8-GHz portion of the spectrum is designated for the FSS. From a practical standpoint, X band can provide service quality on par with C band; however, commercial users will find equipment costs to be substantially higher due to the thinner market. Also, military-type Earth stations are inherently expensive due to need for rugged design and secure operation. We find that some countries have filed for X band as an expansion band, hoping to exploit it for commercial applications like VSAT networks and DTH services. However, at the time of this writing, military usage still dominates. X Band is likewise shared with terrestrial microwave systems, somewhat complicating frequency coordination.

2.4.5 Ku Band

Ku-band spectrum allocations are somewhat more plentiful than C band, comprising 750 MHz for FSS and another 800 MHz for the BSS. Again, we can use dual polarization and satellites positions two deg apart. Closer spacings are not feasible because users prefer to install yet smaller antennas, which have the same or wider beamwidth than the corresponding antennas for C-band service. Geographic separation is possible and in fact could produce multiplication by a factor of up to 10. The maximum available spectrum could therefore amount to over 4 THz.

Exploiting the lack of frequency sharing and the application of higher power in space, digital DTH services like DIRECTV, USSB, and Primestar in

North America usher in the age of low-cost and user-friendly home satellite TV. The United Kingdom, continental Western Europe, and Japan have enjoyed the earlier version of DTH for almost ten years at the time of this writing. As a result of these developments, Ku band has become a household fixture (if not a household word).

The more permissive regulations at Ku band also favor its use for two-way interactive services like voice and data communication. Low-cost VSAT networks typify this exploitation of the band and the regulations. Being above C band, the Ku-band VSATs and DTH receivers must anticipate more rain attenuation. A decrease in capacity can be countered by increasing satellite EIRP. Thin route applications for telephony and data, discussed in Chapters 7 and 8, benefit from the lack of terrestrial microwave radios, allowing VSATs to be placed in urban and suburban sites.

2.4.6 Ka Band

Ka-band spectrum is relatively abundant and therefore attractive for services that cannot find room at the lower frequencies. There is 2 GHz of uplink and downlink spectrum available on a worldwide basis. In addition, the fact that ground antenna beamwidths are between one-half to one-quarter the values that correspond at Ku and C bands means that more satellites could conceivably be accommodated. Conversely, with enough downlink EIRP, smaller antennas will still be compatible with 2-deg spacing. Another facet of Ka band is that small spot beams can be generated on board the satellite with the same antenna aperture, thus improving space division frequency reuse.

The Ka-band region of the spectrum is perhaps the last to be exploited for commercial satellite communications. Research organizations in the United States, Western Europe, and Japan have spent significant sums of money on experimental satellites and network application tests. However, true commercial adoption of this piece of the spectrum is still yet to come. (We exclude from consideration here the use of 18-GHz uplinks to BSS satellites.)

From a technical standpoint, Ka band has many challenges, the biggest being the much greater attenuation for a given amount of rainfall. This can, of course, be overcome by increasing the transmitted power or receiver sensitivity (e.g., antenna diameter) to gain link margin. Some other techniques that could be applied in addition to or in place of these include (1) reducing the data rate during rainfall, (2) transferring the transmission to a lower frequency such as Ku or C band, and (3) using multiple-site diversity to sidestep heavy raincells. Consideration of Ka band for an application will involve finding the most optimum combination of these techniques.

The advent of the information superhighway has encouraged several organizations to consider Ka band as an effective means to reach the individual subscriber. *Ultrasmall aperture terminals* (USATs) capable of providing two-way high-speed data, in the range of 384 Kbps to 2 Mbps, are entirely feasible. Hughes Electronics filed with the FCC in 1993 for a two-satellite system called Spaceway that would support such low-cost terminals. In 1994, they extended this application to include up to an additional 15 satellites to extend the service worldwide. The timetable for Spaceway is toward the latter part of the twentieth century. Almost at the same time, several strong backers introduced another proposal called Teledesic, which would employ the same Ka band from LEO satellites—up to 840 in number. While this sounds rather incredible, strong support from Craig McCaw, cofounder of McCaw Cellular (now part of AT&T), has lent considerable credibility to Teledesic. Now, it would seem that Ka band has a reasonably good prospect of becoming valuable.

References

[1] Morgan, W. L., and G. D. Gordon, *Communications Satellite Handbook*, New York: John Wiley and Sons, 1989.

[2] Fthenakis, E., *Manual of Satellite Communications*, New York: McGraw-Hill Book Company, 1984.

[3] Elbert, B. R., *Introduction to Satellite Communication*, Norwood, MA: Artech House, 1987.

[4] Elbert, B. R., and B. Martyna, *Client/Server Computing—Architecture, Applications, and Distributed Systems Management*, Norwood, MA: Artech House, 1994.

Issues in Satellite System Management

3

With the wide adoption of the technology, satellite communications has brought with it a number of issues that must be addressed before an application can be implemented. Satellite capacity is only available if the right satellites are placed in service and cover the region of interest. Considering the complexity of a satellite and its supporting network, applications can be expensive to install and manage. However, if the issues are addressed correctly, the economic and functional needs of the application will be satisfied.

A viable satellite communications business is built on a solid technical foundation along the lines discussed in the previous two chapters. Such features as the frequency band, capacity and bandwidths, constellation and orbit strategy, and the network topology all have a direct bearing on the attractiveness of the service offerings. The capability of the application is directly influenced by the specific capabilities of the communication payload of the spacecraft. The satellite operator must make the decision whether to launch a satellite with one frequency band or to combine payloads for multiple-frequency operation (called a hybrid satellite). We define and review the options for payload implementation, including the conventional bent pipe as well as evolving concepts like the digital processing repeater. As the satellite payload becomes more unique, the demands on the market and supporting technologies increase. Further to this, the business and operation should consider and properly address all of the issues that this chapter raises.

We review the current state of the art in bus design as it has a bearing on payload power and flexibility. Most satellites today are three-axis designs, but the spinner continues to satisfy mission requirements which are at the lower end of the power range. The power of the combined repeater has the biggest impact on the spacecraft and the user network.

The remainder of this chapter goes into contingency planning from the perspective of the operator and the user. The excellent reliability and flexibility of satellite applications cannot be assured without thorough analysis and proper implementation. For example, a satellite operator must implement a system with multiple satellites so that no single event can terminate vital service to users. Users, on the other hand, must approach satellite communications with an open mind and open eyes. They might arrange for backup transponder capacity for use in the event of some type of failure. Both parties will also need to obtain insurance to reduce financial loss.

The information that follows provides detailed points and approaches for these issues. It provides background on some of the more critical areas that often hamper the introduction and smooth operation of effective systems. Readers should also consider how other potential problems not identified here could adversely impact their services and plan accordingly.

3.1 SATELLITE SYSTEM IMPLEMENTATION

Many of the issues that must be considered by the operators of terrestrial telephone, television, and cellular networks must also be faced by providers of satellite applications. The most basic type of space segment, shown in Figure 3.1, employs one or more GEO satellites and a TT&C ground station. The associated ground segment can contain a large quantity of Earth stations, the specific number and size, depending on the application and business. The ground segment is very diverse because the Earth stations are installed by and operated by a variety of organizations (including, more recently, individuals). Importantly, we have moved out of the era when the space and ground segments are owned and operated by one company.

Due to the size of the investment and the complexity of the work, the satellite operator is usually a tightly organized company with the requisite financial and technical resources. It engages in the business of providing satellite capacity to the user community within the area of coverage. While the majority of commercial satellite operators focus on a single country, there is a new trend toward regional systems that compete directly with the global consortium, INTELSAT. Capacity can be offered on a wholesale basis, which means that complete transponders or major portions thereof (even the entire operating satellite, in some cases) are marketed and sold at a negotiated price. Each deal is different, considering the factors of price (lease or buy), backup provisions, and the term. The retail case comes into play where the satellite serves the public directly, such as in MSS and BSS networks. We consider such business issues in detail in Chapter 11.

To create the space segment, the satellite operator contracts with a spacecraft manufacturer for many of the elements needed for implementation. Historically, most operators took responsibility for putting the satellite into operation, including the purchase and insurance of the launch itself. More recently, contracts have required in-orbit delivery of the satellite, which reduces the technical

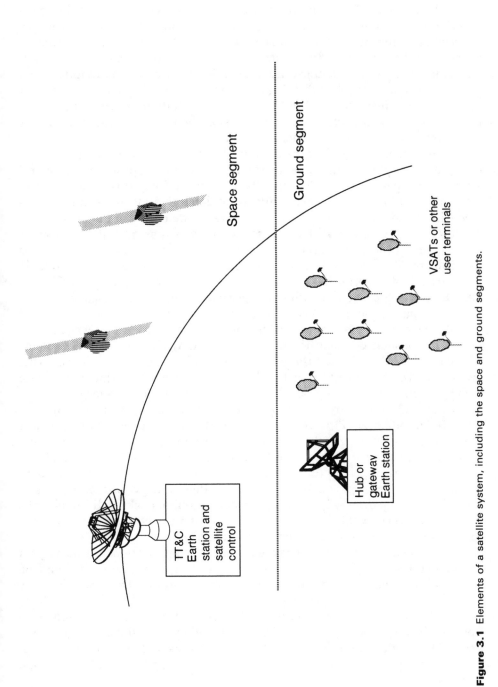

Figure 3.1 Elements of a satellite system, including the space and ground segments.

demand and some of risk on the satellite purchaser. Organizations that have done it this way include the U.S. Navy, PanAmSat, and Optus Communications. However, satellite buyers still need a competent staff to monitor the construction of the satellites and ground facilities. This can be accomplished with consultants, the quality of which depends more on the experience of individuals than on the cost or size of the consulting organization. The experienced spacecraft consultants include COMSAT and Telesat Canada. Individuals, such as retirees from spacecraft manufacturers, can provide excellent assistance at much lower cost. However, they can be difficult to find.

The capacity demands of cable TV and DTH systems are pushing us toward operating multiple satellites in and around the same orbit position. Successful satellite TV operators like SES and HCI have been doing this for some time, developing and improving the required orbit determination and control strategies. This considers the need for more accurately determining the range of the satellite, since we are talking about separating satellites by tenths of degrees instead of multiple degrees. A few of the smaller operators of domestic satellites like Telenor, Thaicom, and NHK have found that it is more effective to double the capacity of an orbit slot by operating two smaller satellites rather than launching a single satellite with the larger combined payload capacity.

Implementation of the Earth station network can follow a wide variety of paths. One approach is to purchase the network as a turn-key package from a manufacturer such as Scientific Atlanta (Atlanta, GA), Hughes Network Systems (Germantown, MD), Alcatel (Paris, France), or NEC (Yokohama, Japan). This gives good assurance that the network will work as a whole since a common technical architecture will probably be followed. There are systems integration specialists in the field, including Satellite Transmission Systems (Hauppauge, NY), ETS (Melbourne, FL), and IDB Systems (Dallas, TX), who purchase most of the elements from a variety of manufacturers and perform all of the installation and integration work, again on a package basis. The application developer may take on a significant portion of implementation responsibility, depending on its technical strengths and resources. Another strategy for the buyer is to form a strategic partnership with one or more suppliers, who collectively take on technical responsibility as well as some of the financial risk in exchange for a share of revenue or a guarantee of future sales.

The operations and maintenance phase of the application falls heavily on the service provider and in many cases the user as well. The service may be delivered and managed through a large hub or gateway Earth station. This facility may have to be supported by competent technical staff on a 24-hour per day, seven-day per week basis (called 24×7). Such a facility might be operated by the integrator or supplier and shared by several users or groups of users. This is a common practice in VSAT networks and cable TV uplinking. Inexpensive user terminals, whether receive-only or transmit and receive, are designed for unattended operation and would be controlled from the hub. The systems integrator can operate portions or all of the network, including maintenance and repair of equipment. A properly written contract with a competent supplier can result in savings and additional functional advantages for the buyer, such as backup services and protection from technical obsolescence.

The satellite communications industry keeps evolving as satellite operators discover how to enter the businesses of their users and as users experiment with becoming satellite operators. In the case of the former, Hughes Communications created the DIRECTV service that is expected to produce much more revenue than would be possible through the wholesale lease of the required Ku-band satellite capacity. On the other hand, PanAmSat was started up by the management of the Spanish International Network, which was the leading Spanish language network in the United States. As these companies have discovered, their counterpart's business is quite different in the nature of the respective investment.

A basic issue on the space segment side is the degree to which the satellite design should be tailored to the application. Historically, C- and Ku-band satellites in the FSS are designed for maximum flexibility so that a variety of customers needs can be met. A typical FSS transponder may support any one of the following: an analog TV channel, four to eight digital TV channels, a single 60-Mbps data signal such as would come from a wideband TDMA network, or an interactive data network of 2,000 Ku-band VSATs. The satellite operator may have little direct involvement in these applications. Alternatively, they may invest in these facilities to provide value-added services.

The alternative is to design the satellite repeater to meet the requirements for a specific type of signal, tailored for one application. The latest generation of BSS satellites provide high levels of RF power to deliver the signal to very small antennas. These satellites are very effective for broadcasting but would be less suitable for two-way communications from VSATs. With regard to the MSS systems being operated, the antennas and transponder electronics are specifically designed to receive the very low power signals coming from mobile telephones and ultimately handheld phones similar to cellular phones. The downlink transmit power, on the other hand, is much higher per voice channel than is typical of an FSS design. With all of this, the satellite must provide a high degree of frequency reuse since the bandwidth available at L band is only of the order of 30 MHz. As a consequence of specialization, systems like DIRECTV and AMSC are inflexible when it comes to adapting to substantially different applications. On the other hand, these tailored satellites deliver precisely the service that is intended and do it efficiently.

The trend is clearly toward more highly tailored satellites, which will accelerate as the transponder itself becomes digital in nature. Digital processing gives many benefits, such as much greater flexibility in channel routing and antenna beam control. The tradeoff is that once the repeater is made of silicon, the signal format may remain nearly constant throughout the life of the satellite. Looking ahead a decade or more, satellite technology will evolve so that the flexibility of today's fixed bandwidth analog transponder might be available in the digital repeater mode.

The ground segment is also undergoing an evolutionary process, where the first designs were versatile, multipurpose (voice, TV, and data), and expensive. Over time, the Earth station has become more specific to the purpose and lower in price. Digital implementation of analog functions like the modem now permit very sophisticated small Earth stations that can sell for about the price of a good camcorder.

The choice of level of integration of the ground and space segment activities is a strategic decision of the satellite operator and application developer. There are some operators who have launched spacecraft with no specific thought to how their space segment services will function with the ground segment of prospective users. In contrast, other operators have simulated their hypothetical communication network performance at every step from when the concept is defined to factory tests to in-orbit. These results have been fed back into the design of critical elements of the satellite and ground user equipment. If the integration between these is very tight, the such efforts are mandatory. The choice somewhat depends on how unique the spacecraft payload and frequency band are in relation to alternative systems and services.

We begin the discussion of satellite system issues with a review of some of the most important tradeoffs in satellite design. The two halves of the satellite—the communications payload and the spacecraft bus—are reviewed separately. The breakdown of the main elements in the payload and bus are as follow.

- Communications subsystem:
  - Receivers,
  - Multiplexers,
  - High-power amplifiers,
  - Switching,
  - Digital processing;
- Antenna subsystem:
  - Reflector,
  - Feed,
  - Feed network;
- Spacecraft bus:
  - Tracking, telemetry, and command,
  - Solar panel,
  - Batteries,

- Reaction control and propulsion,
- Spececraft control processing,
- Attitude control,
- Thermal control,
- Structure.

3.2 COMMUNICATIONS PAYLOAD CONFIGURATIONS

Communications payloads are increasing in capability and power, getting more complex as time passes. Examples include the newer DTH missions, which are designed to maintain maximum EIRP and digital throughput in the downlink. Likewise, state-of-the-art MSS satellites have large, deployable antennas to allow handheld phones to operate directly over the satellite path. Onboard digital processors are likewise included as a means to improve the capacity and flexibility of the network.

Some traditional issues are still with us. One of the most basic is the selection of the frequency band, which was addressed in Chapter 2. Suppose that the operator wants to be able to address the widest range of applications and has consequently decided to implement both C- and Ku-band satellites. With today's range of spacecraft designs, one can launch independent C- and Ku-band satellites. This was done in the first generation of the Ku-band SBS and C-band Galaxy systems in the United States. Each satellite design can be optimized for its particular service. The alternative is to build a much larger spacecraft that can carry both payloads at the same time. The first such hybrids were purchased by INTELSAT (Intelsat V), Telesat Canada (Anik B), and Southern Pacific Railway (Spacenet).

A third overall design issue deals with the coverage footprint, which has a direct bearing on the design of the spacecraft antenna system. The two basic alternatives are to either create a single footprint that covers the selected service area or to divide the coverage up into regions that each get their own spot beam. These two approaches have significant differences in terms of capacity, operational flexibility, and technical complexity.

3.2.1 Bandwidth/Power Levels

As satellite applications are targeted more toward end users, the demand increases for smaller ground antennas and, as a consequence, higher satellite power. Satellite operators tend to seek a marketing advantage by having greater EIRP in the newest generation of spacecraft. A key parameter for spacecraft design in this environment is the efficiency of conversion from DC (supplied by the solar panels and batteries) to RF (power amplifier output).

An overall comparison of the two basic types of power amplifiers is provided in Table 3.1. Traveling-wave tubes tend to have the highest efficiency and have been rediscovered for broadcasting and digital information distribution. TWTs above 200W have challenged developers because of a lack of adequate on-orbit experience. In comparison, 100W to 150W amplifiers are viewed as being dependable, and experience with the newer generation launched in the early 1990s has been very good. Higher power levels are obtained by paralleling pairs of amplifiers. As we approach the year 2000, TWT power will move to the 200W level in response to application requirements.

Solid-state power amplifiers have become popular for power up to about 20W. SSPAs promise longer life because they do not contain a clear-cut wearout mechanism. On the other hand, TWTs demonstrate a 20-year lifetime through extensive on-orbit experience, even though they have a well-known wearout mechanism in the cathode. High-power GaAs FET devices are delicate and must be maintained at a relatively cool temperature over life. SSPAs operate at low voltage and high current and can fail randomly due to design or manufacturing defects (particularly where leads are bonded to substrates). A given SSPA uses many FETs arranged in parallel combinations, requiring that all FETs be functioning in order to provide the rated power output.

TWTAs have evolved to stay up with SSPAs because they function as a kind of linear electric motor that can be inherently very efficient if the energy of the moving element (the electron beam) can be conserved by recycling (via multiple collectors). State-of-the-art for space TWTs is now four collectors in a direct-radiating case, giving DC to RF efficiencies over 60% at Ku band. The

Table 3.1
Comparison of Spacecraft Power Amplifiers Using Typical Values, Without Linearization

Characteristic	TWTA	SSPA
Frequency bands	L through Ka	L through Ku
Power output	10W to 200W	5W to 30W
Bandwidth	20%	5%
Gain	40 dB to 60 dB	6 dB to 20 dB (input driver amplifier can be added)
Efficiency	40 to 60%	30 to 40%
Weight	1.5 kg to 5.5 kg	1 kg to 2 kg
Linearity		
C/IM at maximum power	12 dB	16 dB
AM to PM conversion	5 deg/dB	2.5 deg/dB
Maximum phaseshift	40 deg	20 deg
Operating temperature	60°C	30°C

high-voltage power supplies of TWTs continue to be a source of concern due to their complexity and potential for high-voltage failure. SSPAs are tuned to specific frequency channels to maximize power output and efficiency, resulting in a bandwidth that is substantially less than what can routinely be obtained from TWTs. These differences complicate the choice of the type of amplifier, which cannot be made until the full requirements of the mission are understood and documented.

Table 3.1 also contains a number of secondary parameters that may or may not be significant in a particular application. Power level, bandwidth, gain, and efficiency have direct consequences for the service and spacecraft design. The weight of the amplifier, including its power supply, is important to the size and cost of the satellite itself. In some missions, the quantity of amplifiers may indeed be limited by weight. Another consideration is the weight of the thermal control elements needed to remove heat and thereby control amplifier temperature (particularly important for SSPAs and high-power TWTAs). SSPAs and low-power TWTAs may be conduction cooled through their mounting surfaces. Heat pipe technology improves performance by conducting heat to external radiating surfaces. As the power of the tube increases, direct radiation becomes an option. When exposed to cold space, the collector end of the high-power TWT will emit a substantial fraction of the internally generated heat.

Linearity specifications are very important in high-speed digital transmission and multiple-carrier applications for VSATs and mobile communications. This would favor the SSPA, but it is possible to incorporate linearization as part of a TWTA transponder design. A linearizer places a complimentary nonlinear transfer characteristic in front of the amplifier and thereby compensates for part of the distortion.

With all of the effort that goes into designing and building a powerful and efficient amplifier, there exists the vital importance of minimizing the RF loss between antenna and the amplifier. As indicated in Table 2.1, an output loss of 2 dB results in a diversion of 58% of the amplifier power into resistive heating. This has a direct bearing on the satellite as a whole, because an efficiency improvement means that less total DC power needs to be provided by the bus. A lot more attention has been focused in recent years on minimizing this loss and maximizing the radiated power. Techniques like shaped reflectors with single-feed horns were introduced with this in mind.

Transponder bandwidth is next in importance to users. For a given application, the signal bandwidth determines the minimum transponder bandwidth; anything more than that usually cannot be used effectively. Analog video applications generally require bandwidths in the range of 23 MHz to 36 MHz. A VSAT network or other SCPC application could require less bandwidth and hence can share a transponder with other users.

The most common approach in spacecraft design is first to determine the minimum acceptable bandwidth for any service that can be anticipated. Then, the designer calculates the maximum number of transponders that can physically be carried by the spacecraft and yet meet the minimum requirement for EIRP per channel. Working backward, the designer divides the available bandwidth (typically 500 MHz) by the number of transponders that can be carried (usually in the range of 4 to 16). This yields the maximum bandwidth per transponder.

A more detailed kind of optimization has been performed in recent years using advanced computer analysis and simulation techniques. Using an engineering workstation and sophisticated signal analysis software like the *Signal Processing Workbench* (SPW), the satellite engineer can determine which transponder parameters have a significant impact on the overall link and spacecraft design. There are two types of analyses, namely, the analog/digital approach, which looks at the signal as it passes through the linear and nonlinear elements of the repeater; and the discrete time model, which generates simulated traffic (telephone calls or data messages, as appropriate). Either or both might be used in a given system. The critical step in using computer analysis is first to calibrate the model on a real-world system such as an operating satellite link or a laboratory table-top model. Once calibrated, the computer model is useful for testing various equipment arrangements and technical specifications. The approach is very powerful because you can theoretically include everything that can possibly impact performance and thereby adjust relevant characteristics. During the construction of the system or even after it is in operation, the application engineer can reanalyze the links to pull out more capability or troubleshoot problems that arise over time.

3.2.2 Single-Frequency Payload

The single-frequency payload represents the most focused approach to satellite design. The concept was first introduced by Early Bird (Intelsat I) in 1965 and advanced in the 24-transponder Satcom spacecraft built by RCA Astrospace (now a division of Lockheed Martin). Up to that point, Hughes Aircraft Company held the record with 12 C-band transponders on a spacecraft of about half the size and weight. Similar satellites were introduced by Ford Aerospace (now Space Systems/Loral), Aerospatiale of France, and Matra-Marconi Space of France and the United Kingdom.

In the 1990s, single-frequency payloads have become targeted toward specific applications in TV and mobile communications. The TV marketplace is dominated by cable TV and DTH, where the quantity of available TV channels at the same orbit position becomes important. The majority of the cable TV satellites for the United States, including the Galaxy and GE series, are single-frequency designs, optimized to the requirements of the cable TV networks. This considers all of the technical, operational, and financial factors in providing service. From a technical prospective, the transponder gain and power is made to match the Earth stations used to uplink and receive the signals. What is more important to this class of customer is that the capacity must be there when needed. With about 65% of U.S. families depending on satellite-delivered programming, the cable TV networks put a very high value on the reliability of getting the capacity to orbit and operating once it gets there.

The satellite operators that address cable markets therefore need excellent plans for launch and on-orbit backup (discussed further at the end of this chapter). Experience has shown that the best way to do this is to construct a series of identical satellites and launch them according to a well-orchestrated plan. This must consider how the capacity is sold to the cable programmers as well as the strategy for replacing existing satellites that reach end of life. A satellite operator commits a serious mistake by not having a well-thought out plan for satellite replacement. The single-frequency satellite can fit well into such a plan. Matters are more complicated when an operator wishes to replace individual C- and Ku-band satellites with a dual-frequency hybrid satellite. The benefit of doing this is reduced investment cost per transponder and simpler operation, but the common timing and orbit slot required might be incompatible with the satellite pair reaching the end of life.

A second area of the TV market where single-frequency satellites are preferred is in DTH. Spacecraft for Astra, DIRECTV, and PrimeStar are single-frequency designs that are tailored to the specific requirements of the respective DTH networks. This considers the quantity of transponders, the size of the receiving antenna (and therefore the satellite EIRP), the signal format (which determines the channel capacity and quality) and the coverage area. Taken together, these factors have enormous leverage on the economics and attractiveness of the service, second only to the programming. The delivery of the signal to millions of small dishes demands the highest EIRP that is feasible with a given state of the art. DBS satellites tend to push the limits on power, as opposed to weight. This leaves little left over for C- or L-band repeater elements, which if included could force a compromise of some type. (An exception, for smaller markets, is considered in the next section.)

The cost of a high-power DTH satellite is often more than that of, say, a C-band satellite that serves cable TV. This should not be a concern because of the much larger quantity of receiving antennas. In fact, in [1] we demonstrate that to achieve an optimum G/T consistent with today's class of DTH receiver, the satellite EIRP must be approximately 50 dBW. By optimum, we mean that the total cost of the satellite and all of the receiving Earth stations is at a minimum. Lower EIRP means that satellite dishes must be larger, raising the cost of the ground segment faster than the saving in the satellite. Going in the direction of increasing satellite EIRP is likewise unattractive since the increase in satellite cost outweighs the savings in receive dishes. Figure 6.7, in fact, demonstrates that the high investment cost of the satellite is ultimately very economical on a cost-per-user basis.

3.2.3 Multiple-Frequency Hybrid Payloads

Hybrid satellites were first introduced by INTELSAT at C and Ku bands with the launch of Intelsat V. A third L-band payload was added to Intelsat V-A for use by Inmarsat. The first domestic hybrid, Anik B, was operated by Telesat Canada in the late 1970s; and two American companies—Sprint Communications and American Satellite Corporation (both since merged into GTE and the satellites subsequently sold to GE Americom)—also were early adopters. The idea behind the use of the hybrid was to address both the C- and Ku-band marketplaces at a reduced cost per transponder. These satellites were pathfinders in the field but largely failed in the marketplace. The reason for this is that neither the C- nor the Ku-band payloads had sufficient EIRP to compete with the single-band satellites already serving customers. The early hybrids had sacrificed performance for low cost. What was needed was a much larger spacecraft bus that could support competitive payloads. The first satellite to do this was Galaxy VII, a C and Ku-band hybrid on the Hughes HS-601 platform.

The exception to that rule that DBS and MSS satellites should be single-frequency designs is where the satellite is to serve a relatively small market such as a medium-sized country like Mexico or Indonesia. The demand for any single-frequency band in this type of market is less than what would support a single-frequency satellite.

3.2.4 Shaped Versus Spot Beam Antennas

The coverage pattern of the satellite determines the addressable market and the flexibility of extending services. For a constant transponder output power, the EIRP varies inversely with the beam area. This directly results from the same inverse relationship between antenna gain and area of coverage. In fact, the basic relationship between these two parameters is

$$G = 27000/\theta^2 \tag{3.1}$$

where G is the gain as a ratio and θ the average diameter of a circular coverage area, measured from GEO in degrees.

This is further illustrated in Figure 3.2 for two differing coverage areas: the country of Colombia and the continent of South America. The Colombian

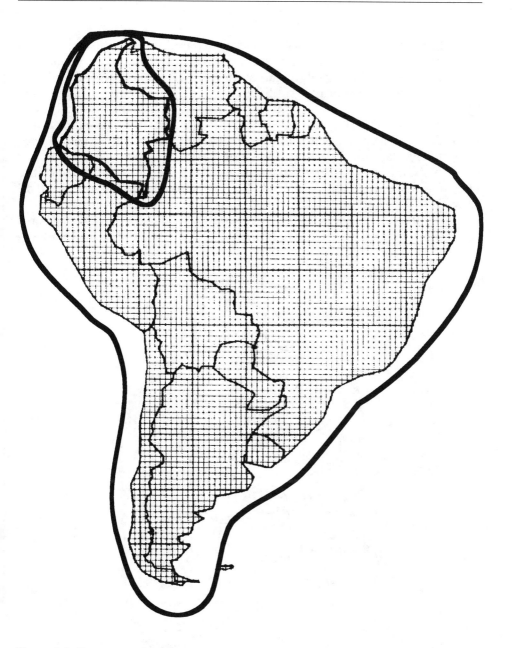

Figure 3.2 Coverage options for a domestic or regional satellite in South America. Satellite is in GEO at 65 deg WL.

market would be served with a national beam that is directed exclusively toward this country, delivering high gain and no direct frequency reuse (other than through cross-polarization). From an orbit position of 65 deg WL, the picture of the land area shows heavy grid lines that are 1 deg apart. The landmass of Colombia covers approximately 2.8 deg^2, while the example of a national coverage contour is slightly larger at 4.6 deg^2. In comparison, the landmass and example antenna coverage of South America are approximately 40 and 52 deg^2, respectively. Because the area of the South American beam is 10 times that of Colombia's, the gain over the entire continent is a full 10 dB less. One could, of course, maintain the same level of EIRP by increasing TWT power by 10 dB as well.

An alternative that is shown in Figure 3.3 subdivides the coverage area many times over using small spot beams. Assuming that each beam is 0.4 deg

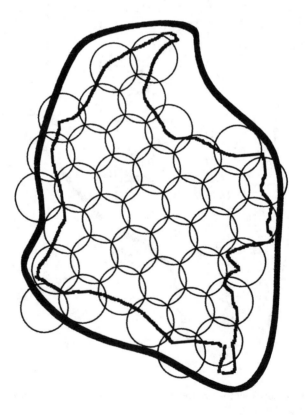

Figure 3.3 Comparison of area coverage and spot beam coverage of the landmass of Colombia, South America. There are 38 spot beams, each 0.4 deg in diameter (0.126 deg^2).

in diameter, it will take approximately 38 such spots to provide the full national coverage. As the size of the spot is only 0.126 deg^2, the gain increases substantially by approximately 15 dB compared to the area beam. The 28 spot beams are arranged in a 7-beam reuse pattern with one-seventh of the allocated spectrum assigned into each spot. Spots that reuse the same piece of spectrum are separated by two adjacent spots that are noninterfering. To use these beams effectively, the satellite must have an onboard beam-routing scheme.

The general relationship among the coverage area, beam size, and number of beams is indicated in Figures 3.4 and 3.5. The first graph allows us to estimate the directivity of a beam of a given area, measured in square degrees. For the case of Colombia, which extends approximately 2.8 deg^2 as viewed from GEO, the satellite can deliver about 40-dBi gain. A doubling of the area to include, say, Venezuela will reduce gain by 3 dB to about 37 dBi. Extending further to cover all of South America in one 50-deg^2 beam pushes the gain down all the way to 27 dBi. A word of caution about the directivity numbers. These are approximate values that do not include the relevant losses in a real antenna system. Also, we have not evaluated the actual beam shaping that would be provided during the design of the antenna system. These numbers are intended to provide a general feel for the relationships.

The second figure plots the number of beams required to cover either Colombia or all of South America. This is the concept behind Figure 3.3, which indicates that it would take about 38 spots of 0.4 deg each to fully cover the country. Beam area is plotted along the x-axis to be consistent with the previous figure. The information tells us that it can take an inordinate number of beams to cover a large landmass. On the other hand, the gain of such beams is substantially higher and the potential for frequency reuse much greater when considering a high density of small spot beams.

There are two application areas where the multiple-beam approach appears to be the most appropriate: L-band MSS networks to serve handheld phones, and Ka-band FSS networks for advanced broadband communications to inexpensive personal VSATs. As discussed in Chapter 2, L-band spectrum is very limited and we must incorporate as much frequency-reuse as possible. This, coupled with the difficult requirement of serving low-power handheld phones, demands a large reflector antenna with many small spot beams. The bandwidth is more ample at Ka band, so the major concern is with delivering high digital bandwidths (up to 2 Mbps) to an antenna of less than one meter.

The choice among the coverage alternatives depends on the interaction of the technical and business factors that confront the satellite operator (who may also be the application provider). From a pure marketing perspective, the single area coverage approach is the most flexible since you can deliver both individual services and broadcast services as well. The wide beamwidth produces a relatively low antenna gain, and the only frequency reuse is from

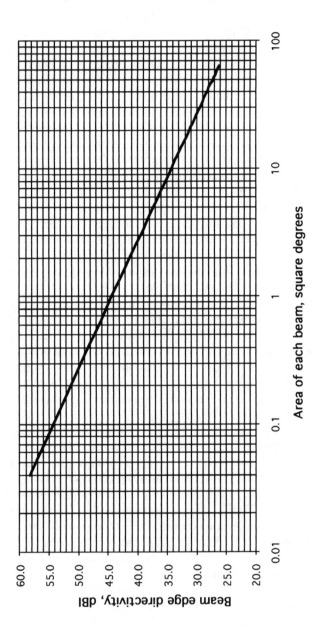

Beam edge directivity, dBl

Area of each beam, square degrees

Figure 3.4 An estimate of the directivity versus the coverage area.

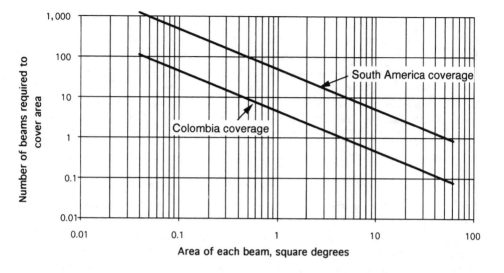

Figure 3.5 Approximate number of beams required to cover the country of Colombia or the entire South American landmass.

cross polarization. Moving toward multiple spot beams can greatly improve the attractiveness of the service to large quantities of simple, inexpensive Earth terminals. The ultimate example is the handheld satellite phone, which demands the greatest quantity of beams and their corresponding high gain. When we move in this direction, we restrict the range of services that can be delivered to the user community. Hopefully, this still represents a large and financially attractive market.

3.2.5 Repeater Design (Analog and Digital)

The repeater is that portion of the communications payload that transfers communication carriers from the uplink antenna to the downlink antenna of the spacecraft. In established C- and Ku-band satellite systems, the repeater is divided into transponders, each of which can transmit a predefined amount of bandwidth and downlink power. It is common practice to call a repeater a transponder and vice versa, although repeater is the more general term. As we extend the satellite design by increasing the number of beams and introducing digital processing technology, the simple transponder designs of the past are no longer appropriate.

  In the following paragraphs, we review the traditional type of transponder, called the bent pipe, along with newer concepts such as the *satellite-switched*

TDMA repeater (SS/TDMA), the full *demodulation-to-remodulation* repeater (demod-remod), and the *digital signal processing repeater* (digital repeater). Each represents an increasing degree of complexity and sophistication. As one moves toward increasing levels of complexity, the satellite becomes more and more a part of the overall network of ground stations and is inseparable from it. This tends to increase performance and effectiveness for a specific network implementation but renders the satellite less flexible in terms of its ability to support different traffic types not considered prior to launch.

3.2.5.1 Bent Pipe Repeater

Each transponder of a bent pipe repeater receives and retransmits a fixed-bandwidth segment to a common service area. The example of a six-transponder design in Figure 3.6 has a single wideband receiver that takes the entire uplink frequency band, typically 500 MHz, amplifies, and transfers it to the correspond-ing downlink band. The bank of input filters, labeled F1 through F6, subdivides the total bandwidth into 72-MHz segments, each amplified to a high level by a dedicated power amplifier. The combined output of all six amplifiers (each on a different frequency) is applied to a common downlink antenna with an output multiplexer composed of six reactively coupled waveguide filters.

The engineering design of the transponder channel is a high art because we must worry about dozens of parameters. Repeater parameters like receive *G/T*, transmit EIRP, transponder bandwidth, and nonlinear distortion have a large impact on users. These should be specified for every application. A multi-tude of others, like gain flatness, delay distortion, and AM-to-PM conversion,

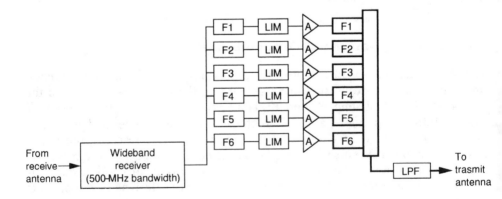

Figure 3.6 A simple bent pipe with six transponder channels, each transmitting approximately 72 MHz.

are often of less concern to real-world applications. Wideband digital transmission at 60 Mbps in a 36-MHz transponder is an exception because these distortions can significantly reduce throughput or increase the EIRP requirement for the same throughput. The limiter/amplifier (LIM) in Figure 3.6 provides a degree of control over data transfer by adjusting the input power and possibly correcting some of the nonlinear distortion. Typical transponder characteristics for the bent pipe design are listed in Table 3.2. Actual values will vary from design to design, in response to the type of amplifier, the frequency of operation, and design choices for the intended service.

Some examples of how repeater parameters can be related to particular signal types are shown in Table 3.3. To do this properly, the designer must fully understand the signals being transferred and the distortions to those signals caused by the various elements of the transponder. With the advent of engineering workstation computers and signal analysis software, this optimization has become relatively routine. However, the issue remains about whether it is wise to design the transponder for a particular signal and corresponding application. A useful alternative is to utilize a multifunctional design where a compromise is struck among a variety of expected signal types. This is actually how the first transponders were designed.

If the signal format does not change during the lifetime of the satellite, then the transponder is eligible for optimization. Consider first if you will operate the transponder with only a single carrier over a wide bandwidth or if you will carry multiple carriers in the same transponder. The DTH transponder has characteristics that are determined first by the frequency assignments in

Table 3.2
Typical Transponder Characteristics in a Bent-Pipe Repeater

Characteristic	Typical Value
Gain (saturation flux density)	−80.0 dBW/m^2
Linearity for multiple carriers (C/IM)	12 dB at saturation
Noise power ratio (NPR)	16 dB at 4 dB output backoff
Nonlinear phase shift (AM to PM conversion)	3 deg/dB
Amplitude frequency response (gain flatness)	±0.25 dB over useful bandwidth
Out-of-band attenuation (input channel separation)	−30 dB in adjacent channel
Cross-polarization isolation (XPOL)	30 dB (linear)
Multipath performance	dependent on overall design
Frequency tolerance and stability	10^{-6} in the translation frequency
Gain stability	±1.5 dB over lifetime

Table 3.3
Specific Signal Types and the Transponder Characteristics That
Can Be Optimized for Improved Performance

Signal Type	Transponder Characteristic
FM Video	TWT AM/PM and filter group delay characteristics
Wideband digital data	Gain slope and nonlinear phase shift
SCPC	Gain flatness and linearity (NPR)
Wideband TDMA	Linearity, gain flatness, and transient response of the power supply
VSAT Operations	Gain flatness/linearity and frequency stability
Mobile SCPC	Frequency stability and NPR

the ITU BSS Plan (see Chapter 7) and second by the type of signal format (analog or digital).

More advanced repeaters that use digital signal processing offer a great deal of flexibility in routing traffic and permitting very inexpensive user terminals to gain access to a range of mobile and fixed services. Alternatively, the processor function can be implemented with analog components that nevertheless have a degree of flexibility. The MSAT satellites operated by AMSC and Telesat Mobile contain *surface acoustic wave* (SAW) filters that, when combined with commandable down/up converters and switching, allow the ground network operator to alter the bandwidths and routing of SCPC bandwidth segments. This controls the balance of traffic and permits the operator to isolate sources of interference. Both the analog and digital processors are powerful tools in the hands of the satellite operator.

The selection of the bent pipe transponder has implications for users of satellite capacity. Bent pipes are nearly transparent to the user and can be subdivided in power and bandwidth, as discussed previously in this section. Moving toward the more sophisticated designs, the satellite becomes integrated with the network and transparency is lost. One of the first specialized commercial repeaters was carried aboard the Spacenet IIIR satellite. This was the Geostar payload, which introduced an L-band vehicular position determination service in the United States. Geostar had to contract with the spacecraft manufacturer to have the equipment installed on the satellite and then had to pay Spacenet for the operation and use of the payload. Subsequently, Geostar failed as a business. The special transponder was vacated and could not be reassigned to some other revenue producing application.

3.2.5.2 Satellite-Switched TDMA

Satellite communications engineers long ago recognized that digital TDMA offers the greatest multiple-access throughput. In the bent pipe design, a wideband TDMA network supports a large quantity of Earth stations while at the same time employs all of the available power and bandwidth. It is not unusual for the throughput to be 95%. Stations must be synchronized and their burst transmissions time-sequenced to prevent any overlap. In the TDMA mesh network, each station is capable of receiving the bursts from all other stations, including that of the reference stations as well as its own. The latter point simplifies the initial access and acquisition process.

The multibeam satellites first introduced by INTELSAT complicated the TDMA network because stations could not receive their own bursts. The solution was to provide feedback over the satellite. Connecting traffic between beams is conceptually simple, provided you have a means to switch bursts one by one. The technique of *satellite-switched TDMA* (SS-TDMA), shown in Figure 3.7, uses a matrix of PIN diode switching elements to route bursts between beams. Used on the first generation of *tracking and data relay satellite* (TDRS), SS-TDMA is conceptually simple and relatively easy to implement. Each beam gets its own wideband receiver and power amplifier. The bursts are transmitted at speeds up to the maximum capacity of the channel, which can amount

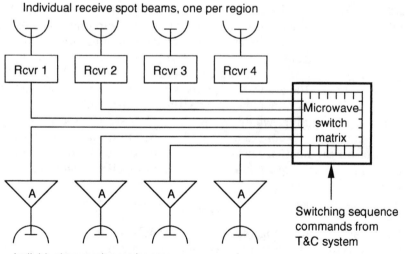

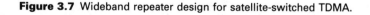

Figure 3.7 Wideband repeater design for satellite-switched TDMA.

to 250 Mbps or more. The switching elements themselves do not constrain bandwidth and must be capable of changing state in one-tenth of a microsecond (0.1 μs). This is not particularly challenging from a technical standpoint. The switching pattern is similar to what is followed in standard terrestrial *time division multiplex* (TDM) equipment or in a common device called the *digital access and cross connect system* (DACS) used by telephone companies to transfer individual 64-Kbps voice channels. The actual switching pattern is uplinked from the TT&C Earth station and then held in an onboard memory. A new pattern can be uploaded and subsequently activated in response to changing demand.

The first commercial use of SS-TDMA was on the Intelsat VI satellites, of which five were built and put into service. The payload of Intelsat VI contains a considerable quantity of C- and Ku-band bent-pipe transponders. The satellite must satisfy the basic transmission needs of literally hundreds of Earth stations within the three ocean regions. One of the main constraints in SS-TDMA is that each Earth station must be capable of transmitting at the maximum bit rate, which is in excess of 100 Mbps. The cost of the baseband and RF equipment needed to do this may be high in relation to the value of the traffic that can be carried, as compared to the more attractive medium of undersea digital fiber. INTELSAT has not included SS-TDMA on its subsequent satellites, choosing instead the bent pipe approach.

The pioneering efforts of Western Union with the first flight SS-TDMA payload on TDRS and subsequently by INTELSAT with the Intelsat VI series showed clearly that a sophisticated time division switching repeater can be depended upon from a technical and operational standpoint. Business considerations have stood in front of making SS-TDMA viable in the near term. One of the big problems for the operators of these satellites (currently NASA in the case of TDRS) is that the SS-TDMA payload elements cannot be allocated over more attractive applications after the satellite is launched.

From this beginning, we evolve to the digital signal processing repeater with its greater capacity for frequency channelization and time slot routing. As discussed in the next two sections, these designs offer some performance advantages for the coming generations of communications satellites.

3.2.5.3 Digital Signal Processing Repeater

The digital signal processing repeater is a significant advancement from the analog versions that ε ritch and route user traffic onboard the satellite. A *digital signal processor* (DSP) is a computing device that converts (transforms) an information signal from one form into another unique form. Historically, the DSP was programmed on a multipurpose digital computer as a way to save the time and energy of doing the transform mathematically with integral calculus.

The most popular DSP process is the *fast Fourier transform* (FFT), which is related to both the Fourier transform and Fourier series taught to all electrical engineering students. It takes a signal in the time domain (i.e., a waveform) and converts it into a collection frequencies (i.e., a frequency spectrum). The inverse FFT does just the opposite—transforming a frequency spectrum into a time waveform. When in either digital format, we can multiply, filter, and modulate the signals to produce a variety of alternate signal types. In this manner, a digital processor can perform the same functions in software that would have to be done with physical hardware elements like mixers, filters, and modulators. Modern DSP chips and systems can operate over many megahertz of bandwidth, which is what we need to build an effective digital repeater.

A block diagram of a hypothetical digital processing repeater is shown in Figure 3.8. The antenna and wideband receivers perform their traditional analog functions, while filtering and switching occur in the digitized sections of the repeater. This is indicated within the box at the center of the figure. The first function at the input of the processor is to convert the incoming frequency spectrum into a digital data stream. This is accomplished using *pulse code modulation* (PCM) at a high speed. The selection of the number of quantization levels and hence the number of bits per sample determines the amount of degradation to signal quality attributable to the processor. The digitized chan-

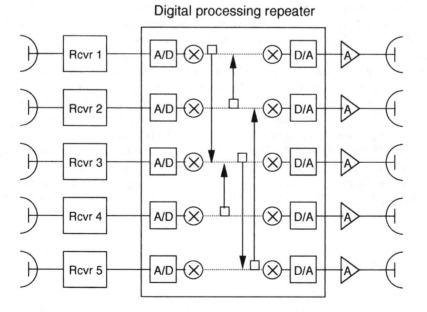

Figure 3.8 Channel routing in a digital processing repeater.

nels can be routed either as time slots or as narrow frequency bands. Other functions that are possible include automatic gain control, phase adjustment (as part of a phased array antenna system), and channel multiplexing. After all of the processing is complete, the data is reconverted back into its analog form (i.e., D/A conversion).

While the functions are shown as discrete components, they are actually performed mathematically by the processor and memory chips that implement the desired functionality. This means that the processor is designed for a specific purpose. To convert to the transmit band, the output of each channel is fed to a hardware upconverter that translates analog signals from an intermediate frequency range around 100 MHz to the downlink band. From this point, the signals are amplified in a conventional power amplifier and applied to the appropriate transmit antenna of the satellite.

The onboard digital processor is controlled from the ground and is capable of all of the functions described in the previous section for SS-TDMA. The state of the art at the time of this writing restricts the input/output bandwidth per channel of the processor to 100 MHz, which is less than can be provided by the IF type of satellite switch. As development continues, it will be possible to transfer the entire uplink/downlink bandwidth through each port on the processor. This only requires greater processor speed, something that one expects to see as the technology is improved over time.

The digital processor repeater is first being applied commercially in L-band satellites, which provide mobile communication services. DSP-based repeaters have also been proposed for Ka-band satellites that could reach orbit before the end of the century. High-capacity and sophistication for the processing function translate into the size, weight, and power consumption of the processor itself. In comparison to the typical microprocessor found in a personal computer, the digital repeater processor needs many more specialized DSP chips and supporting circuitry. These tend to run at a much higher speed and therefore consume more power. Another consideration is the degree of redundancy that needs to be included. Certain common functions, like clocks, memory, and power supplies, can be made redundant. But the actual channel processing elements would normally be single string. Redundancy must then be provided by including extra strings such that excess capacity may be reallocated in case of a partial failure.

3.2.5.4 Demodulation/Remodulation

The last type of repeater is not really new as it has been applied in nearly every commercial and government satellite launched to date. This is because satellites use a demodulator and remodulator chain to perform the ranging function associated with determining the orbit of the satellite. Up until now, demod/remod was not applied for revenue-bearing communication signals themselves.

As shown in Figure 3.9, a demod/remod repeater looks nearly identical to the bent-pipe, but with a demodulator and modulator added to each transponder channel. The minimum function of this combination is to prevent the direct addition of uplink noise to the downlink noise. Instead, uplink RF noise is transferred to the baseband of the signal where it causes a specific amount of impairment such as increased error rate. The uplink will threshold at a point determined by the demodulator onboard the satellite, while the downlink will threshold at a point determined by the demodulator in the receiving Earth station. Another benefit is that the downlink EIRP will be stable because the carrier that is applied to it is generated in the satellite modulator and not the uplinking Earth station. This same effect is produced by a limiter on the input side of the TWT.

Some missions might suffice with this capability alone. For example, we could build a very effective satellite that broadcasts data to millions of receivers. The only variation in downlink received power will be that caused by fading along the path between the satellite and the receiving Earth station. Uplink RF noise will introduce errors in the satellite demodulator, which will be transferred directly to the downlink. For example, if the uplink produces an error rate of 10^{-7} and the downlink produces an error rate of 10^{-6}, then the combined error rate is $1.1 \cdot 10^{-6}$. This condition might correspond to the uplink C/N being only 1 dB greater than the downlink. The increase in error rate of 10% is almost immeasurable in the recovered data. In comparison, without demod/remod, a

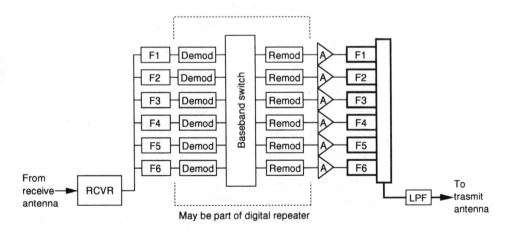

Figure 3.9 Modification of the bent-pipe repeater to provide modulation/remodulation. The demod/baseband/remod features could be provided by a digital processor.

1-dB difference would reduce the downlink C/N by 2 dB. The error rate for the received data would now be two orders of magnitude less than the downlink by itself, or 10^{-4}. Data communication throughput could be reduced due to the requirement for retransmissions between end-user devices.

Once we have recovered the original data in the satellite, it is likely that we would want to do some additional digital processing and switching. Figure 3.8 shows a baseband switch inside the repeater, similar in function to a digital telephone exchange that switches 64-Kbps channels. We can set up and connect telephone calls using the switch in the sky. Telephone and switched data can be offered directly to the public, allowing users to pay only for the services that they use. In contrast, bent pipe transponder capacity cannot easily be sold on a per-minute basis because of the difficulty of managing access. Incidentally, the demod-switch-remod functions can also be part of the digital processor repeater. Another possible application is in a multibeam satellite, as illustrated in Figure 3.8.

3.2.6 Additional Payload Issues

This section discusses a number of additional issues in the design or implementation of the communication payload. While not exhaustive, we provide some ideas and food for thought when planning a spacecraft or major application design.

Satellites operating at higher frequencies like Ku and Ka band might be fitted with a reference beacon for reception by communication Earth stations. This provides a reference for determining the amount of rain attenuation being experienced on the link. Another use is as an independent control channel for onboard communication functions such as the digital repeater discussed earlier in this chapter.

The command link must function at all times, which means that the command receiver must be permanently on and physically connected to the antenna. No switches or other interaction with the communication part of the repeater should be allowed. Command encryption might have to be considered for very secure operation, but this also should not interfere with safe operation in the case of an emergency.

Generally, the uplink coverage footprint should be as nearly identical to the downlink as possible. This allows transmitting Earth stations to be located anywhere in the entire area of coverage. However, there are systems like DTH and MSS with only a few ground transmitters in the fixed uplink part of the spectrum, so consideration should be given to restricting the uplink coverage area. This provides an improvement in spacecraft G/T and SFD, which in turn can improve link quality and availability. Alternatively, smaller uplink antennas can be used, which is a consideration at Ka band where large antennas are expensive and more complex to operate.

Another uplink issue is the appropriate use of *uplink power control* (UPC) to maintain carrier power at the satellite during heavy rain. UPC has proven effective in Ku-band VSAT hubs and Ka-band video uplinks, where an entire network is dependent on the reception of a strong and stable broadcast channel. There are concerns for the accuracy and responsiveness of the UPC control loop because any error will translate into a potential for network instability and loss of service. To this end, the UPC should be thoroughly tested prior to operation and maintained in proper working order for as long as it must function. Level control in individual user terminals is also an option, but here it is the service to the single user that is the consideration. This is particularly important in MSS, where mobile terminals will experience deep fading due to terrain blockage. The reaction time of the UPC will impact service performance and, on an aggregate basis, network capacity.

There is a high degree of interaction between the main elements of the payload, particularly the antenna and repeater. Components, subsystems, and the entire payload will need to be tested and retested on the ground and in space. The approach for sparing the amplifiers and other critical devices needs careful consideration as well, in addition to the overall requirement to isolate and remove any potential single-point-failure mode in the spacecraft or ground segment.

Tracking, telemetry and command (TT&C) requirements must be considered at the same time as the communication payload is designed and optimized. Most satellites include T&C frequencies at the upper or lower edge of the communication band. However, this may not be appropriate if existing ground tracking stations cannot employ the same frequencies. In this case, another band, such as C or Ku, may have to be employed and the requisite equipment included on the satellite. The correct number of telemetry points and commands need to be determined and compared to what the system can provide. In commercial satellites, telemetry is needed to assure reliable operation and to permit troubleshooting in the event of a problem. While it is not needed for deep engineering study, adequate telemetry is nevertheless vital to the long-life operation of the entire network.

3.3 SPACECRAFT BUS CONFIGURATIONS

The spacecraft vehicle, more particularly the bus, must fulfill the requirements of the communication mission, typically lasting 15 years in the case of GEO. Hughes pioneered the spin stabilization technique for GEO, while European and other U.S. manufacturers pursued the three-axis early on, gaining experience with the nonspinner. Spinners are simpler than three-axis; hence, they tend to be troublefree. However, three-axis can deliver more power at less overall launch weight and, hence, are preferred for high-power missions (like DTH and LMSS). Even Hughes, the source for spinners, offers three-axis designs.

Regardless of spacecraft configuration, operations personnel and users have to contend with on-orbit problems that result from flaws in the design, manufacture, or operational procedures. The despun section of a spinner contains the communication payload. Total spin-up can result in loss of the satellite, that is, a flat spin. This is a rare event, and in fact only had one inadvertent occurrence in United States for an experimental satellite called Tacsat.

Loss of control of a three-axis spacecraft is somewhat more likely because of the low inertia of the vehicle and the complexity that comes about from the need for many thrusters. There are fundamentally two means for maintaining an inertial reference and degree of control. There are the zero-momentum systems that use three reaction wheels, each along a different orthogonal axis, to allow correction in any of the three directions. Each wheel can spin in either direction, thus affording forward and reverse control. The other, more common approach is the momentum bias system where a single wheel provides all of the reference and some net momentum along a virtual spin axis. A pair of such wheels are needed to assure a redundant system, which is still only a small fraction of what a spinner can provide. An emergency in either occurs from loss of pitch lock, where the body slowly rotates around the north/south direction (perpendicular to the equatorial plane). As long as the wheel is kept within certain speed limits, the spacecraft will tend to remain erect and stable.

Problems in three-axis recovery from loss of lock have occurred in the past due to (1) improper use of thrusters (they should have been kept off), (2) inappropriate onboard autonomous protection software that switched the wheel(s) off, and (3) lack of sensor (or memory-stored) information so that an autonomously recovering spacecraft does not know where to turn to expect the sun. Depending on where the sun sensors are mounted and how much risk the operator takes, the recovery can take up to nearly 24 hours, during which communication service is almost surely lost. The newer designs are more intelligent and robust, and therefore these problems should become a thing of the past. In any case, it is possible to override such undesirable modes by ground command; using these commands properly is also a key to the safe operation of any spacecraft.

Having a stable platform that does not succumb to spinups and loss of attitude control is paramount for commercial satellite communication applications. Users expect their satellite operator to provide dependable service; in fact, they would prefer not to have to worry or even think about it. Selecting the particular bus design and the safeguards at all steps of the manufacture and operation are essential to achieving this result. For this reason, most operators prefer not to be the first in line for a new bus design.

3.3.1 Spinner Characteristics

The spinning satellite design has the longest track record of successful operation and is recognized as the most reliable configuration. Over the years, its performance has improved greatly, but even still, it is generally favored for low- to medium-power missions. Spinners like Galaxy I and Telstar 3 performed their TV transmission functions nearly flawlessly for 10 years. The complexity level of the payload and antenna system might be limited because of constraints on physical attachment area and volume. The following are some of the more recognized features of this type of design:

- High on-orbit stability due to gyroscopic stiffness;
- Simple control methodology using only two propulsion thrusters;
- Uniform thermal distribution due to natural rotation of the body;
- One moving part required—the bearing and power transfer assembly;
- Proven record of long-life performance (some have operated for over 15 years);
- One current manufacturer (Hughes);
- Solar panel area limited by launch dimensions, which can be extended by telescoping an outer panel.

At the time of this writing, Hughes still produced the HS-376 model of spinner, as shown in Figure 1.4. More than 50 of these satellites have been launched, and nearly 15 are in operation. They continue to be purchased for domestic satellite systems like Galaxy (United States), AP-Star (China and Southeast Asia), Measat (Malaysia), and Brasilsat (Brazil). The maximum payload power available from current models is in the range of 1 kW to 2 kW. The operational advantages of the spinner, in terms of simplicity and reliability, have been achieved by recent three-axis designs, discussed in the next section. This means that as users demand more capability from their satellites, the corresponding demand for spinners will decrease (unless Hughes or some other company devises a way to significantly increase the power and weight capability of the spinner without a substantial increase in its cost of manufacture).

3.3.2 Three-Axis Characteristics

Great strides have been made in establishing the three-axis spacecraft as a solid provider of satellite communication services to commercial users. After nearly 20 years of experimentation and development, the three-axis is quite literally "ready for prime time." The Hughes HS-601 spacecraft is illustrated in Figure 1.5 as an example. All of the current spacecraft manufacturers produce the three-axis type of design, each vying for the high ground of offering the greatest capability in terms of DC power, payload weight, and flexibility of antenna attachment. Purchasers, who are satellite operators, are also concerned about stability, reliability, and ease of operation.

The primary features of the three-axis design are summarized as:

- Substantially more capability to produce DC power—up to 8 kW at the time of this writing;
- Greater room for antennas, including complex deployed structures of over 10 meters in diameter;
- Reduced angular momentum due to the use of a momentum wheel instead of a spinning body (reaction-wheel systems with zero-momentum are also applied);
- Complex attitude control, requiring an onboard computer and array of thrusters;
- Thermal conditions vary in daily and seasonal cycles, similar to the despun payload platform of the spinner;
- Thermal control through use of heaters, louvers, heatpipes, and other devices;
- More moving parts such as momentum wheels and solar panel positioners;
- Better efficiency in terms of weight and power;
- Eclipse performance typically better because it represents a smaller departure from normal conditions.

This list suggests that there are tradeoffs in the use of a three-axis spacecraft. You can obtain almost any power level that you could require with significantly better efficiency than provided by the spinner. The other side of the coin is that the three-axis is considerably more complex and potentially somewhat less reliable than spinners that have been launched in the recent past. However, this is counterbalanced by years of experience coupled with many important improvements that manufacturers have included in modern three-axis satellites. Today, the control methodology of the three-axis is well understood and embodied in the onboard spacecraft control processor. This is a general-purpose computer that is able to control the spacecraft during its operating lifetime and includes many backup and fault-protection modes.

A cautionary note is needed. In order for this optimistic view to apply, you must be sure that the three-axis spacecraft that you are using is the latest in the series and incorporates the control features and redundancy that you expect. This can be verified by an experienced spacecraft systems engineer who has operational experience with recent three-axis satellites.

3.3.3 Spacecraft Power Constraints

The demand for more downlink RF power has consequences for the design, size, and weight of the spacecraft bus. This is in addition to any impact on the communication hardware, such as power amplifiers and waveguides, which resulted from the use of a high power level. We review in qualitative terms the most significant impacts on the spacecraft bus. Readers who need to quantify the impact should work closely with the spacecraft designer or manufacturer, who is in the best position to evaluate new requirements and their impact on the spacecraft.

3.3.3.1 Power System Limit (Panel Area and Weight)

The most direct impact of increased RF power is the requirement for greater DC power from the electrical power system onboard the satellite. The spinner and three-axis designs approach the generation of prime power in a slightly different manner. Generally speaking, flat solar panels on the three-axis can be extended in length and power almost without limit, while the cylindrical panels on the spinner cannot exceed the dimensions and volume available within the launch vehicle shroud. With the state of the art in 1996, the spinner can generate approximately 2,300W of DC power, of which about 2000W are available for the payload. The 300W of difference between the two values is needed to power the bus. The comparable numbers for the three-axis are approximately 9,000W and 8,000W for total prime power and available payload power, respectively. Three-axis bus designs that are currently on the drawing board promise up to 16 kW of prime power.

The beginning-of-life power output of the solar panel can be measured on the ground with reasonably good accuracy. In orbit, the solar cells in the panel are exposed to the space radiation environment, which consists of free electrons and protons. Most of this charged flux emanates from the sun and follows the 11-year solar cycle. Panel designers use projections of the radiation flux that have been compiled by respected organizations like the U.S. *National Aeronautics and Space Administration* (NASA) and the *European Space Agency* (ESA). However, no model can predict with certainty what the environment will be for a particular mission. Instead, the approach is to use a reasonably conservative estimate of the environment over the solar cycle in question. What actually happens will probably be less than the projection. In the last cycle (peaking around 1990), there were a number of unexpectedly large proton events, the worst of which caused as much as a 1% drop in panel power output. Satellites constructed during the 1980s used silicon solar cells, which delivered powers up to a maximum of about 4 kW on three axis buses. Newer, more efficient cells were introduced in the 1990s, using gallium arsenide (GaAs)-on-silicon semiconductor material. This increases power by up to a factor of two, making 8-kW systems feasible.

3.3.3.2 Battery System Limit (Volume and Weight)

Assuming that the payload must operate in eclipse as well in sunlight, the power required by the payload would also demand battery capacity. There are two technologies available today for commercial spacecraft, namely, nickel cadmium (NiCd) and nickel hydrogen (NiH). NiCd batteries are similar in concept to standard rechargeable NiCd batteries found in home electronic equipment. However, the particular configuration employed in space is designed for much longer life, low weight, and greater discharge capacity. The other technology, NiH, has also been used for many years. Provided that the eclipse demand is above a certain minimum value, one can obtain 50% to 100% greater charge capacity for the same weight as NiCd. Up until recently, NiH was only found on the larger satellites but is now available on the smaller spinners as well. The choice has nothing to do with the spacecraft configuration but rather depends on the total power required and the specific capability of the spacecraft manufacturer to produce and integrate NiH batteries on a cost/effective basis.

Mobile communications services could well employ a satellite with reduced eclipse capacity. This is because MSS is really a telephone network in space, where subscribers make calls according to their own particular usage patterns. On an aggregate basis, telephone calling follows a predictable pattern of rising in the morning and hitting a peak, called the busy hour, some time in the early to mid-afternoon. Not having all of the power available at 2 A.M. would be an acceptable compromise, provided that the payload and power system are designed for this kind of variability.

3.3.3.3 Thermal System Capacity

We mentioned in Section 3.2.5 that the thermal control subsystem of the spacecraft must remove excess heat so that internal temperatures do not exceed design limits at any time during the life of the satellite. Any increase in input and output power will introduce more heat that the thermal system must reject.

With the growth in power demand and packing density of electronics, spacecraft manufacturers have adopted the heat pipe as an effective means to move heat from the source to the external radiating surface. A heat pipe is a long tube containing a substance that can transition between liquid and gaseous states. As with any classical refrigeration cycle, the liquid is converted to gas at the hot end, flowing to the cool end that is exposed to space. The gas condenses, releases its heat, and returns through the pipe to the hot end. High-power three-axis satellites employ braces of heat pipes to move heat from points of high concentration, such as underneath high-power TWTAs, to external surfaces that can radiate the heat to space. If we generate more heat, then more heat pipe capacity is needed as well as possibly greater radiation surface. This is not a trivial change because it could require extensive rearrangement of the spacecraft structural system and additional weight to provide more thermal control capability. In some cases, physical volume constraints could preclude adding the needed thermal control facilities. One alternative is to design all electronic and mechanical equipment to operate over a wider temperature range.

If these steps are not taken correctly, the result will be a hotter spacecraft and a potential reduction in lifetime and performance. Techniques to control the degradation of external control surfaces have been suggested since the hottest temperatures occur at end of life. To date, however, no commercial satellite has incorporated anything that would tend to reduce temperature as the spacecraft gets older and more contaminated. This whole subject will be getting more attention in coming years as power is increased. For users, there is little that one can do except to ask the right questions of the satellite operator or manufacturer. They should be able to demonstrate by analysis and measurement how their spacecraft will fare over the operating life.

3.3.3.4 Fuel Capacity and Loading

Along with the increase of the spacecraft power and thermal control support, the satellite engineer must consider the additional fuel load to maintain lifetime. For a given type of propulsion system and fuel, any increase in dry weight of the spacecraft will require the same proportional increase in fuel weight. We typically want to keep the satellite within a north/south and east/west box that is 0.2 deg on each side. Most of the stationkeeping fuel is required to control the inclination of the satellite, that is, north/south stationkeeping. Obviously, if the need for north/south stationkeeping can be eliminated, then substantially less fuel will be required. This can work for MSS missions where users employ broadbeam antennas but is probably not feasible for fixed applications with fixed dish antennas like DTH and VSAT networks.

It is possible to improve thrust performance without adding fuel by increasing the *specific impulse* (I_{sp}), which measures thrust per unit weight (in units of seconds) of the fuel. In typical GEO and non-GEO spacecraft designs, I_{sp} is in the range of 170s to 300s. The conventional *reaction control system* (RCS) employs hydrazine as a single fuel and uses a blow-down system with a simple gas pressurant. Moving to a bipropellant system with monomethyl hydrazine and nitrogen tetroxide oxidizer, along with a regulated system to maintain constant pressure, will increase the I_{sp} by 50% or more. This type of operation is acceptable for long burns used to raise the orbit or in north/south stationkeeping. Short thrust pulses bring the performance way down because the thruster does not have time to reach full operating temperature.

Of recent interest are various forms of ion propulsion. Hughes has introduced its *Xenon ion propulsion system* (XIPS) into the HS-601HP spacecraft product line. Other electrical thrust systems like the arc-jet are becoming available as well. These state-of-the-art propulsion technologies promise values of I_{sp} in excess of 1000s. The issue with any of these new concepts (some of which have been around for decades) is that of life expectancy. This is being addressed through lifetest programs and, in conjunction with early on-orbit experience, will prove the dependability of XIPS and the arc-jet.

The trend in GEO satellites has been to reach as long a life as possible, extending past 10 years to 15 years or more. If we were to move back to 10 years or even 8 years, it would be possible to reduce fuel consumption. This requires a proper analysis of the mission, which is beyond the scope of this book. However, it is possible to have the satellite operator or manufacturer determine the amount of the resultant fuel. A shorter lifetime often has little impact on profitability of the investment because of the effect of discounting the later years of revenue.

It might be possible to increase lifetime by using a different launch vehicle to place the satellite into transfer orbit. In most missions, some of the RCS fuel is reserved for orbit corrections and even for the perigee kick function. If the rocket can be depended on to carry out more of these functions (and with improved accuracy), then RCS fuel can be saved for stationkeeping. Evaluation of this also requires a thorough analysis by the launch vehicle provider.

If in the final analysis you determine that there must be an increase in fuel load, then there still is the consideration of fuel tank capacity. If the fuel demand is going to be significantly greater than current designs, then larger tanks can be installed. The first impact is the difficulty of integrating the larger tanks. The second impact is the structural loading during launch of the larger and heavier (full) tanks. Such changes can have a significant impact on the overall design, including its ability to qualify for launch on a particular rocket. Therefore, a thorough evaluation of such a proposed change is justified.

In summary, increasing RF power will have a significant and possibly major impact on the spacecraft. For this reason, it has to be carefully considered. To do this right, you will have to involve many parties—the satellite operator, the spacecraft manufacturer, and possibly the launch service provider. This is not an impossible task provided it is dealt with in a thorough manner.

3.4 CONTINGENCY PLANNING

Satellite operators and users must engage in contingency planning, which involves making arrangements for backup satellite capacity and succession when operating satellites reach end of life. For operators, this is a matter of maintaining the business in the face of possible launch and on-orbit failures. Users of these satellites share that concern and would probably will leave a given satellite system if capacity is not available in the event of a failure.

Providing the backup and replacement capacity is costly and if done wrong can lead to a disastrous result for all parties. For all of these reasons, operators and users can participate in the solution to providing continuity of orbital service.

3.4.1 Risks in Satellite Operation

The following subsections identify the most important risks that affect the delivery of space segment service to users. We offer some basic approaches to the resolution of each of these risks. However, this is not a substitute for a detailed plan that is compiled for the unique circumstances of the particular operator and/or user.

3.4.1.1 Launch Failure

The satellite operator and user must make provision for the distinct possibility that a given launch will not be successful. Spacecraft manufacturers provide a variety of services that take account of the probability of approximately 10% that the satellite will not reach its specified orbit and provide service. For example, the contract for the satellite might include a provision for a second spacecraft to be ready for backup launch within a specified period after the failure. The contract might even provide for delivery in orbit by a specified date, which implies that the spacecraft manufacturer will have to go through the (expensive) steps that would otherwise fall upon the operator. In the end, however, the user pays the costs of covering the risk.

The rest of the steps that the operator can take include having an on-orbit satellite available to maintain the service during the period between the failure and the next launch. This is covered later in this chapter, under the topic of succession strategy. On the user side, some form of contingency plan must be put together. This could involve contracting with another satellite operator to have backup transponders available in the event that the new satellite does not go into operation on time.

3.4.1.2 Reduced Technical Capability

Any organization that is engaged in a high-technology activity is exposed to the risk that it will not be able to maintain a sufficient level of technical competence. This depends on the people who work for the company and includes the number of qualified people and their level of training. Historically, companies and government agencies have attempted to build competence through inhouse education programs and on-the-job training. There has been a trend in recent years to require that new people come to the company already trained, either because they worked for another organization in the same or a similar line of business or because of their individual educational experiences. This reduces the training burden on companies but increases the risk from *poaching*, the tendency of companies to lure qualified people away from each other with attractive offers of employment.

In satellite operation and application, loss of technical capability has not been a problem in developed countries, probably because of the outflow of engineers and technicians from the defense and aerospace industries. People are also becoming available as large telecommunications companies downsize to become more competitive. Eventually, this overhang will diminish as more and more people reach retirement age. We will, at that point, depend on the production level of new engineering and other technical graduates. As a consequence, satellite communication organizations may find it more and more difficult after the year 2000 to maintain adequate staff.

A related aspect of this problem is that the technical demand on an organization can increase as new classes of satellite systems and communications technologies are introduced on a large scale. The transfer from analog to digital video along with the popularity of DTH is putting pressure on the satellite job market. Experienced people in the industry now have demonstrated flexibility, possibly because of the range of services available on satellite networks worldwide. Furthermore, newer satellites are larger and more complex than the bent pipe systems of the 1980s and early 1990s. Digital cellular networks also represent a technical challenge to traditional types of radio technologies. The impact of MSS becoming a mass market has yet to be felt. For a lot of us, the fact that the business is getting more complicated makes it that much more interesting and challenging.

3.4.1.3 Loss of On-Orbit Lifetime

Newcomers to satellite communication may have a somewhat obsolete view of satellite operations, possibly driven by highly visible launch vehicles or LEO satellites that re-entered the atmosphere and crashed to the ground. The actual experience is that the great majority of satellites live out their life expectancies and can be counted upon to provide service for a duration of 10 to 15 years. There are exceptions where some kind of catastrophic failure ended the satellite's life prematurely, but these are very few and far between.

An important but often overlooked task of the satellite operator is the proper and efficient maintenance of orbit control. Many GEO systems enter service using a single TT & C Earth station with one antenna. This has adequate ranging and control accuracy if the satellite is to be controlled to 0.2 deg on each side of the stationkeeping box. As more satellites are added to the same orbit position, improved accuracy will become a requirement. A second TT&C station is usually needed to provide a second set of range data to the orbit determination process. This allows the software to come up with an accurate orbit more quickly. The situation with regard to non-GEO satellites is still open to question because none of the planned systems are yet in service. The need to maintain multiple satellites in non-GEO orbits and to coordinate the arrangement of multiple orbits to assure continuous service has yet to be practiced in a live system.

Even with the excellent experience to date, the risk of loss is so great that operators and users must have contingency plans. The approaches for this are precisely those that we covered earlier in this chapter. If the risk of reduced lifetime could be anticipated by a few years, then the parties can simply plan on launching the replacement ahead of the originally planned date. This introduces very little disruption in the normal planning cycle and will not cost much extra. On the other hand, planning for an unexpected loss of life, such as that experienced when a satellite abruptly loses a significant fraction of its transponder capacity, means having an extra satellite available in orbit. This on-orbit spare would not be sitting idle but would be employed for preemptible services that can be discontinued when and if the satellite is needed for its service restoral role. Preemptible services will produce revenues that help offset the investment and operating cost of the spare.

3.4.1.4 Loss of Ground Facilities

Ground facilities tend to be less reliable than the satellites that they support. Being on the ground, they are exposed to many environmental risks, such as flood, Earthquake, fire, wind, theft, and civil unrest. In addition, ground facilities are dependent on external support to keep them running. Some of this can be countered through backup means, such as an *uninterruptable power supply* (UPS), local water storage or supply, and storage of large quantities of supplies and spare equipment. At some point, the ground facility will not be able to fulfill its role either as a control point for the satellite or as a communication node.

Assuming that we have taken appropriate measures to strengthen a particular facility against the expected hazards, the only thing that remains is to provide an independent backup. For a satellite operator, this means having a backup TT&C station and satellite control center. This type of strategy provides a very high degree of confidence that service will be maintained even if the primary site goes out of service. The physical facilities can probably be more easily replaced than the people who operate them. As stated under the section on reduced technical capability, having qualified people available can become a challenge. If we routinely have one operating site to control the satellites, it would be quite a burden to try to maintain a backup site with qualified staff as well. Most of the time, this staff will have little to do and therefore might not be as experienced as those who work at the main operating site. This can be countered by providing routine training and exercises for the staff and by rotating qualified people between the two locations.

For satellite users and their communication Earth station facilities, the trend has been for the sites to be unattended. The trained staff would normally be located at the network control center and at distributed maintenance facilities. The key here is to have enough staff deployed at different locations so that there is inherent diversity in the operation. A concern is with a single network control point and the possibility that it will be knocked out. The best approach here is to have at least two such facilities, each supporting half of the network. In the event of an outage, the other facility takes over management of the entire network.

3.4.1.5 Harmful Interference

Any radiocommunication service is potentially a victim of harmful radio frequency interference, which can be either accidental or intentional. By harmful we mean that authorized services are disrupted or rendered unsatisfactory to users. This is different from unacceptable interference, which is a term in frequency coordination to indicate that the calculated interference level is above some detection threshold (discussed in Chapter 10). The vast majority of harmful interference events are accidental in nature, resulting from an error in operation or an equipment failure of some type. This means that whatever the cause, the interference will be found and corrected as a matter of course because the error or failure produces a direct loss of performance for the unknowing perpetrator. Intentional interference is rare and often quite notorious. In many countries, particularly in the developed world, intentionally causing harmful radio frequency interference is a crime. This has been an effective deterrent, mainly because the lawful operators want it that way.

Satellite communication is particularly vulnerable because any transmitter on the ground that is within the satellite footprint can be a source of harmful interference. In area-coverage systems, it is difficult but not impossible to locate the source. Most of our efforts are expended on monitoring all of the transponders in the downlink so that interference can be observed as soon as it appears. This is augmented with good direct telephone communication with users, which are usually the first to notice an interference event. The key to controlling and eliminating harmful interference is to constantly maintain this type of vigilance over the system. As soon as any interference is detected, the operations staff must move quickly to identify the source and demand correction. In the vast majority of the cases, this is effective in a matter of minutes. The remaining cases take longer to correct, sometimes hours or days if the source cannot be isolated quickly. The approach here is to reduce the impact of the interference by moving users to different frequencies or different transponders. This allows the problem to be studied more carefully without the pressure of having to maintain service.

Intentional interference is a source of anxiety among satellite operators and users alike. There is always the possibility that a radio pirate might either take over an existing legitimate Earth station or build one for the express purpose of causing some kind of abuse (harm or theft). In the rare cases where this has happened (only two that this author can recall), the perpetrator was identified and prosecuted. It turns out that the type of person who would do this sort of thing has emotional problems. This allows the police and other authorities to track down the individual. In the meantime, people in the industry are given the opportunity to think about how this kind of disruption can be detected more quickly and how to prepare for the next episode. It provides an opportunity for all to increase the level of vigilance.

3.4.1.6 Sabotage

Another source of intentional disruption is the physical type, which we call sabotage. Since the satellite is controlled from the ground, it is conceivable that someone might attempt to vandalize an operating TT&C station. Any high-power Earth station used for TV uplinking might also be used to jam the command frequency or even take control, given the proper command encoding equipment. The newer generation of commercial satellites tend to have secure command systems to make a takeover a very remote possibility.

Most Earth stations that are capable of causing sabotage to the satellite are protected with security perimeters. The amount of this type of physical security will depend on the risk. In the United States, it is normal practice to provide security fences, doors, and even guards. Facilities in remote areas might have less physical security, but some minimum amount is still justified. Recall the old adage that most locks are designed to keep an honest person honest.

Satellite control facilities that provide government communications services are often protected to the fullest. This is a special case and is really beyond the normal scope of commercial affairs. However, the thought processes that the government applies can be useful to protect high-value installations. It is always prudent to think about what type of attack might be possible and what could be done to minimize the risk or impact. Some time spent in anticipation of this kind of event is well worth the effort. For example, the only thing that may need to be done is to use the physical security that is already in place but is currently not being used. For example, security doors with TV monitors might be installed but are deactivated as a matter of convenience. If the risk increases for some reason, then all you might need to do is to reinstate the use of these doors and TV monitors. Computer systems are very capable of providing greater security than is used on a routine basis. Tightening security might only be a matter of using the access control mechanisms (including passwords) already provided in the operating systems and data communications network.

3.4.1.7 Compromise of Information

Finally, there is the risk that proprietary technical or business information could be released to the outside and used by a competitor of some type. Any organization or individual who could use this information to our disadvantage would be viewed as a competitor. Satellite operators are privy to a lot of this type of information from the satellite manufacturer, who would not want detailed design information to leak to another manufacturer. This is usually covered in the contract and can bring stiff penalties. Other proprietary information is developed within the organization, covering the unique way that business is organized and conducted. It turns out that the sum total of all business information and documentation represents a barrier to entry into the business because of the high cost that would be incurred by a potential competitor to recreate it.

The approach to protecting information from compromise is usually to maintain physical security for the area where the information is stored. Proprietary information is indicated with a label or stamp that states the information is proprietary to the particular organization. Further steps involve putting terms and conditions in contracts with external parties, requiring them to protect this type of information once it has been identified as proprietary.

In the end, the best protection against loss of proprietary information is to keep the organization moving forward. This means that any particular piece of information only has value for a limited period of time. If it is compromised, then the loss is minimal, possibly only amounting to the nuisance of having to change over to the new procedure and technique in the worst case. Also, any one piece of information or technical approach by itself probably is not that valuable. Rather, the value is in the collection of this information and the way in which the organization applies it.

3.4.2 Available Insurance Coverage

The policies and procedure described in the previous section deal with the operational impact of risk. There is always the financial impact of loss, for which insurance is an effective preventive measure. We consider some of the more common types of insurance that can be purchased by satellite operators and users.

3.4.2.1 Launch Insurance

A newly launched satellite stands between an 80 and 90% chance on the average of successfully reaching orbit (GEO, MEO, and LEO) and being capable of a planned start of service. Some launch vehicles and supporting services have achieved the higher end of the range, including Arianespace's Ariane 3 and 4 rockets and McDonnell Douglas's Delta series. Lockheed Martin's Titan and, through a recent acquisition from General Dynamics, Atlas Centaur, have nearly as good a record as the leaders. The rockets available from China Great Wall Industry Corporation of the People's Republic of China are good performers, but the record to date is still advancing from the low end of the scale. And lastly, the tried-and-true Russian rockets left over from the Soviet space program are finding buyers in the commercial marketplace.

With the exception of the first successful launch of an Orbcom satellite on the Pegasus rocket, all launches of commercial communications satellites prior to the time of this writing have been for GEO service. These involved at most two satellites at a time on the Ariane 4 and 5, or, prior to the change in U.S. Government policy, the Space Shuttle. Planning for GEO systems is on the basis of launching one or two satellites at a time and amounts to betting on the toss of dice. As we shall see later in the chapter, building a multiple-satellite GEO system is on a step-by-step basis and can proceed more or less in a serial manner. As we move into the era of the global LEO and MEO systems, the risk of launching satellites takes on a different tone. Global systems like Iridium and Globalstar, discussed in Chapter 9, cannot start service with a single satellite but require an initial operating constellation of dozens of working satellites. On a relative basis, one launch by itself does not pose as much of a financial risk as with GEO systems. On the other hand, a serious problem with the design or manufacturing process of the rocket used to create the LEO or MEO constellation can halt implementation and the start of service by between six months and one year.

The simple fact is that launching satellites is a risky business and demands every possible step to assure the financial and operational viability of the user and the satellite operator. Spacecraft manufacturers may or may not bear part of the risk, depending on the nature of the particular contract. They, too, need to consider how to insure their financial exposure.

Launch insurance is generally available to the parties who stand to lose in the event of a failure to reach orbit. The satellite operator can purchase coverage equal to the purchase price of the spacecraft and launch vehicle. This is typically increased to assure that all of the expected cost of a replacement is included. To do this, the number is adjusted upward for inflation between the contracted price (which is probably two to three years old) and the time when a negotiation for the new spacecraft and launch would happen. Alternatively, the operator may use option prices that were previously negotiated with the manufacturer and launch vehicle provider. The last item to be included is the cost of the next launch insurance policy.

Operators of LEO and MEO networks take an entirely different stance with respect to launch insurance. Some may prefer to self-insure, meaning that no specific launch insurance with be purchased. By purchasing sufficient extra spacecraft and launch vehicles, they will provide the needed "insurance" against the expected failure rate of the launch vehicle systems employed. This will not protect them against a systematic problem with a particular system, but the operator could reduce risk by selecting a second source early in the program.

Major users who purchase transponders for the life of the satellite also purchase launch insurance. In the 1980s, cable TV networks like HBO and Turner Broadcasting purchased such insurance from the same sources as the operators. The issue here is that there could be a very substantial financial risk placed on a single launch, representing such a large loss as to be uninsurable. The same applies to a multiple launch where two satellites are insured. The simple answer to this kind of problem is to stay within the limits imposed by the marketplace. If the risk associated with a particular launch exceeds what the insurers will cover, then the insured might get together and divide their assumed risk.

The entities that require the insurance have multiple sources for coverage, depending on their country of origin. Often an insured party will work with its existing underwriter. In previous years, much of the coverage found its way to the largest insurance market in the world for high-risk activities—Lloyds of London. As many readers know, Lloyds is not an insurance company but rather a coordinator. They represent literally thousands of insurers, called *names*. These are companies and even individuals who attempt to make money by betting against disaster. Unfortunately, there have been more disasters in the satellite, shipping, air, and other industries to make the business of being a name rather unattractive. The launch insurance game is now in the hands of a new breed of underwriter who tends to take less risk by charging high premiums in the range of 20% to 30%.

Other insurance coverage is typically bundled in with the purchase price of the satellite and launch vehicle. There is risk of financial and human life loss due to some kind of catastrophe at the launch site. This is a rare, but not unknown, type of loss. The providers of hardware and services typically insure against these losses. Failure to launch on time is typically not a risk borne by theses providers, except possibly that the spacecraft manufacturer could, under contract, be held to such a claim. This depends on the particular arrangements made ahead of time.

3.4.2.2 On-Orbit Life Insurance

Commencing with the initiation of service, satellite operators usually insure their operating satellites against loss of lifetime. The price of this coverage is proportional to the value of the satellite reduced by the number of years already expended in orbit. A direct analogy is the kind of warranty that automobile tire manufacturers provide, which is reduced by either the years remaining or the consumed tread.

The cost of life insurance has been in the range of 1.5% to 2.5% per year. Owners of transponders can also purchase life insurance, or, alternatively, it could be provided as part of the transponder purchase agreement (i.e., similar to the tire warranty). Users who rent their satellite capacity have no direct need to insure the remaining life because they simply do not have to pay if the capacity is not available due to a satellite failure. Their situation could be difficult, however, if they have not made other provisions for replacement service.

3.4.2.3 General Liability Coverage

There is a wide variety of other insurance coverages that are valuable to those engaged in the satellite communications field. Some examples include standard workman's compensation insurance, insurance for loss during transportation of equipment, patent liability coverage, insurance to provide replacement of lost facilities or services, and liability insurance to cover the intentional and unintentional actions of employees and management. There is likely to be a need for insurance against liability for injury or damage that results from a launch failure or the possibility that a satellite may reenter the atmosphere before it reaches its final orbit.

3.4.3 Space Development—Estimating Lead Time

Communication spacecraft used in GEO, MEO, and LEO networks require a considerable time for the design and manufacturing cycles. These last from as long as 36 months for a new design to as little as 12 months for a very mature design with some existing inventory of parts or subsystems. A typical spacecraft of standard design will take about 24 months to deliver to the launch site from the time that the manufacturer is authorized to proceed with construction. The launch service provider also will require lead time to arrange for construction of the rocket and to reserve the launch site. The resulting waiting time to launch could be as long as 30 months. This means that the developer of a new application or system must allow sufficient lead time.

An overall timeline for a typical spacecraft development program is shown in Figure 3.10. This takes the perspective of the satellite operator or developer of an application that is dependent on the availability of a new satellite type. It allows for a precontract period of about six months to collect business and technical requirements and to prepare technical specifications. The period could be shortened if the requirements are standard and no new development is required, such as for a "plain vanilla" C-band satellite for video distribution. On the other hand, if we are talking about a new concept for which no precursor exists, the precontract period could last one or two years.

The satellite operator will normally procure the spacecraft according to a competitive process to provide some confidence that the best terms have been obtained. This considers the technical, cost, and schedule requirements for the project. Such a procurement may take anywhere from three to six months or more, depending on the same circumstances mentioned in the previous paragraph. We assume in Figure 3.10 that a decision has been made during the pre-contract period on the supplier and specifications. The supplier will proceed with the design engineering portion of the program, culminating with a *preliminary design review* (PDR) some time around the sixth month. The PDR will be evaluated by management from the supplier as well as the satellite operator and its technical and business staff. Once completed and progress affirmed, the spacecraft supplier will move into the detailed design and manufacturing phase. At some point, perhaps 12 months from start, a *critical design review* (CDR) will be held for the same reviewers with the objective of ratifying that the spacecraft program is proceeding correctly.

Units may now be integrated into systems and tested for compliance with the end-to-end requirements of the satellite. During the *integration and test* (I&T) phase, the subsystems are installed in the spacecraft and tested both in ambient air and in a chamber that simulates the space environment. The objectives here include:

- Build a spacecraft that is capable of surviving the launch and on-orbit environments;
- Demonstrate that the satellite will meet its performance specifications on an end-to-end basis;
- Demonstrate that the satellite can be operated effectively by radio control through the TT&C system.

After the spacecraft completes the I&T phase, it is ready to be shipped to the launch site where it is again checked for integrity and operability and placed on top of the launch vehicle to be tested once again. The launch site preparations take between one and two months, depending on the type of vehicle and the number of spacecraft to be launched at the same time. Also to be arranged are

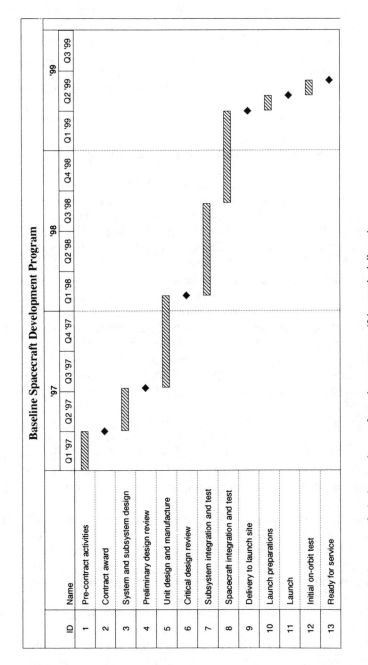

Figure 3.10 A typical spacecraft design and manufacturing program (24-month delivery).

the tracking stations and services that allow the satellite manufacturer and operator to conduct transfer orbit operations and initial on-station testing. Once launched, the satellite is tracked through the various phases and put into service. It can then be handed over to the satellite operator and service initiated to users. If this is a new type of network service, then users may need to conduct network checkout testing to demonstrate that everything is in working order and ready for commercial service.

3.4.4 Satellite Backup and Replacement Strategy

Under the assumption that an operator's satellites will work as planned, one must still plan for replacement of the satellites at end of life. This can be a complex process because of the time needed to design and manufacture the replacement satellite (not to mention the time it takes to figure out what kind of satellite to buy). The operating lifetime of a particular satellite is only known within something of the order of a plus or minus three months accuracy, so there is risk associated with launch delays.

An example of a replacement strategy for a hypothetical satellite system consisting of three orbit positions is shown in Figure 3.11. As this suggests, the best and simplest approach is to start with the current orbital arrangement and build a series of timelines (arrayed from the top to the bottom of the page). The satellite operator in this example starts in 1996 with three operating satellites: F1 and F2, launched in 1985, and F3, launched in 1990. This particular situation might have come about because F1 and F2 were launched within six months of each other to provide a reliable system of two satellites; since both reached orbit successfully, the third satellite, a launch spare, could be delayed until demand materialized. The operator chose to place F3 into service in 1990 as an on-orbit spare and use it for occasional video and other preemptible services. This provides high confidence that at least two satellites will be available. We assume here that the operating lifetime of each satellite is approximately 12 years.

The satellite operator purchased two replacement spacecraft (F1R and F2R) for delivery and launch in 1998 and 1999. This will ensure continuity of service, provided that both launches are successful. Figure 3.11 indicates that in 1998, F3 will be taken out of service and drifted over to F2's orbit position. This will allow F3 to take over for F2 when its lifetime runs out. Next, the replacement for F1, called F1R, will be launched in 1999 so that services can be transferred to it in a timely manner. The year 2000 sees F2R launched and placed into F3's old orbit position, which will have been vacant for about a year. This scenario provides high confidence that at lease two orbit positions will be maintained during the entire transition. If there had been a launch

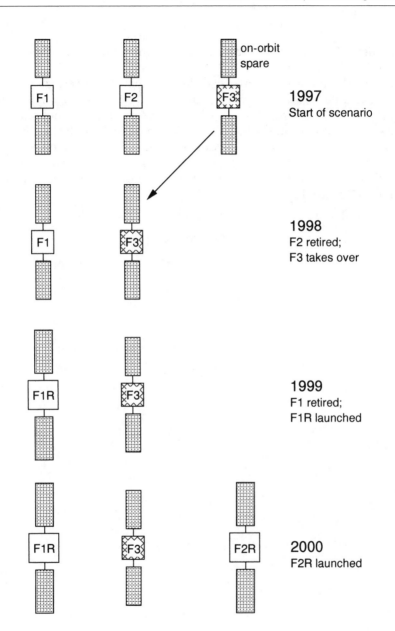

Figure 3.11 A typical satellite replacement scenario.

failure, then F3 would have lasted long enough to permit another spacecraft to be built and launched.

Obviously, there are many possible replacement scenarios and it makes sense to examine as many as can be imagined. There are financial as well as regulatory considerations. While this is done, users of the system must be kept informed and made to understand how the operator is replacing the satellite that makes their respective businesses possible. This is very critical because experience has shown that an operator who practices good replacement planning will tend to have a better reception in the marketplace. The same goes for the application developer and user.

Reference

[1] Elbert, B. R., *Introduction to Satellite Communication*, Norwood, MA: Artech House, 1987.

Part II
Television Distribution
and Broadcasting

Television Applications and Standards

4

As the principle application area for commercial communications satellites, television is very suited to the characteristics of the medium. The point-to-multipoint nature of video communications fits the wide-area broadcast feature of the satellite link. The ability of a satellite to serve a particular TV market is simply determined by the coverage area footprint. Thus, the Galaxy satellites are optimized to serve the fifty United States, while AsiaSat satellites cover east and southern Asia. The TV signals themselves must be consistent with the technical and content characteristics of the region served, aiming for one or more particular user segments. The important segments include network broadcasting to local over-the-air TV stations, cable TV (CATV) systems, and direct-to-home (DTH) subscribers who own their own dishes. A complete discussion of the DTH segment is provided in Chapter 6.

Figure 4.1 provides a framework for the discussions in this chapter. Of fundamental importance are the standards used in the creation, organization, and distribution of the programming product. In the analog domain, the same format is used during each stage of preparation and delivery. This imposes tight specifications on the transmission performance of the channel, particularly the video *signal-to-noise ratio* (*S/N*) and various impairments that distort the picture. These are covered at the end of this chapter. Digital formats, which are more tolerant of noise and distortion, are described in the next chapter. During the preparation of the product, the general view is that no impairment should be introduced. In digital terms, this means that the highest data rate possible should be used. Distribution of the signal to the consumer can be with the lowest data rate that is consistent with an adequate perceived quality.

The attractive nature of modern TV programming is a tribute to the technology and skill used to put the "product" together. The other vital aspect of this medium is the range of possible applications to which the product is put to use. Network broadcast is the broad category that includes the various forms of entertainment television. Once carried out as a local service, network broad-

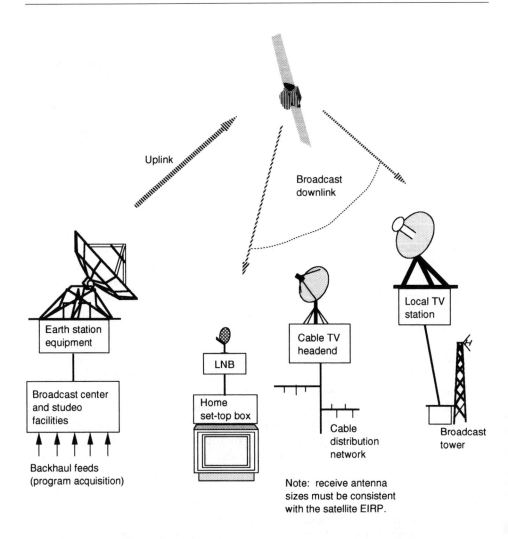

Uplink

Broadcast
downlink

Earth station
equipment

Broadcast center
and studeo
facilities

Backhaul feeds
(program acquisition)

LNB

Home
set-top box

Cable TV
headend

Cable
distribution
network

Local TV
station

Broadcast
tower

Note: receive antenna
sizes must be consistent
with the satellite EIRP.

Figure 4.1 Framework for television applications.

casting is now a national or even international medium. Businesses also exploit
the live-action nature of television through private broadcasting and two-way
interactive video teleconferencing. In the case of the latter, the medium provides
the means to engage in a new form of interpersonal communication where the
content is also supplied in real time.

The functional requirements and standards for various TV applications
fall into a number of categories. In a very comprehensive study of the satellite
communications industry, the consulting firm of Booz, Allen, and Hamilton

has identified two broad categories: programming and business TV [1]. Included under programming are a variety of uses of the TV medium to provide entertainment and information to the public.

The availability of the signal is determined by the equipment used to receive it coupled with the mechanism used to pay for it. With DTH systems, individuals can receive the signal directly with appropriate equipment, providing the most direct connection between the programming supplier and the public. On the other hand, the supplier will often want to restrict distribution to those whom have paid for it or to those for whom it is intended for a variety of other reasons. Restrictions on delivery can be based on geography or association with a group. In the following subsections, we review the categories of satellite video applications as a means to better define the requirements for each.

4.1 ENTERTAINMENT PROGRAMMING

The ultimate consumer of TV programming is the household. In modern times, people have more time available for recreation and therefore are looking for the best value [2]. Beyond a doubt, this is from the home TV set. There are approximately 100 million homes with television sets in the United States. The average time spent viewing TV in each of these homes is about seven hours. In advanced economies, the quality of this programming is quite high and everybody can find something that they enjoy. The price of doing this, after you have purchased the set, is nearly zero. Adding more variety usually means spending more disposable income. For any other form, such as movies on the screen, sporting events, and theatrical events, the price per hour is much higher. But programming is still going to be the best overall buy, even when obtaining it through pay services over cable and satellite.

The nature of TV programming is quite sublime, where its value is determined by the size and type of audience that it draws. By far, the most valuable programming is that which is developed for and used in the major TV networks that broadcast their signals through local transmitters and over cable TV systems. The total revenue of the U.S. TV networks is about $32 billion, which is comparable to what newspapers collect from their advertisers. About one-third of this revenue goes to the national networks themselves, with the rest divided up among the local TV stations and the cable TV networks. The most attractive type of programming is that which appeals to a broad cross section of the public, which can be divided into certain key age groups. Behind this is the money that comes from manufacturers, retailers, and service providers, who advertise what they sell through the TV medium.

The following discussion suggests what the nature is of this most valuable segment of programming, based on the line-up of TV shows and their relative popularity [3]. The most popular program on the ABC network in the United States is *Home Improvement*, a situation comedy (sitcom) that stars Tim Allen. This particular program appeals to the most sought-after age group by advertisers; that is, 18 to 45 years. It has been on top at ABC for three years. One might ask if its popularity is due to the story line or to the star, Tim Allen, who also makes movies. Two other very good draws are *Roseanne*, which was number 4 in its sixth season on TV, starring Roseanne, and *Coach*, which was number 8 in its sixth year. ABC in fact had the most shows in the top ten of all of the networks. Perhaps a common thread that exists among its top shows is that they feature stand-up comedians in their lead roles (stand-up comics who also happen to be good actors). Two other successful shows follow a totally different line. The ABC *Monday Night Football* program carries a regular professional football game but at a time when working men and women can relax after a day at the office or factory. The other is *NYPD Blue*, a racy police drama that has attracted both a solid audience and some public criticism. The top national news magazine shows, like *60 Minutes* and *20/20*, can also place very well in the standings.

One could spend a lot of time studying the different programs, their makeup, and source of popularity. An example of a listing of the top shows in the week of May 8–14, 1995, is given in Table 4.1 [4]. This was taken about one year after that of the previous paragraphs, indicating the trendy nature of popularity of national TV shows. The business is very dependent on the creativity of the writers, producers, directors and actors—just like the movies. Also, what is popular this year can lose appeal in the passage of time. All popular shows have substantial value as reruns on cable TV networks and independent TV stations. *Law and Order*, for example, is very successful as a rerun on the A&E cable network at the same time that new shows in the series are appearing on CBS. The same can be said of the venerable *Murder, She Wrote* drama series, starring Angela Lansbury.

The national network TV programs discussed in the previous paragraphs and listed in Table 4.1 are produced and first released in the United States. This results from the fact that the United States still represents the best market in the world for such English-language programming and because the networks and their program producers are United States–based. Even still, many of these programs are distributed throughout the world, resulting in substantial additional revenues for their producers.

In Europe, the quality of satellite-delivered programming tends to be higher, in commercial value, than what passes through the terrestrial networks. This evolved probably because TV stations were largely installed and operated by governments as a public service and not as a business. The same could be said of TV in Asia and other parts of the world. For this reason, services from Sky of the United Kingdom, RTL of Luxemboug, and Canal Plus of France have

Table 4.1
Top-Rated Shows for the U.S. TV Networks, May 8–14, 1995,
According to Nielsen Ratings

Rating	Show	Network
1	*E.R.*	NBC
2	*Friends*	NBC
3	*Seinfeld*	NBC
4	*Saturday Night Movie*	ABC
5	*Home Improvement*	ABC
6	*Sunday Night Movie*	NBC
7	*NYPD Blue*	ABC
8	*Ellen*	ABC
9	*Monday Night Movie*	NBC
10	*Grace Under Fire*	ABC
11	*Murder, She Wrote*	CBS
12	*Coach*	ABC
13	*Frasier*	NBC
14	*Primetime Live*	ABC
15	*Murphy Brown*	CBS
16	*60 Minutes*	CBS
17	*Monday Night Movie*	ABC
18	*Mad About You*	NBC
19	*Thunder Bay*	ABC
20	*Roseanne*	ABC

become powerhouses and mainstays in the programming field. The programming mix on these systems includes locally developed shows and, more particularly, many of the most popular entertainment from the U.S. networks and studios. An example of the most popular satellite-delivered programs is presented in Table 4.2. The Sky services, in particular, are offered by British Sky Broadcasting (BSkyB), formed from the merger of Sky, controlled by News Corp., and British Satellite Broadcasting. The combined company has reached 6 million subscribers in 1996 through a combination of cable and DTH access, split roughly 33 and 66%, respectively.

In Asia, the Japanese enjoy probably the most extensive mix of commercial television broadcasts and networks. This is because of the economic status of the country and the great appetite that the people have for entertainment and information. There is an extensive infrastructure of microwave and fiber, as well as several TV stations in each major city. Japan, then, is more like the United States than any other country. Much of the programming is locally produced, but there is a love of American TV shows as well. Hong Kong and Singapore also have excellent local broadcasting services in multiple languages

Table 4.2
The Top 25 Satellite Programs in 1995 (*Cable and Satellite Europe*, February 1996)

Program	Channel
Bigtime Boxing	Sky Sports
Super Sunday Match	Sky Sports
Jurassic Park	Movie Channel
Football	Sky Sports
FA Cup	Sky Sports
Demolition Man	Movie Channel
Fugitive	Movie Channel
Bushido	Sky Sports
Mrs. Doubtfire	Sky Movies
Review	Sky Sports
Jungle Book	Disney Channel

to reflect the population. Some of these stations, particularly TVB in Hong Kong, are extending themselves throughout Asia in response to the rapidly growing demand for TV.

More recently, the Star TV services, initiated by Hutchison Wampoa and sold in 1993 to News Corp., are attracting audiences through C-band DTH and cable TV. This supplements and in some cases substitutes for local TV in areas where the broadcasting is weak or nonexistent. Star began its operation at the start of the 1990s using the relaunched Westar 6 satellite and a second satellite, AsiaSat 2 went into service in 1996. The coverage of these satellites extends throughout Asia and all the way to the Middle East, making the programming services available to over two-thirds of the world's population. Programming is a mix of U.S.- and European-derived entertainment, news, music, and regional movies. That U.S. programming maintains and even increases its popularity in Europe and Asia is a source of concern to the national governments who may have a policy of encouraging local content and culture. On the other hand, the satellites that deliver these services, including the Astra series in Europe and AsiaSat in Asia, are the recognized hot birds where ground antennas remain focused.

Programming can be distributed through a variety of systems. Through the 1980s and the 1990s, satellites have been the principle carrier of programming material. When satellites first became popular for TV transmission, it was a simple matter to format the signal for distribution. As discussed in Section 4.4, the analog signal format that comes from the camera is essentially the same format used in production and transmission and for viewing by the home receiver as well. With the introduction of digital TV, new formats are being evaluated and exploited in certain situations. The typical low-cost home TV receiver itself is still the same. In fact, television itself was first demonstrated at the 1939 World's Fair, where large numbers of people saw it. Early receivers became available in 1940s for limited broadcast reception. According to Paul Resch, an industry observer and practitioner, one of the original receivers would still take the signal off the air today—except of course for Channel 1, which has been reassigned to mobile radio. Interestingly, the same principle applies to the first telephones in that you could receive a call over most telephone lines.

Program distribution systems throughout the industry employ all or some of the architecture shown in Figure 4.2. There are many options as to how this architecture can be implemented, but the most economical approach is to combine these elements into a single facility. This is unique to the satellite industry, where one organization can create, distribute, and sell a product from one location. Network broadcasters like CBS and Fox in the United States, Star TV in Hong Kong, and Tokyo Broadcasting in Japan exploit the efficiency of this arrangement. The basic functions provided are:

- Program origination in a studio;
- Display of prerecorded material, on film or tape;
- Program acquisition by terrestrial links to remote studios and other venues;
- Program acquisition by satellite links (typically C band) from syndication sources;
- Reception of material using electronic news gathering (terrestrial microwave);
- Reception of material using Ku-band *satellite new gathering* (SNG);
- Editing of these various inputs into the actual program to be broadcast;
- Relay of programming captured from satellite links;
- Transmission from the studio to a satellite uplink (either C or Ku band);
- Transmission from the studio to a local broadcast transmitter and tower;
- Control and switching of input and output video and audio signals to prepare program material for recording and distribution.

The selection and use of combinations of these functions depend on the type of programming and service involved. For example, some programmers produce all of their own material, including sports event coverage, news, and movies, and therefore require the most extensive and reliable facilities. NBC, for example, has redundant broadcast centers in New York City, NY, and Burbank, CA. A smaller operation like the Disney Channel in Singapore will emphasize the local replay of tape and the retransmission of programming that is received over a Pacific Ocean satellite. The least impressive arrangement that

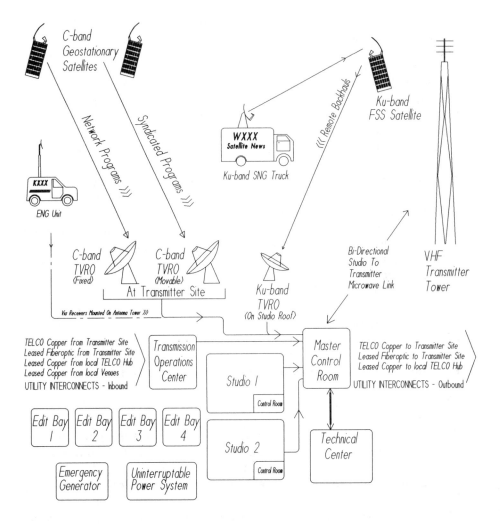

Figure 4.2 Main broadcast center architecture for program development, acquisition, and distribution to affiliate TV stations (courtesy of Paul Resch).

this author has seen was a pair of tape machines connected to a video uplink that represented the complete program distribution system for a start-up cable TV subscription move channel.

One of the most important features of modern broadcast centers is the use of backhaul circuits, many of which are provided over other satellites. As indicated in Figure 4.2, there are several backhaul inputs available at the facility.

The satellite capacity that is used for this could be from a pool of available transponders that is maintained by the network. In markets like the United States and Europe, occasional video services from larger satellite operators provide the temporary transponders when needed at a substantially lower cost. The TV network or station reserves the transponder ahead of time and arranges for an uplink for use during the programming event. More recently, backhaul has become available over fiber optic networks. A company called Vyvyx in the United States uses the fiber optic network of WilTel to make occasional connections between networks like Fox Television with football stadiums around the country. In Malaysia, the second carrier, Maxis, is likewise extending its fiber to locations to afford this type of point-to-point video transmission.

The truck in the center of figure indicates how SNG service is introduced so that the network may move into a site no matter where it is located. TV networks and stations own these trucks as they provide a great deal of versatility and flexibility in covering remote venues and events. The typical SNG truck includes a 2.4m antenna and 400W Ku-band transmitter, along with a limited quantity of production equipment to permit some local editing. The one condition in using SNG and occasional video service, of course, is that the location have access to a satellite that can also be received at the broadcast center. This could be a problem with coverage of very remote places, particularly in other countries. A double-hop arrangement would then have to be made with an international service provider such as INTELSAT or PanAmSat.

4.1.1 Network Broadcast

The network broadcast to TV stations originates from the video uplink shown in Figure 4.1 (this particular feature is not shown in Figure 4.2 but could be added to one of the program acquisition antennas). According to Booz · Allen, the network broadcast category of TV programming includes commercial and noncommercial services. A typical network *feed* can last from a few hours per day, during evening *prime time*, to a full 24-hour service. The basic characteristic is that the feed contains networkwide programs intended for local rebroadcast by TV stations. In North America, each TV station is separately owned and managed. There are also groups of stations under common ownership, including stations in major cities like New York, Chicago, Los Angeles, and San Francisco that are owned by the national networks themselves. In the United States, there is currently a limit on the total size of audience that can be addressed by stations that are owned by the same network. Other companies and individuals can own multiple stations (another restriction is maintained for newspapers so that they cannot dominate the supply of news in a given city).

Network broadcasting in free-market economies has generally been supported by advertisers who pay to have their message put in front of the mass market. Total revenue from television advertising was $14.4 billion in 1990 and was experiencing an 8% compound growth rate at that time. Well-known brands like Coca-Cola, Sony, Nestlé, and Texaco got that way through the medium of commercial television. As far as the sponsor is concerned, the most important piece of material to be transmitted is the advertising message. The rest is used as a draw for the audience, and the larger the audience the better. This is not meant as a condemnation, because all of us can enjoy much of the programming that is created with advertising dollars. Rather, we need to understand that the economics of network programming in the main are driven by the need to influence buying behavior.

National brands can be advertised via the network feed, but much of the programming and advertising can be generated at the local level. In the United States, there is a government mandate to encourage local control of broadcast TV. This means that the local TV station must include facilities to create as well as distribute program content, including news and commercials. Local sports events are popular, so the station needs to be able to collect sports feeds. This includes the situation when, say, the Los Angeles Dodgers go to New York to play the New York Mets baseball team. The LA sports announcers follow the team to New York where they occupy a broadcast booth at the stadium. Their coverage of the away-from-home game is carried by satellite (under special arrangement or as a network feed) back to the hometown, where it can be broadcast locally.

Local stations also receive compensation from the network with which they are affiliated. This source of revenue has declined as the networks have sought to reduce costs in the face of competition from cable TV and DTH. There have been a number of new startup networks in the United States, most notably the now established Fox TV network and more recently UPN, which is backed by Paramount. The money appears to be there in national advertising, but getting an adequate share of it is becoming more of a challenge.

In Chapter 5, we review developments in various forms of digital TV, including HDTV. U.S. TV networks have been participating in the technical development and political debate that concerns HDTV. It is the policy of the FCC that HDTV will be made available to the viewing public within a time frame of less than 10 years from the writing of this book. By way of contrast, NHK of Japan has been broadcasting HDTV programs using their hybrid analog/digital MUSE format over the Japanese DTH network. To cause this to happen in the United States, the FCC will require that TV sets that appear on the market after a certain date must contain HDTV decoders, in addition to the standard NTSC capability. Dick Smith, chief engineer of the FCC, stated, "If we do not include HDTV at the beginning—and did not require the receivers to decode that—it would not be possible to add it later on. We want to keep our options open and be able to do all of the above. With that, I think there's a great future for terrestrial (network) broadcasting" [5].

At the same time that the FCC is pushing hard on the networks to implement the HDTV standard once it is adopted, they are maintaining some of the privileges that the networks have enjoyed over the decades since TV was introduced. The complicated rules that govern how the networks can control the affiliate stations that broadcast the network shows to the public are changing slowly. They are also very complicated. One such rule, called the *right to reject*, permits affiliates to refuse network programming. The territorial-exclusivity rule prohibits agreements that would not allow a station from broadcasting a network show that the local affiliate does not broadcast (note the triple negative). In 1995, the FCC announced its intent to maintain this rule under the restriction that it solely cannot be applied on financial grounds (the fourth negative). On the one hand, members of the FCC are asking if the FCC should be in the business of regulating the relationship between the networks and their affiliates; yet on the other hand, it continues to do just that.

In much of Europe, Latin America, and Asia, stations within a country can be controlled by the national network. Their primary support in the past has been from government funding, the motivation for which can be quite varied. As the premier government broadcaster, the BBC is acclaimed for the intelligence put into the material that is presented. They have been free to use their judgment to decide what is best for the viewing public. Some of it makes its way to the United States to be enjoyed on public television and some of the cable networks. But today, the BBC is under more scrutiny to show that it continues to deserve subsidy from the U.K. populace. A similar situation exists in the United States with respect to PBS, which could not exist without U.S. government support. PBS is the national programming exchange and distributor for programming that is largely created by public TV stations. Ironically, PBS was the first to adopt satellite distribution at a time when the commercial network broadcasters relied on terrestrial microwave.

4.1.2 Cable TV

The cable TV industry was created by the simple need to improve TV reception in remote areas. Distant VHF and UHF broadcasts are received by high-gain antennas on mountain tops or tall towers, which are shared by residents of a given community. In fact, the original name of the service was *community-antenna TV* (CATV), an acronym that is still in use. This type of business did not become an industry as it is today until the introduction of pay TV services delivered to the cable TV system by geostationary satellite. The availability of a wider array of programming has made cable TV attractive to urban residents.

Now, with a single cable access, we receive the local TV channels (including some that are just out of range or from nearby cities), national cable TV networks, *pay-per-view* (PPV) movies and events, and international programming as well.

4.1.2.1 Cable System Architecture

An example of the design and layout of a typical cable TV system is shown in Figure 4.3 [6]. The signals are distributed by a branching network composed of 75-Ω coaxial cable and wideband distribution amplifiers, which provide acceptable signal quality to homes. The topology of the network must be tailored for maximum efficiency of distribution and lowest investment and maintenance costs. The five major parts of this type of distribution network include the following.

Headends are receiving antennas and equipment that provide the TV signals to be distributed. Cable TV headends include VHF and UHF antennas and C- and Ku-band satellite receiving dishes, along with the required number of TV receivers to demodulate the channels and transfer them to the cable network.

Trunk cables bring the collection of channels from the headend to the neighborhood to be served, usually on a point-to-point basis. Amplifiers are introduced at appropriate distances to maintain adequate signal strength and *S/N*. If the distance to be covered is extremely long, it may be preferable to use a terrestrial line-of-sight microwave link or fiber optic cable. Another use of trunk cable is to allow multiple headend sites to connect to a common equipment room that houses the video receivers and distribution amplifiers.

Distribution cables (or feeder cables) branch out from the terminating point of the trunk cable and run past the homes in the neighborhood. There would be many distribution cables needed to cover a large community. This tends to maintain a better signal quality in terms of *S/N* and distortion because the video signals do not have to pass through as many individual amplifiers. Also, the failure of one feeder line or amplifier only affects a relatively small group of homes. Problems and power outages along the distribution cables are the principle causes of service degradation and interruption.

Drop cable to the home connects from the distribution cable to a tie point akin to the telephone box or electric power meter connection. From the home connection, internal coaxial cable is used to bring the signals to the TV set top box.

A *set-top box* (or consumer electronics unit) allows the subscriber to select particular channels for viewing and to control which channels are available under any particular service plan or option. The basic set-top box merely contains a downconverter that can translate one channel from the available spectrum to, say, channel 3 or 4, where it can be picked up by a standard TV set.

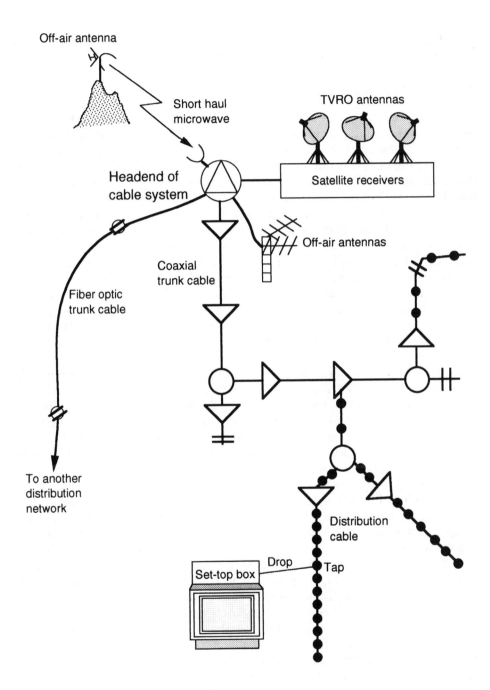

Figure 4.3 Typical layout of a cable TV system.

This duplicates what goes in a TV tuner and VCR but supports a more efficient spectrum arrangement than is used over the air.

4.1.2.2 Set-Top Box

The set-top box was originally introduced to provide a convenient interface between the set (and, more recently, the VCR) and cable drop. In addition, the box improves quality through more effective channel tuning and filtering. This is because the channels are adjacent to each other on the cable, whereas over-the-air broadcasts allow more guardband between channels within a common broadcast area. This scheme was adhered to in the United States so that the selectivity of the standard TV tuner could be relaxed [7]. The box also screens out over-the-air radiation that could interfere with cable signals. These are typically delayed in time due to the lower speed of propagation over coax as compared to the air medium. Eventually, the set-top box added new capabilities to extend the normal frequency range for greater channel capacity. Functionality for descrambling and conditional access, discussed in Chapter 6, are incorporated in later models. Data can be delivered over the cable to the set-top box, either as a separate channel or through the vertical blanking signal of the video. The latter is limited as to the speed and capacity that can be delivered to the home.

The design and function of the coming generation of set-top boxes is shown in Figure 4.4 [7]. This model is not in active use on cable but is similar to the *Digital Satellite System* (DSS) used by DIRECTV in the United States and the *European Digital Video Broadcast* (DVB) standard. The video is digitized at the source and distributed over the satellite or terrestrial network. The electronics that are needed to perform the decoding and demultiplexing functions are coming available through wide adoption of the MPEG-2 standard (discussed in Chapter 5). Two services such as data communication and telephone are being introduced using a second (return) cable to allow up-stream communications. The additional functions in Figure 4.4 include the on-screen program guide and menus that allow the subscriber to customize the service. Out-of-band signaling, which is the name for a separate frequency channel used to distributed high-speed data to subscribers, can deliver an array of teletext and information services. CD-quality audio can be offered in the same manner, with the number of channels limited only by the availability of source material (one video channel is capable of supporting 50 stereo audio channels).

The technology to design and manufacture this type of set-top box is already here. The difficulty that the cable TV industry faces is to select the particular set of features and resulting services that they wish to make available to subscribers. There is an open question as to the commercial viability of interactivity, which is something that will add significantly to the cost of the

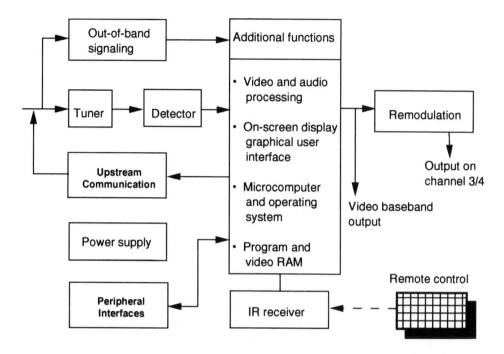

Figure 4.4 The next generation of TV set-top box, including addressable features and digital decompression. An interface to an external device such as a PC is provided.

box and the cable plant as well. Data distribution, which can mimic interactive service through the use of teletext (i.e., repeating the information stream at a reasonably fast interval), is already in operation but is hardly universal. Conversion to the MPEG-2 digital format is a strong possibility in North America and Europe. Therefore, the next generation of set-top box will likely appear as a more user-friendly but conventional *tuner* for existing cable network programs.

There is another view of interactivity that responds to the greatly increased interest in the Internet and the World Wide Web. The current telephone network is generally inadequate for comfortable use of many Internet and WWW services, particularly file and image transfer. With the possible introduction of two-way data over the cable at speeds well above the 28.8-Kbps capability of analog telephone lines, there is a good opportunity for cable TV to leapfrog the telephone companies in providing Internet access. This feeling has begun to take hold in the U.S. cable TV industry, where many more conventional interactive TV projects have been placed on hold or canceled entirely [8]. U.S. regional Bell telephone companies had planned to rebuild local networks for interactive cable service, but these are being slowed or halted because of the concern that consumers may not be willing to support the massive investment involved. Instead of using the TV set, cable operators are adopting the PC as the preferred platform for providing interactivity.

In 1996, only two major interactive TV projects remained in the United States, namely, *Time Warner's Full Service Network* (FSN), which was launched in 1994, and Videotron's unlaunched *University, Bidirectionality, Interactivity* (UBI) consortium in Quebec, Canada. FSN was launched with *video on demand* (VOD) with 36 movies, nine stores offering products in an electronic shopping mall (the same concept also appears via America Online and CompuServe as well as the Web), six card games, and three network games. A total of 50 subscribers have access to this experimental offering. Time Warner continues to experiment with the system by adding on-demand services, more moves, and games and changing the pricing basis. The UBI approach is perhaps ahead of its time by offering electronic catalogs and reservation and ticketing services but no VOD. This was intentional because Videotron determined that the consumer would not be able to bear the cost of the necessary infrastructure. The service was supposed to have been introduced in 1994 but has not reached the consumer as of the date of this writing. Their difficulty has been the development and delivery of the appropriate set-top boxes, which would seem to perform many of the functions of a PC running the Mosaic or Netscape software to access the Web.

4.1.2.3 Cable System Design

Cable TV systems differ in the number of channels that they are capable of carrying. This depends, of course, on the bandwidth of the cabling plant and amplifiers as well as the tuning capability of the set-top box. The lowest capacity systems deliver 12 to 22 channels as they only operate in a 170-MHz bandwidth, between 50 MHz and 220 MHz. These systems are still found in rural areas and in small cluster communities. In developing countries, such a capacity would be viewed as substantial, so a design with this limitation would still be very attractive on a near-term basis. However, the clear trend is toward systems with 50 to 60 channels of capacity over a single cable, occupying a bandwidth of 500 MHz. Doubling the capacity to 120 channels is done by adding a second cable so that each home is served with a pair of cables. What is attractive about the dual cable is that it can accommodate a future upgrade to two-way communications. Otherwise, bidirectional transmission would reduce capacity by requiring segmentation of the band as well as the use of bidirectional amplifiers.

In more modern plant designs, the coax is being replaced by fiber optic cables and optical repeaters, as this technology is more reliable, has greater bandwidth, is lower in loss, and is more cost/effective on a long-term basis.

Fiber can be used on the trunk, distribution, and drop cable lines. The initial step being followed is called *hybrid fiber coax* (HFC), by which the video from the headend is digitized and routed around the metropolitan area on the fiber backbone. The actual distribution to neighborhoods would continue to use analog coax. The HFC approach allows the telephone companies to use existing fiber plant, although the video codecs, which are very costly, would have to be added. The next step in applying fiber optics is called "fiber to the curb," where the fiber extends through the distribution cable but not to or into the home. The transmission over the fiber distribution cable could either be analog or digital. This is a natural extension that leaves the expensive and difficult last step to new construction and future upgrades. A converter at each tie point transfers the channels from lightwave to the loop coax cable. This uses many of the benefits of fiber without having to replace existing loops and set-top boxes.

Signal quality at the subscriber end is intended to be better than what reaches the typical home through over-the-air broadcasting. This, of course, depends on how well the home would be able to receive the TV signal with a roof antenna of some kind. The NTSC system has been around for a long time, and standards for quality are reasonably well understood. In 1959, the *Television Allocations Study Organization* (TASO) devised a five-point scale to allow viewers to rate the quality of different levels of noise in the picture. The scale is described in Table 4.3, indicating a six-level scale that covers the range of (1) Excellent to (6) Unusable. In the context of cable TV, viewers expect to experience a level of fine to excellent, that is, TASO grades 2 and 1. To meet

Table 4.3
The TASO Picture Quality Rating Scale Used in Subjective Evaluation of Video Signal-to-Noise Ratio

Number	Name	Description
1	Excellent	The picture is of extremely high quality—as good as you can desire.
2	Fine	The picture is of high quality, providing enjoyable viewing; interference is perceptible.
3	Passable	The picture is of acceptable quality; interference is not objectionable.
4	Marginal	The picture is poor in quality, and one wishes they could improve it; interference is somewhat objectionable.
5	Inferior	The picture is very poor but watchable; definitely objectionable interference is present.
6	Unusable	The picture is so bad, one could not watch it.

a TASO grade of 2, the cable must deliver a *S/N* of 46 dB to 47 dB. The empirical data presented in Table 4.4 would suggest that this level of *S/N* would meet TASO 1 or 2, meaning that the picture is fine to excellent. This would be the value that arrives at the home receiver and therefore includes all contributions from the source material and origination processing, satellite link, Earth station, and cable plant.

Distortion must be within acceptable limits, which is one of the more challenging aspects of cable design. This is because the amplifiers along the cable each add intermodulation noise to the total distortion budget. A summary of noise and distortion requirements for analog cable plant design is provided in Table 4.5.

These analog parameters make sense when the satellite and cable distribution systems provide an analog transmission channel. When we move to the digital format for everything but the TV receiver itself, it no longer makes any sense to view the problem in this manner. As discussed in Chapter 5, the process of digitizing and compressing the analog TV signal introduces impairments that cannot be removed. The viewing quality, however, is at least equivalent to what the standard analog cable system is cable of delivering. With the digital set-

Table 4.4
The Relationship Between Measured *S/N* of
the Video Signal and the Viewing Quality

Video S/N, dB (weighted)	Viewing quality
40	Snow objectionable in picture
42	Snow clearly visible
44	Snow just perceptible
46	No snow visible

Table 4.5
Video Quality Objectives for Analog Cable Plant Design
(NTSC Video Standard)

Parameter	Abbreviation	Value
Signal-to-noise ratio	S/N	46.0 dB
Composite second-order distortion	CSO	−53.0 dB
Composite triple-beat interference	CTB	−53 dB
Signal into TV receiver		1 mV
		0 dBmV

top box, all of the impairment is produced at the sending end and the resulting picture would be the same even if the encoder were directly connected without the satellite link and cable in the middle. The key parameter for the link itself is BER as opposed to S/N. If the BER is maintained above the minimum requirement, then picture quality will be completely stable. Below this operating point, the picture rapidly becomes unusable as opposed to being degraded.

Security in conventional cable TV systems is needed to prevent unauthorized viewing. This comes into play when the cable-operating company offers different levels of service: a basic level, which includes local TV stations plus five to ten basic cable networks like CNN, TBS, ESPN, and the Weather Channel. The next level might add ten more basic services, like The Discovery Channel, Arts and Entertainment (A&E), and MTV. There might even be a third level with basic services that adds a regional sports channel and a comedy network like Comedy Central. Higher charges are incurred if the subscriber takes the premium (i.e., pay TV) channels, which do not carry advertising and therefore are paid for exclusively with subscriber revenues. The cable operator must be able to restrict the watching within the same home to allow only the paid-for channels to be available.

There are two security techniques in popular use, both of which are very low in cost. The cheapest approach is to insert a jamming waveform on the particular frequency channel and provide the paid subscriber with the ability to remove the signal within the set-top box. This type of protection is very poor since the means to evade it is very simple in electronic terms. The next in cost is the use of a negative trap, which is a band stop filter that removes or distorts a specific channel or a contiguous band of channels. The negative trap filter affects all TV sets in the same household since it is placed in the incoming line.

The direction that the industry is taking in security is to use signal scrambling and addressability of the set-top box. This is the same technique pioneered on satellite TV networks. The scrambling is accomplished either by (1) suppressing the synchronization signal, which is needed in the TV set to lock on to the scanning waveform; or (2) baseband scrambling, which modifies the video and audio in a pseudorandom manner, requiring the set-top box to have prior knowledge of the random pattern of scrambling. Both of these approaches are combined with addressability, which allows the cable operator to activate a given channel on the set-top box. Upon customer order, the cable operator transmits a unique code over the cable that causes the set-top box to descramble the channel in question.

4.1.2.4 Cable Service and Programming

Cable subscribers pay a monthly access fee for use of the infrastructure. Rights to provide this type of service are usually granted by the local community government as a monopoly franchise. In recent years, the monopoly privilege has come under attack as new entrants wish to offer a wider array of services, including interactive video and data. This aspect will be covered later in this chapter.

There are two classes of cable TV channels: advertiser-supported and pay (also called subscription or premium). A breakdown of the leading United States–based advertiser-supported cable TV channels, also referred to as cable networks, is provided in Table 4.6. These channels were selected based on their viewership and general appeal. An advantage of an advertiser-supported channel is that, in theory, the channel can be transmitted in the clear without scrambling. The only issue is to provide a measure of the audience size, which will satisfy potential advertisers. In reality, advertiser-supported channels must be scrambled to protect the copyrights of producers and owners of the programs. The producer charges the programmer such money for the right to distribute the movie to a specific market (say a country or even a city). Reception in a nonauthorized region must be curtailed. Also, some ad-supported channels, such as CNN and ESPN, charge small but significant subscriber fees to defray some of the expense of producing programs. Sports and news channels are less costly to produce because the original material is self-generating.

Typical advertiser-supported channel formats include:

- *News (24-hour):* either a general-interest service like CNN and Sky News, or one the focuses on a particular aspect such as the Weather Channel or CNBC.

Table 4.6
Typical Cable TV Channels From the United States (1995)

Name	Ownership	Availability
ESPN	Disney/ABC	Global
CNN	Turner/TCI	Global
CNBC	NBC	United States
MTV	Viacom	Global
VH-1	Viacom	United States
Nickelodeon	Viacom	Global
The Discovery Channel		United States
USA Network	HBO	United States
Lifetime		United States
Home Shopping Network	Silver King	United States
The Nashville Network		United States
A&E		United States
The Weather Channel		United States

- *Sports:* may be 24-hour like the news channels or limited in time to correspond to when certain types of sporting events are held. ESPN is by far the most popular sports channel in the world and happens to have the largest viewership in the United States as well. European sports networks include Sky Sports and Eurosport.
- *Super stations:* Originated by Turner Broadcasting as WTBS, a super station is principally a local TV station that is redistributed by satellite throughout the country or region. This is more of a novelty in the United States where TV stations are usually restricted to local service. In most countries, there are national stations that rebroadcast the same signal nationwide and, hence by this definition, are already super stations.
- *Movie channel (old release):* Theatrical movies have long been a popular form of TV programming. The satellite-delivered cable channel dedicated to movies is popular as well. Through advertiser support, the subscriber does not pay an additional fee for the service but must endure the commercials none the less. Examples of 24-hour movie channels that are advertiser-supported include TNT, which is general interest, and American Movie Classics (AMC), which chooses to display older films that have enduring appeal. Similar services are available in Europe through Sky and Super and in Asia as part of the StarTV package.
- *Science and other special interest:* Cable TV has long had the promise of encouraging channel formats directed at selected audiences. This particular category is potentially of general interest but would not survive in over-the-air broadcasting except possibly in government-supported educational programs. The Discovery Channel (TDC) first appeared in the United States in 1984 and was an instant success. It selected program material that had originally appeared on the U.S. PBS. However, the difference here is TDC is advertiser-supported while PBS is not. Another immensely popular special interest channel is Arts and Entertainment (A&E), a cable network offering cultural programming from the BBC, PBS, and internal production sources, augmented by reruns of popular TV network series like *Murder, She Wrote* and *Law and Order.*

The advertiser-supported cable TV channel represents an interesting opportunity as a new venture. This is because there is literally an unlimited number of possible formats and subsequent subscriber markets. The cable medium allows a new channel to reach a reasonably large viewer base that would be hungry for new entertainment. There are a number of issues that a new entrant must address before attempting to get started in what has become a very competitive field. The programming idea itself is not the most difficult part. Rather, the first problem is start-up capital to acquire the transponder space, uplink facility, and studio. This must be available 24 hours a day. The staff to run these facilities must be familiar with the business and able to manage this type of operation on a very professional basis. Viewers will not accept anything less than the high production values that are standard in cable and over-the-air broadcasting. The other side of the coin is that conventional TV channels have more general appeal and can draw very large audiences. Their greater access and viewership means that they obtain higher advertising revenues from national advertisers like Coca-Cola, General Motors, and McDonalds. This then provides the money needed to produce the most attractive programming that, not coincidentally, makes the channel more attractive than cable to advertisers.

Narrowly focused channels can exist on cable because some metropolitan areas have relatively high concentrations of specific viewers. The Chinese Channel, for example, is viable within major cities like Los Angeles, New York, and San Francisco, where the Mandarin-speaking population totals over one million. Cable systems that serve these communities would offer the Chinese Channel, which delivers potential customers to a specialized type of advertiser.

Another major challenge that the new entrant faces is gaining access to the subscriber base. Cable subscribers cannot on their own cause a new channel to appear on their cable. Rather, it is the cable system operator who controls technical access and the purse strings as well. In many developed countries, cable TV systems are often massed under a national operator or holding company, called a *multiple systems operator* (MSO) in North America. This means that the national operator controls a substantial quantity of potential subscribers and hence can make or break the new programmer. For this reason, a new service will have to make it very attractive to the MSO by doing such things as allowing them to keep any subscription fees that are collected plus a portion of the ad revenue. It is even not without precedent for the fledgling programmer to have to give up some of its equity in order to reach the subscribers. Having gained access, the programmer can then approach the advertisers to start the flow of ad revenue.

Premium or pay channels are not advertiser-supported but rather are paid for entirely by the subscriber through a monthly fee. The amount varies from as little as $5 to as much as $15 per month, depending on the quality of the programming and the ability of the subscriber base to pay. Delivery of the channel must be controlled through one of the security techniques described earlier in this chapter. The need for this has been verified many times over, the most recent experience being the development of the backyard dish "industry" in the United States. In that particular case, homeowners were encouraged to spend between $3,000 and $10,000 for a C-band receiving system because they could receive free cable channels, including the pay services like HBO and Disney. This was possible in the mid-1980s because none of the services were scrambled. After HBO and others began scrambling, backyard dish sales nearly halted and this new industry almost collapsed.

Movie channels are clearly the most viable form of the premium channels as they can attract a large audience. For HBO, the world market leader, premium service means gaining a significant percentage of cable subscribers. Not all subscribers, however, will choose to pay the significantly higher cost of this service, but for those who do, it delivers a continuous flow of movies that include at least one new hit a week. Movie channels cost a lot of money to produce because of the high cost of acquiring the material and the need to collect revenues from every watcher on a monthly basis. The cable provides a degree of security itself with the cable operator taking responsibility for collecting the revenue. They keep a percentage of what the subscriber pays for the cable network service.

The cost of acquiring and/or developing programming for a pay TV channel far exceeds anything else, including the investment or operating cost of satellite transponders. The purchase of recent-release movies dictates whether the channel can survive or even make money. In the early 1990s when *British Satellite Broadcasting* (BSB) was competing with Sky TV for market share and for movie material as well, the cost of programming skyrocketed. This severely weakened BSB, who subsequently merged with Sky to form *British Sky Broadcasting* (BSkyB), which is controlled by News Corp. (the force behind Sky).

Leaders in the premium channel business also produce their own movies and specials. HBO, for example, is a credible movie studio of its own, having produced many movies for its cable services and for the box office as well. They are also in the business of producing special events such as musical concerts featuring famous stars like Diana Ross and Madonna. These are shown on the premium channel and find their way to other outlets such as theatrical movies and advertiser-supported cable channels.

The Walt Disney Company entered the subscription cable TV business in the early 1980s and achieved success with their Disney Channel. This concentrates on children's entertainment and family-oriented (Disney) movies. It is a premium service without advertising, something that many parents appreciate because their children are not bombarded with ads for toys and junk food. The Disney Channel is successful but has not reached the market size of HBO. Other Disney activities in TV include the production of TV series, game shows, and movies, as well as the sale of box office hits on video cassette. In 1996, Disney made its move into main-stream TV broadcasting by acquiring ABC with its network, TV stations in leading markets, and ESPN.

The final category of premium service is PPV, a system for delivering single events to the subscriber for a specified one-time charge. PPV was introduced as a form of closed-circuit transmission of sporting events, particularly professional boxing. To view the event, customers purchase a ticket and then went to a designated viewing location such as a theater or bar to witness the event live on a screen or large monitor. Over time, the cable networks and cable TV systems figured out how to use the cable distribution system and the set-top box to permit PPV in the home.

Recent release movies were added in the mid-1980s over the first PPV network, Viewer's Choice. The issue at the time was the release "window" that defines the delay, measured in months, between when the film is first shown in the United States in regular theaters and when it is made available through cable's PPV facility. Tied to this is the release window for video cassettes. While the movie studios and distributors do not make this kind of information readily available, the sequence seems to be the theatrical release in the United States and other global markets, followed within three to six months by video cassettes, followed by PPV. All three of these media provide the studios with a nice boost of revenue, where the ultimate viewer pays by the increment and the studio gets a discernible share. The total period is approximately nine months, give or take a few months. Not long after the U.S. release, the film is distributed to theaters around the world. Non-U.S. exhibition of films has become an important source of revenue to the studios, which may explain why the cassette and PPV release windows can be delayed as much as another six months. The key point in all of this is that the studios and distributors optimize the release timing to obtain the maximum amount of revenue from a given movie in the shortest possible period of time.

PPV networks, then, must compete with cassette rental and purchase, as well as the successive showing of the same movies over the premium cable channels, the latter obtaining their movies perhaps two months later. All movies do not find their way to PPV for reasons that only the studios know. The result of all of this is that PPV is a niche in the overall picture, while premium channels tend to be a mainstream revenue producer for the cable system and cable network operators.

There are a number of variants of PPV that are generally recognized. In its most primitive form, the subscriber must reserve the particular program well in advance and the service provider then provides a special access device such as a descrambler or inverse trap to be connected prior to viewing. An improved form results with an addressable set-top box, which allows the cable operator to activate the show remotely. The subscriber must still make an advance reservation over the telephone, anywhere between one month and one hour in advance of the event or show. The ordering and setup of program delivery is entirely manual.

The final variant is called impulse PPV (IPPV), which is made possible by the modern set-top box and some type of interactive connection between the subscriber and the cable system operator. The simplest scheme uses a telephone hookup whereby the box automatically dials the operator to request the PPV event. This can be done literally within minutes of the event. Interactive cable eliminates the need for the telephone call because the request is made using a data communication network. Alternatively, the set-top box can authorize and then descramble the show without going back to the cable operator. Subsequently, the box calls back to report an aggregate number of PPV viewings. This particular technique is used in DTH-delivered PPV, which is discussed in Chapter 6.

Vendors of cable equipment have found it difficult to implement true IPPV in many systems and so are offering variants that come under the category of "near" IPPV. The problem with this is, what exactly is near IPPV? One approach is to broadcast the same movie on several channels at the same time but to stagger the start times by between twenty minutes and one hour, a technique called "multiplexing." For a 90-minute movie, for example, this would require either three or four channels. The viewer would only have to wait a maximum of twenty or thirty minutes to the start of the movie (the average wait time would be half this amount). Access control to this can be done by the set-top box without intervention from the cable operator. The movies could be delivered all the way from the cable TV network, there only being the requirement that several simultaneous channels be used over the satellite. As we move to digital compression and transmission, this multiplexed form of IPPV will no doubt become more affordable and popular.

The tape play equipment needed to originate the movies can be located at the cable headend or studio. This could be attractive for a large system where the usage justifies the investment and operating cost. The more common form is to have the origination point at the uplink to the program distribution satellite.

4.2 EDUCATIONAL TV

Satellite delivery of educational TV programs and courses is a relatively small niche in the overall business of video distribution. It got its beginning through the terrestrial medium of microwave transmission.

4.2.1 University Distance Education

Standford University developed the first closed-circuit education TV network using the *instructional television fixed service* (ITFS) frequency assignment at S band [9]. This special allocation by the FCC allowed the Stanford Instructional Television Network to serve working professionals in the San Francisco area. Similar networks were created by UCLA, USC, and the *California State University* (CSU) system. Several universities in the CSU system followed suit, as *California State University at Chico* (Chico State) began to serve Northern California and California State Long Beach concentrated on Los Angeles and Orange Counties in Southern California.

Chico State introduced satellite delivery of its computer science classes in September 1984, to extend the reach to the multiplicity of high-technology companies in the San Francisco area, particularly Silicon Valley. A student could complete an M.S. in computer science without ever attending a single class in Chico, some 200 km to the northeast of San Francisco. By 1987, Chico State was broadcasting five courses a semester to nine corporate participants, including HP, Texas Instruments, General Dynamics, Alcoa Laboratories, Pacific Telesis, and MCI. The network extended across 20 locations, as far east as Pennsylvania and as far north as Washington State. From a technical standpoint, any receiving dish within the U.S. footprint could watch as well. Similar networks were established in the Commonwealth of Virginia (the formal name of the state of Virginia) in 1985 to allow graduate engineers and business administration students to attend classes at convenient locations.

The universities that employ ITFS and satellite-delivered educational broadcasting use these media as extension mechanisms. In other words, their main mission is still to serve the on-campus student population through live classes where the instructor is physically present. Most satellite-delivered education is one way, without an interactive feature that allows students and instructor to ask questions of each other. Also, examinations are more difficult to administer, particularly when you consider the opportunity for cheating. For this reason, most degreed programs on satellite required the students to be examined at a special location or on campus.

Adding the interactive feature is feasible and, in fact, is done in a number of installations. The simplest and least expensive approach is to use a dial-up telephone connection and what amounts to a speaker phone. The incoming calls to the studio are bridged in a conferencing unit to allow all sites to hear the question and answer. Since the instructor is literally blind as to what is going on in any particular remote classroom, it is useful to include an indicator light to show where a question or comment might originate. More sophisticated conferencing systems include a data channel, which allows several useful features. One or more of the following facilities will have value, depending on the nature of the instruction:

- A readout that indicates the source of a question or comment;
- A mechanism to collect responses to multiple-choice questions from remote classrooms, useful for measuring the effectiveness of the teaching;
- Ability to open up (or close) the return sound channel on an individual classroom basis;
- A forward and return graphics capability to allow students to present their ideas (instead of a return video channel, which is usually inconvenient and

prohibitively expensive as well); when necessary, this can be accomplished with standard fax machines;

- A computer networking function to allow exchange of text or files.

Another education network is operated by an organization called the *National Technical University* (NTU), which has neither a campus nor its own instructors. Instead, all of its classes are drawn from existing universities around the United States. Professors from 24 member universities offer courses leading to the M.S. degree in computer science, electrical engineering, engineering management, and manufacturing systems engineering. Fifteen of the universities that support NTU have studios and uplinks, and over 450 courses are offered each semester. The 2,000 students who attend NTU are provided with prepared materials ahead of class, and many of the classes are recorded on tape for closer study and first-time viewing when classes are unavoidably missed.

4.2.2 Corporate Education and Interactive Networks

One of the most sophisticated satellite education facilities ever implemented was the *Interactive Satellite Education Network* (ISEN), operated for IBM by Hughes Communications between 1983 and 1993. It contained all of the features in the previous list, employing the satellite for both directions of transmission. As shown in Figure 4.5, it included four instructor studios and 20 classroom sites around the continental United States. A typical video receive site (which can transmit voice and low-speed data as well) can present all four classes at the same time. A typical classroom can hold 16 attendees and has two monitors to allow the instructor to display his or her face along with a transparency (or, alternatively, any combination of these and a 35-mm slide, computer graphic, or video tape). The basic arrangement of these two facilities is shown in Figure 4.6. Attendee access to the return channel is through a *student response unit* (SRU) on each desk, which contains the microphone, activation switch, indicator light (showing if the instructor has put this position "on the air"), and a set of five *radio buttons* to allow each student to indicate a selection to a multiple-choice question from the instructor. The SRU approach has been adopted by many other networks and can be applied even if the return channel is over a terrestrial network rather than the satellite.

A point worth mentioning is the role of the administrator, such as that employed at each remote classroom location of ISEN. In general, the success of each class and the network as a whole depend on how well the service is organized and the resulting impression this makes on students. The IBM approach was to have a qualified ISEN specialist at each location who would assist the students with administration of the class, local problems, and equip-

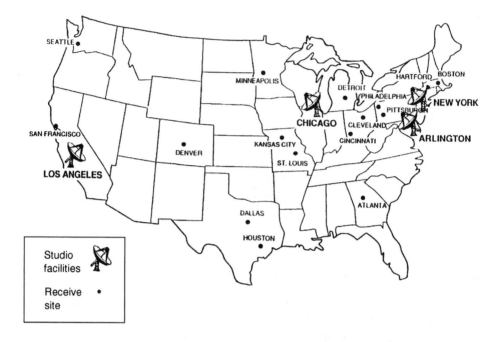

Figure 4.5 Geographic locations of the studio and receiving sites in the IBM Interactive Satellite Education Network (graphic provided by Hughes Communications, Inc.).

ment operation. Depending on the frequency of use of a particular location, the administrator's role could be a full-time job, a part-time job, or an additional assignment. Management of the instructor studio location is even more critical, as will be discussed later in this section.

Regarding the ISEN technical design, IBM set a very high objective for class availability. It was the view of management that an outage at one site in 20 would cause a delay or cancellation of the entire class. To minimize this possibility, C band was selected for its lower rain fades. One 36-MHz transponder on the Galaxy 2 satellite, as indicated in the spectrum plot in Figure 4.7, was sufficient to carry the four video carriers from the studios and the 20 audio/data carriers from the remote sites. The audio/data carriers time-share the frequency slots (shown by the thin carrier lines) while the video carriers, each of which transfers two TV channels at 1.5 Mbps each, are constant. The display also shows the "humps" of intermodulation noise in the transponder, which resulted from operating the transponder as close to saturation as possible.

ISEN met all of its technical and operational requirements during its lifetime. Several of the Earth stations had to be located away from the classroom

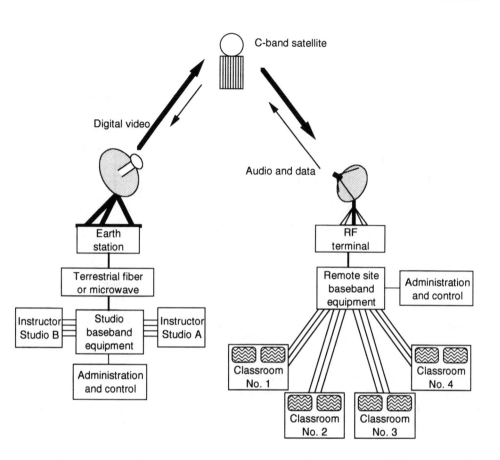

Figure 4.6 The configurations of the studio uplink site and one remote classroom receive site. Classroom receive sites can transmit FM SCPC audio and data back to the studio providing the instruction.

site in order to prevent terrestrial interference. Most of this transmission was obtained from the local telephone company in the form of multiple T1 circuits. The network achieved very high availability, usually in excess of 99.95% for class delivery, because of the excellent propagation characteristics of C band. In fact, the majority of outages were due to equipment failures and interruptions of the terrestrial links between the classroom sites and Earth stations.

4.2.3 Digital Compression in Education

All of the university satellite networks began their operation with conventional analog transmission technology in the same manner as the commercial broad-

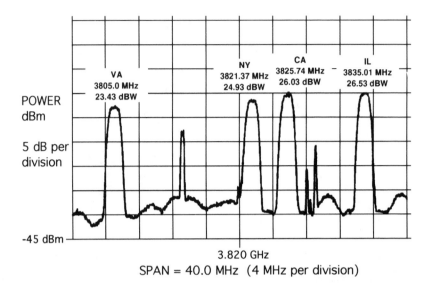

POWER
dBm

5 dB per
division

-45 dBm

VA
3805.0 MHz
23.43 dBW

NY
3821.37 MHz
24.93 dBW

CA
3825.74 MHz
26.03 dBW

IL
3835.01 MHz
26.53 dBW

3.820 GHz
SPAN = 40.0 MHz (4 MHz per division)

Figure 4.7 Spectrum analyzer display of the C-band FDMA carrier loading of the Galaxy 4 transponder used for the Interactive Satellite Education Network.

casters and cable networks. Digital video compression technology had already been developed and used on satellites as early as 1982. ISEN, which went into service around that same time as the Chico network, employed digital compression right from the start, which allowed IBM to squeeze all eight video channels into the same transponder. However, the equipment to accomplish that was, at the time, very costly and represented long-term financial commitment to the service. The university networks had to start on a low budget, using as much existing equipment and services as possible. This meant that analog systems and established uplinks and satellites were the proper approach. The cost per hour per class would be higher on the satellite, but the cost of the ground equipment would be substantially less. Also, adding a new downlink is very modest when analog systems are installed.

NTU, like the other university networks, began using analog TV transmission but chose to operate at Ku band rather that C band. Medium-power Ku band was selected as a strategy to reduce the size and cost of the receiving system. Also, there are fewer constraints on locating a Ku-band video uplink since coordination with terrestrial microwave services is not required. Unfortunately, Ku-band satellites were few in number in the early 1980s and NTU had to consider how to acquire adequate channel capacity for the long term. Consequently, they converted to compressed digital transmission to permit at least four TV signals to share one transponder. Their timing in 1990 of going this way was excellent because *Compression Laboratories, Inc.* (CLI) of Palo Alto, CA, had just introduced its moderately priced Spectrum Saver product line. This system can match the needs of educational networks, which require that a common uplink TV signal be received by many downlinks. The *one-to-many* connectivity of this application system is achieved with a relatively low-cost digital IRD and a significantly more costly encoder.

Satellite-delivered education has been extended to elementary and second-ary (high) schools in the United States. Some of these networks are commercial ventures intended to return a profit in exchange for improving the classroom experience in what is a very difficult educational challenge. The reason for this is that students are confronted with a wide variety of entertainment on home TV sets and many have trouble paying attention during six hours in class. The first systems to be introduced employed Ku band to reduce dish size and installation cost and analog transmission to make the indoor electronics as inexpensive as possible. This is because most of the primary and secondary school education networks require that the school purchase the equipment and then pay for the service. With thousands of receivers around the country, the cost per class of full-transponder rental is small relative to that of the studio production itself.

4.2.4 Guidelines for Effective Distance Learning

Creating the education network is relatively straightforward. However, more challenges lie in employing the technology effectively. The following guidelines are suggested by CLI to enhance the effectiveness of distance learning. These provide good advice for any organization that is considering using or has already committed to using satellite TV as the means to provide an acceptable classroom experience to students.

- Introduce the instructors to the network before they are expected to go "on camera." Provide a training session so that they understand how their presentation style must be modified to the requirements of the new medium.
- Begin the indoctrination early by bringing the instructors together with the administrators who will operate the network. Discuss any special arrangements that are needed on both sides.
- Reassure instructors that they can adapt their teaching styles to meet the new medium of television. The lack of visual feedback from students will give some instructors difficulty, but this can be overcome after a few exposures to educational TV.
- Demonstrate all of the facilities ahead of time so that instructors become comfortable. Give them time to practice with people at the other end who

can provide feedback. An effective way to practice is to include small groups of peer instructors who teach in the same field. This reduces some of the uncertainty in the dialog, which must traverse a considerable distance.

- Plan and schedule the classes so that there are no surprises on the day of the class. Make instructors and administrators aware of the schedule and procedures that must be followed. Typically, the equipment and satellite capacity will need to be scheduled ahead of time and there will be little if any flexibility. Supporting materials must be available at the time of the class and hence will have to be delivered ahead of time (possibly by mail or overnight delivery service).
- Communicate to all participants the expected benefits of using this medium rather than focusing on its limitations. Promote the system and the session so that there is maximum chance for success.

4.3 BUSINESS TV

Entertainment and educational TV provide the foundation for the business application of the video medium. Businesses can employ applications that involve private broadcasting, which relies on the point-to-multipoint nature of the satellite delivery medium, and video teleconferencing (VTC), which uses point-to-point two-way links to add the visual element to the standard interactivity of voice telephony. These techniques are used widely in the United States, Europe, and leading Asian nations, although their growth has been restrained by the cost and complexity of operating the equipment and arranging private broadcasts and VTC events. There is an expectation for more rapid adoption in the coming decade because equipment prices are coming down and the next generation of digital satellite and terrestrial networks will overcome many of the current operational limitations.

4.3.1 Private Broadcasting

Private broadcasting is no different technically from video distribution and one-to-many educational TV. The originator of the program uses a TV studio and uplink to create the broadcast, and the signal is received at multiple sites that have simple TV receive-only antennas and electronics. The broadcast is viewed on TV monitors in conference rooms and, ultimately, the desktop. Private broadcasts are often scheduled and may be employed almost daily in the routine of business. Examples include:

- Announcements of new product introductions and marketing campaigns, in heavily marketing-oriented organizations like Frito-Lay and Microsoft;

- Distribution of financial or critical business news that can impact the company or its customers, which is popular in the financial services industry for stock brokers like Merrill Lynch and large investment banks like CS First Boston;
- Instructions on product application and display for merchandising for retail store chains like Wal-Mart and Sears;
- Public relations–oriented communication between the government and the press;
- Instructional information and product guidance for representatives and dealers, by major manufacturers like General Motors and IBM.

Any of these applications could justify a dedicated private broadcasting network that operates daily or even several times per day. If the need is less frequent, perhaps once per week or month, then the network can be put together on an ad hoc basis by renting the studio. This reduces the capital commitment but increases the operating cost. The only problem with this approach is that, due to the higher operating cost, it becomes a candidate for cutting when times get bad. On the other hand, depreciation and maintenance changes, as well as the cost of acquiring long-term satellite capacity, can be a heavy burden in times of financial need.

Many applications for private broadcasting can be satisfied on an ad hoc basis, that is, without the acquisition of a dedicated studio facility, uplink, and transponder capacity. Downlinks, on the other hand, would have to be installed on a more permanent basis and consideration given to which satellite would be the focus of the ad hoc network. The following are some examples of ad hoc private broadcasts.

- Press conferences of top executives who must inform the public of a major change in strategy or financial performance;
- Announcements by chief executives to the entire employee population across a wide geographical area, which may occur when there is a change of leadership or a major acquisition;
- Interviews with political candidates during a national campaign, for distribution to local TV stations and eventual rebroadcast; the reason why this is private broadcasting and not program origination is that the candidate usually pays for the event from campaign funds;
- Marketing presentations on major new products such as the Microsoft Windows 95 PC operating system or the General Motors's Aurora motor car.

As a private broadcast, there is usually not a requirement for interactivity because of the potentially large number of remote locations and attendees. Also, the person doing the talking is almost always following a written script that does not allow for interruptions. There may be a question-and-answer period at some point in the broadcast, which could involve either people in the studio or call-ins over a return channel (almost always through dialup telephone). The problem here is the unpredictability of call-ins, which can put the presenter at a major disadvantage in front of a potentially large audience that is the target of what is otherwise a well-prepared presentation (of course, it might be better to use precleared questions that the presenter is already prepared to answer).

Private broadcasting received a boost by the rising popularity of VSAT networks in the United States and Europe. As discussed in Chapter 7, a VSAT can be equipped with either an analog or digital *integrated receiver-decoder* (IRD) to receive private TV broadcasts. The cost of this upgrade is small compared to the cost of the VSAT network and represents an excellent way to increase the return on investment. The IRD is connected to the downlink using a simple power splitter. The outputs of the IRD are connected to one or more video monitors located in conference rooms. The quality of reception is not affected in any way by data or voice services that are provided by the VSAT indoor unit. The only consideration is that the antenna be of sufficient size to provide an adequate link C/N, derived by a link budget calculation as discussed in Chapter 2.

There are two options for the uplink for the VSAT private broadcast—either a dedicated facility that is owned and operated by the corporation or a rented facility used occasionally. The uplink can be completely separate from the VSAT hub, using a different antenna and even coming from a totally different part of the country. Some form of scrambling or encryption could be used to secure the proprietary content. The only condition is that the uplink be capable of transmitting to the same satellite where the VSATs are pointed. Transponder capacity for the ad hoc event would be rented from the appropriate satellite operator.

The bottom line in private broadcasting is that it is a traditional TV medium, where the presenter is a star on the screen. Consequently, considerable effort must be placed on the visual impression of the scene and the presenter(s). This should be organized and directed by someone who has experience in TV production.

4.3.2 Video Teleconferencing

Video teleconferencing links and networks were touted in the 1980s as attractive ways to reduce business travel costs and improve organization performance by increasing communication among distant groups. Major U.S. corporations like ARCO and Citibank invested millions of dollars on special VTC-equipped conference rooms, video compression codecs, and Earth stations needed to deliver adequate bandwidth. These pioneers demonstrated that it was feasible to tie people together and that those participating could fulfill many useful purposes. As time progressed, the cost of the rooms and codecs came down, along with increased availability of much cheaper terrestrial communications using the fiber optic networks of the long-distance carriers. The innovator and leader in this field is Sprint Communications, which was first to go totally fiber in its national long-distance network and continues to lead the market in providing connectivity for VTC users. Satellite links were subsequently adopted to the unique needs of VTC when HNS and Spar Telecommunications began to offer mesh networking systems using larger-sized VSATs. The advantage of this approach is that a VTC can be scheduled and activated by central control, even though the sites involved are located elsewhere in the network. The partial transponder bandwidth needed for such a private network would be leased from a satellite operator, perhaps on a long-term basis.

Many of the most popular applications for VTC are summarized as follows.

- *Routine meetings between members of a team that is engaged in a very large project:* Groups in different locations can interact as frequently as daily, which can be vital if the project is moving quickly. Projects of this type are very high valued and often are for a government agency such as the U.S. Defense Department or the national PTT of a country like China.
- *Coordination meetings of a joint venture involving groups in different countries:* In this way, the combined organization can cooperate and collaborate better because they see each other more frequently than they would if face-to-face meetings were relied upon. This tends to build trust and improve communication, which are vital for the success of a joint venture business activity particularly in its formative phase. Likewise, VTCs can deal more effectively during a period of difficulty by allowing issues to be aired and discussed.
- *Routine financial reviews of a multinational corporation that involve many remote locations:* Headquarters financial managers can speak to their counterparts at remote locations, either collectively or one at a time. Not only the operating numbers but their meanings can be discussed. Any new policies or practices would be reviewed and comments collected for consideration.

Those who invest in the rooms and equipment usually provide a satellite and terrestrial access. This increases the versatility of the systems, making possible internal as well as external conferencing. In the case of the latter, the most effective approach is to connect to a terrestrial network on both sides. Sprint, for example, serves the United States, Canada, and many points overseas through connectivity with a large number of counterpart national network operators. This author, for example, has participated in such conferences with points in Japan and Singapore.

The standard arrangement of VTC is for point-to-point connectivity as presented in Figure 4.8 for a typical system. This produces a two-way service where both sides of the conference can see and hear each other. Each end of the connection is equipped with cameras, microphones, monitors or TV projection systems, a digital video compression codec, and a controller. The most expensive item in the system and the one that is most critical to the operation and performance is the codec. The leading suppliers of VTC codecs in 1996 are CLI and PictureTel. The standard VTC codec digitizes and compress the video signal and performs the reverse function as well. In addition, a typical codec provides separate inputs for audio, data, and control. The user can interact with the device using either a separate control box or, more recently, a special type of handheld remote controller and on-screen display. This is advantageous because the older control boxes are not intuitive and usually do not have online instructions available.

The telecommunication aspects of VTC can be provided either by satellite or by a terrestrial digital network. As shown in Figure 4.8, rooms A and B are to be connected by VTC so that a single meeting will involve both locations. The conference table is arranged so that attendees can see the video monitors and can be seen on the other end through the cameras that are mounted on top of the monitors. One camera could provide a wide-area view of all in the room, while a second would point to the speaker. Microphones on the conference table carry speech and activate the appropriate camera. The two ends of the conference would appear either on separate monitors or on a split screen of a single monitor. Another feature is the use of a still projector or computer display to add a graphic capability to the meeting. This could either be substituted for the live picture or sent simultaneously over the data channel that is multiplexed with the digital video.

The cameras and monitor of the VTC in Figure 4.8 are connected to a digital compression codec that performs the processing and multiplexing of all of the inputs and outputs. Overall operation of the room equipment and the telecommunication links is managed by a controller, which could be part of the codec or a separate unit. One function of the controller is to allow the VTC to use either a satellite link or a terrestrial network, depending on what is available and what is the most cost effective for the particular meeting. Both options should generally be provided, although there are fixed and recurring costs associated with each.

Digital compression of video signals is discussed in detail in Chapter 5. Briefly, the video signal is first converted from analog to digital format, resulting in a high bit rate data stream at approximately 100 Mbps. This is substantially higher than the rate to be transmitted over the link. The data is first processed

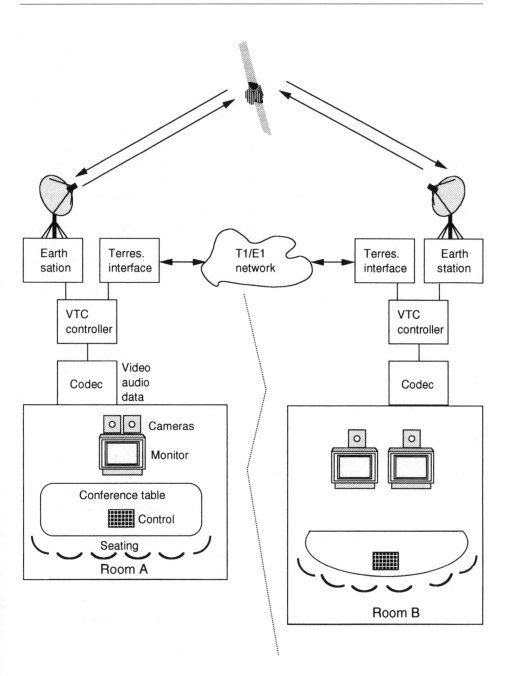

Figure 4.8 Typical arrangement of a two-way video teleconferencing system for use in interactive meetings.

on a single-frame basis; that is, spatial compression is performed on the first frame to reduce the quantity of bits used to represent the image. Frame-to-frame processing (temporal compression) then causes only the changes between adjacent frames to be transmitted. The most popular spatial compression technique is based on the *discrete cosine transform* (DCT), a mathematical conversion algorithm that takes the scanned image and produces a set of coefficients from a corresponding mathematical series. The algorithm is now standardized as part of the H.320 series of ITU-T specifications, which are discussed later in this section. These coefficients are then compressed to further reduce the bandwidth. Instead of sending the picture image, the coefficients are transmitted to the other end where the image can be recreated. The combined effect of spatial compression and frame-to-frame compression produces a moving picture image that can closely approximate the natural motion performance of the original analog TV signal. The amount of naturalness and the ability to track fast movement is directly dependent on the degree of compression. In other words, the more compression, the less natural the resulting TV images but the lower the required transmission speed.

Because of this tradeoff, developers of VTC networks must examine the codecs carefully before committing to a particular brand of equipment and transmission data rate. It is highly desirable to expose prospective users to a typical VTC link of the same design before making this commitment. Otherwise, it is possible that a great deal of time and money could be spent on a network that users find unacceptable for their intended purpose. Vendors often provide demonstrations at expositions and trade shows, but prospective buyers must be sure that the demonstration is for the same arrangement that they intend to purchase, such as monitor size and quality, cameras, codec (including features and standards support), and transmission data rate.

The other part of the process is to digitize the audio, compress the data rate by removing redundancy, and transmit the resulting data stream. The audio data is time division multiplexed with the compressed video data.

A lot of interest has been focused on the use of international VTC standards from the ITU-T. The first published standard was H.261, which defined the compression algorithm. More comprehensive standards have since been defined.

H.320 is the existing family of ITU standards that describes a VTC system, originally referred to as P*64. This simply means that the transmission link employs an integer multiple of 64 Kbps, for example, 64, 128, 256, 384. H.320 is further broken down into more standards, as follows.

H.221 defines the frame structure for audio-visual services at rates ranging from 64 Kbps to 1920 Kbps. The latter corresponds to the international E1 line rate at 2.048 Mbps.

H.242 defines the system for establishing the connection between end VTC codecs or setting up calls between audio-visual terminals using digital channels of up to the E1 line rate. It also defines the data communications protocols used by H.320 codecs when sending compressed video and audio between locations.

H.261 defines the codec processing algorithms for audio-visual service with P*64 transmission speeds, intended to ensure compatibility between countries with different video formats (e.g., NTSC, PAL, and SECAM, as discussed later in this chapter). This is also referred to as the *common intermediate format* (CIF) and allows interoperability.

A fully functional codec has multiple line speeds available and, therefore, can provide different levels of video quality in relation to how much the user is willing to pay for transmission. These can step through 128 Kbps, 256 Kbps, 384 Kbps, 728 Kbps, and 1.544 Mbps. The general reaction to the motion quality, color, and resolution of these various speeds is that 128 Kbps is barely acceptable and that 384 Kbps and 728 Kbps are preferred for typical meetings. If the VTC can transmit full-motion material like movies and TV spots, then 1.544 Mbps or 2.048 Mbps is required. Teleconferencing codecs are designed for meetings and should not be used to transmit fast-action material like horse racing and soccer.

Since the most popular speeds, and the speeds most available on a national and international basis, are 128 Kbps and 384 Kbps, less expensive boxes that fix these rates are appearing. The cost differential between the fully variable and the fixed rate codec is nearly two-to-one. The codec can also provide transmission security using a symmetrical encryption algorithm. This would be important for very private transmissions that would be transmitted over satellite links or public networks.

Satellite-based VTC remains a useful application because it bypasses the remaining limitations of terrestrial networks. Both Hughes Network Systems and Spar Telecommunications offer VTC systems that use the satellite as a common connection point. The HNS approach, called Intellivision, is based on FDMA with each station activated on an assigned channel at the time of transmission. A common control station is used to schedule the conferences and to control the remote equipment. The Spar network uses TDMA so that only one frequency is employed. However, the bandwidth of this channel is proportionately greater because it must support multiple stations in a burst transmission mode. Both systems are relatively user-friendly, allowing a non-technical administrator to arrange and manage teleconferences across the diverse network.

Satellite transmission of two-way VTC raises the interesting possibility of point-to-multipoint communication (e.g., private broadcasting) and true multipoint conferencing (e.g., many-to-many connections). The former is obtained by having only one site transmit and the other sites operate in the receive mode. The audio and control of the transmission would be in one direction as well. There could be audio return either over the satellite using some form of multiple access or by separate telephone dial-up connections. The multipoint conference is an intriguing possibility that has not yet appeared in practice. Through the concept of a video bridge, multiple sites transmit video and audio simultaneously. This means that at a given location, all participants from all locations can be seen and heard. The video bridge display might look like the old TV game show called *Hollywood Squares,* where the screen is divided up into a matrix of boxes with separate pictures, one for each location. You would see each location as a tiny picture among many on the screen. The trouble with this, or course, is that it would be relatively hard to tell who is doing the talking at a given time unless the square containing the active talker is highlighted or expanded in size.

4.4 ANALOG TV STANDARDS

Television standards define the format and quality of video signals that are intended for viewing by the general public. They are applied at the origination point where the picture is acquired, to the studio where programs are prepared, and the link to the broadcast station or cable TV system that transfers the signal to the ultimate viewer. There is little doubt that eventually analog standards will give way to digitally-based processing, transmission and display. The current status of the digital standards is covered in Chapter 5. However, the majority of the existing infrastructure of TV sets, local stations, cable TV systems, Earth stations, and studios are analog in nature. Anyone contemplating a new TV application or network will have to consider this factor when determining how they will reach enough potential viewers to make their venture a success.

Analog TV standards deal with every phase of the process of creating and distributing video programs to the public. We use the following definitions with regard to each of these aspects of program delivery.

Acquisition defines standards for the video format that the camera uses to present the video image. These fall into the three recognized color systems: NTSC, PAL, and SECAM. The standards are defined further, and there are differences that make the specific details differ from country to country, in some cases.

Transmission defines standards that quantify the allowable distortion and degradation due to carrying the signal from the camera or studio to the point of distribution to the viewer. This format is usually not intended for direct reception by the public but is designed for minimum degradation in signal quality. Point-to-point transmission is the normal mode, using fiber optic or coaxial cable, microwave radio or satellite links. Point-to-multipoint transmission via satellite or terrestrial microwave radio towers can be employed to reach the public directly using the same standard, provided that the end user has an appropriate converter box.

Distribution defines standards for the allowable degradation as the signal is carried to the ultimate viewer. Originally, this considers the radiation of the TV channel from local broadcasting stations, directly over the air at VHF and UHF frequencies. More recently, hybrid analog/digital formats like *multiplex analog components* (MAC) and digital standards like *Motion Picture Experts Group* (MPEG) are becoming popular as the system for delivery, where each viewer has a set-top box to convert from the unique distribution format to one of the standard analog formats.

These aspects of the analog TV standards are covered in the following subsections.

4.4.1 Video Format Standards

The 1950s and 1960s saw the adoption of analog color TV standards: *National Television System Committee* (NTSC) in the United States and Japan and *Phase Alternation Line* (PAL) and *Séquentiel Couleur Avec Mémoire* (SECAM) in Europe. These standards are used worldwide, and in some cases the same TV receiver is capable of displaying more than one. For many years, the electronics in receivers employed vacuum tubes, which are relatively expensive and less stable than solid-state equivalents. Therefore, first-generation TV sets were designed for a minimum number of parts. Transistor circuits began to replace tubes in the 1960s, greatly improving both the stability and reliability of home receivers. The first role of digital circuitry in the 1970s was in the form of digital channel display and remote control. Later, integrated circuits were introduced to replace nearly all of the active electronic elements, producing very low cost receivers with far more complexity and sophistication than the original developers might have thought possible. Picture quality is generally felt to be as good as can be obtained with these analog systems, which is still quite acceptable for comfortable viewing of entertainment TV and many business video applications. Use of NTSC, PAL, and SECAM for computer text and image display is generally felt to be unacceptable, except possibly for the current generation of video games.

The analog standards are divided according to two properties: the basic black-and-white signal, also called the luminance, which existed before color was added; and the technique for adding color (chrominance), namely, NTSC, PAL, and SECAM. The luminance creates the black-and-white image during the scanning process over the screen of the picture tube. Color is added by phase modulating a subcarrier frequency that occupies a position within the luminance baseband frequency range. Table 4.7 summarizes the key parameters for worldwide TV systems, where the capital letter indicates the CCIR standard designation for the luminance system. The table is abbreviated, since it requires more than 50 individual technical characteristics to specify each system properly.

4.4.2 Analog Transmission Standards

The purpose of analog transmission standards is to provide television engineers and specialists with standardized performance objectives and measurement methods to determine signal quality. In this instance, we are concerned with

Table 4.7
Summary of Analog TV Standards that Apply Throughout the World, According to the CCIR

Basic TV Standard	M	M	N*	B, D, G, H, N	I	D, K, K1, L
Color system	NTSC	PAL	PAL	PAL	PAL	SECAM
Video bandwidth, MHz	4.2	4.2	4.2	5	5.5	6
Broadcast TV channel bandwidth, MHz	6	6	6	8 (B: 7)	8	8
Field frequency	60 (59.94)	60	50	50	50	50
Line frequency	15,750	15,750	15,625	15,625	15,625	15,625
Sound subcarrier frequency, MHz	4.5	4.5	4.5	5.5	6	6.5
Color subcarrier frequency, MHz	3.579545	3.575611	3.582056	4.433619	4.433619	4.406250 and 4.250000
Video levels (%)						
Blanking	0	0	0	0	0	0
Peak-white	100	100	100	100	100	100
Sych tip	−40	−40	−40	−43	−43	−43
Difference between black and blanking	7.25	7.25	0	0	0	0

the link between the studio and either the local TV station or cable TV head end. Transmission systems that are used for this purpose include microwave radio, satellite links, fiber optic cable (used in an analog manner), and coaxial cable. Digital fiber and microwave includes a video codec that operates at 45 Mbps for nearly lossless transmission. The signal is not normally available to the public along one of these systems and so analog transmission standards are not designed for the minimum cost of reception. Rather, they emphasize the quality of the resulting delivered signal with a minimum of added distortion and interference noise.

An example of a typical transmission system for point-to-point video transfer is shown in Figure 4.9. The studio delivers separate video, audio, and data outputs to a video exciter that is associated with, in this case, a TV uplink Earth station. Point A represents where the video portion of the information is essentially perfect in a technical sense. The transmission system extends from the exciter, which produces a modulated carrier at the IF frequency (typically 70 MHz), containing the three components. Translation to the RF transponder channel is performed by a separate upconverter or as an integral part of the exciter.

The most popular analog technique is to employ *frequency division multiplex* (FDM) to combine the video with the associated audio channels, as shown in Figure 4.10. The video is transferred directly across to the low end of the baseband and stops at frequency f_m. Each audio channel is frequency modulated onto a subcarrier on an upper baseband frequency. This particular example provides two audio channels on separate subcarriers at f_{c1} and f_{c2}, for stereo audio in the primary language. Audio channels for multiple languages can be included by adding subcarriers into the baseband. Up to a total of 10 such subcarriers have been used in practice.

The third subcarrier (at f_{c3}) is for a broadcast data channel to be received at the remote stations or by other downlinks. Some of the possible applications for this data broadcast include:

- Network control and coordination, for automated operation of remote antenna, transmitters, receivers, and studio equipment;
- Program information and verbal instructions to allow the distant stations (in the case of TV broadcasting) to be aware of upcoming events and any special requirements;
- Data services like teletext that can be delivered to the public along with the video or offered as a totally independent service for additional revenue;
- A paging channel to provide nationwide paging services through the facilities of the local broadcast station or an auxiliary transmitting tower.

The composite baseband containing the video and all of the subcarriers is transferred to the IF carrier using FM. Typical baseband and modulation

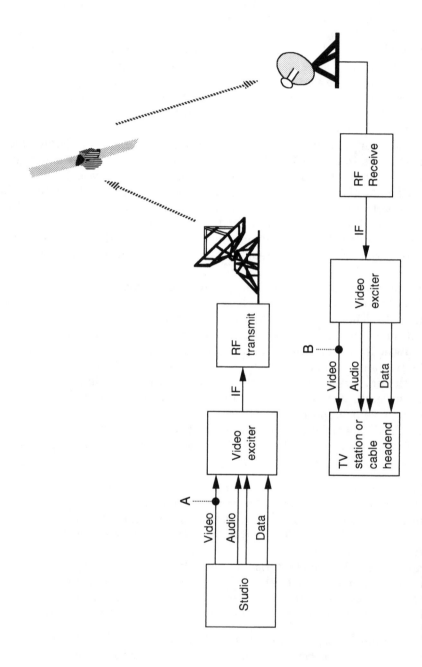

Figure 4.9 Typical interfaces for a video point-to-point or point-to-multipoint transmission system.

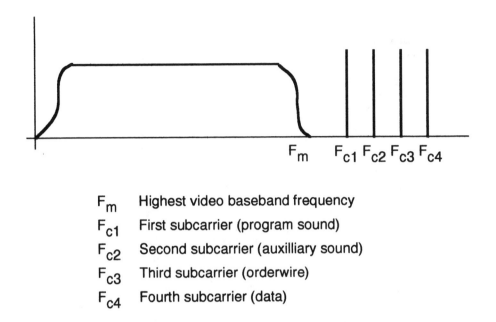

F_m Highest video baseband frequency

F_{c1} First subcarrier (program sound)

F_{c2} Second subcarrier (auxilliary sound)

F_{c3} Third subcarrier (orderwire)

F_{c4} Fourth subcarrier (data)

Figure 4.10 Arrangement of the video baseband used in analog FM transmission over satellite links.

formats are shown in Table 4.8, with separate listings provided for NTSC, PAL, and SECAM. These values are not formal standards but are used in practice around the world on the systems indicated in the first column. They are incorporated in commercial video exciters and receivers from a variety of suppliers in the United States, Europe, and Japan.

Television engineers who work with transmission systems and standards have the intention of not degrading the signal in a significant way, expecting that the eventual distribution process will be less controllable and therefore tend to provide the bulk of the degradation to quality.

The performance requirements for the video signal, as received at point B in Figure 4.9, are contained in widely recognized and supported standards of the U.S. ANSI and the ITU. The *Radiocommunication Bureau of the ITU* (ITU-BR), formerly known as the CCIR, issues recommendations that apply to the international transmission of TV signals over satellite links. While accepted among nations, the ITU-BR recommendations are not detailed enough to assure commercial service quality and therefore must be augmented by the specifications discussed in this section. Other organizations like the *European Broadcasting Union* (EBU) have specific standards that apply within a given country or region. However, the standards that are discussed in the next paragraphs are those that predominate around the world.

Table 4.8
Recommended Baseband Subcarrier Frequencies Used in Satellite Video
Transmission to Convey Program Audio, Stereo, and Data (All Frequencies
in MHz)

Frequency (MHz)	NTSC	PAL	SECAM
Highest baseband frequency, f_m	4.2	5.0	6.0
TV channel width	6.0	6.0	8.0
Primary audio, f_{c1}	6.8	—	—
U.S. cable TV			
U.S. broadcast	6.2/6.8	—	—
SES Astra	—	6.5	—
EUTELSAT	6.6	6.6	6.6
INTELSAT	6.6	6.6	6.6
Stereo audio, f_{c2}	(left/right)		
U.S. cable TV	(encrypted)	—	—
U.S. broadcast	5.94/6.12	—	—
SES Astra	5.94/6.12	—	—
EUTELSAT	—	7.02/7.20	—
INTELSAT	6.65	6.65	6.65

Source: Norman Weinhouse Associates.

4.4.2.1 NTSC Transmission Requirements

The requirements for the transmission of the NTSC signal are specific and detailed. The *Electronic Industries Association* (EIA) and the *Telecommunications Industry Association* (TIA) have produced a well-known standard, that is, EIA/TIA-250-C, "Electrical Performance for Television Systems" [10]. This is an updated version of RS-250-B, which had been the measurement standard for all North American TV transmissions up until 1990 when EIA/TIA-250-C was formerly issued. The latest standard, effectively applied anywhere in the world where the NTSC system is used, is used to evaluate the performance of short-, medium-, and long-haul microwave links, satellite links, and various end-to-end combinations thereof.

The following basic definitions are essential to understanding the role and application of 250-C.

- A *short-haul transmission system* is usually a simple point-to-point transmission link (also called a hop) that is of the order of 30 km in length. These links are used to connect the studio to the broadcasting tower or to a local transmitting Earth station.
- A *medium-haul transmission system* is a microwave or cable relay system consisting of more than one hop over a distance of between 200 km and

4500 km. Such systems were popular before the age of satellite transmission and have reappeared through the introduction of long-haul fiber optic systems.

- A *satellite transmission system* is a single-hop satellite link between a transmitting Earth station and a receiving Earth station through a "bent pipe" satellite repeater. A typical example is shown in Figure 4.9.
- An *end-to-end network* is an interconnection of multiple transmission systems, consisting of, for example, a satellite transmission system with short-haul microwave transmission systems on both ends. This is the typical case, using various combinations of transmission systems that depend on the requirement.
- *IRE units* measure the TV signal, where one IRE unit is 0.01 times the range of the luminance signal. IRE is the abbreviation for the Institute of Radio Engineers, an organization that merged with the American Institute of Electrical Engineers to form the *Institute of Electrical and Electronic Engineers* (IEEE). The typical time waveform in Figure 4.11 displays an NTSC signal showing with the blanking level at 0 IRE units, the maximum (white) video level at 100 IRE, and the synchronization waveform negative pulse extending to –40 IRE. In total, the video signal ranges 1V, peak-to-peak. IRE units are only used for 525-line systems (NTSC), while the

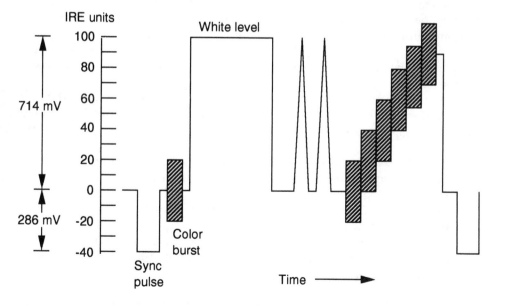

Figure 4.11 The waveform of a typical line of the NTSC signal.

625-line systems (PAL and SECAM) refer either to the percentage of maximum video level (which happens to equal IRE) or to the actual voltage.

- The *average picture level* (APL) is an average taken of the signal level during the active scan time, not including the blanking and synchronization. In other words, it is the integrated average of the picture waveform itself for one horizontal line (e.g., 33.4 ms for NTSC), over the range of 0 to 100 IRE units.

A sample of the specified values for each type of transmission system, summarized from [10], is provided in Table 4.9. This demonstrates the extent and depth of 250-C, which can be difficult to meet unless high-quality equipment (particularly exciters and receivers) is used. Considerable debate erupted during the early 1980s over whether digital video links can or should meet RS 250-B. This was first fueled by the introduction of high-quality, low error rate fiber transmission systems within major U.S. cities like Los Angeles (in fact, it was within LA that the ABC network first experimented with fiber for the 1984 Summer Olympic Games) and later by the availability of compression systems based on the MPEG-2 standard.

In the interest of brevity and clarity, we have limited Table 4.9 to the cases of short-haul terrestrial, satellite, and end-to-end transmission systems. The general trend is that the longer the transmission system and the more components that the signal must traverse, the larger the specification range of allowed performance. The three specifications that are emphasized—differential gain, differential phase, and *S/N*—are the most critical to the visual quality and color of the received analog signal and for that reason they have been adopted by the ITU-BR for their recommendations on television transmission over satellites.

Actual measurement of the specifications is accomplished using standard test waveforms, which are listed in the table and shown in Figures 4.11 through 4.13. The signals work in the following manner.

- A *multiburst signal* (Figure 4.12) measures the frequency response at six discrete frequencies over the video baseband range.
- A *stairstep signal* (Figure 4.13) measures the gain at the color subcarrier frequency at six different brightness levels, from zero (black) to 90% of the maximum white level.
- *Three-level chrominance* (Figure 4.14) detects any change in the phase of the color subcarrier (which produces the color or hue) as a function of the amount of color saturation.

These and other test waveforms are inserted into horizontal lines during the nonvisual portions of the vertical blanking interval. Alternatively, the nor-

Table 4.9
NTSC Television Transmission Performance Requirements (EIA/TIA 250-C, 1990)

Test Parameter	Waveform	Short Haul	Satellite	End-to-end
Baseband frequency response	Multiburst	±2.5 IRE	±7 IRE	±12 IRE
Chrominance to luminance gain inequality	Modulated stairstep	±2 IRE	±4 IRE	±7 IRE
Chrominance to luminance delay inequality	Modulated stairstep	±20 ns	±26 ns	±60 ns
Differential gain	Modulated stairstep	2 IRE (2%)	4 IRE (4%)	10 IRE (10%)
Differential phase	Modulated stairstep	0.7 deg	1.5 deg	3 deg
Luminance nonlinearity	Modulated stairstep	2%	6%	10%
Chrominance to luminance intermodulation	3-level chroma signal	1 IRE	2 IRE	4 IRE
Chrominance nonlinear gain	3-level chroma signal	1 IRE	2 IRE	5 IRE
Chrominance nonlinear phase	3-level chroma signal	1 deg	2 deg	5 deg
Dynamic gain of video signal	Stairstep with variable APL	2 IRE	4 IRE	6 IRE
Dynamic gain of sync signal	Stairstep with variable APL	1.2 IRE	2 IRE	2.8 IRE
Signal-to-noise ratio	10 kHz to 4.2 MHz, weighted	67 dB	56 dB	54 dB
Availability of the video signal at S/N ≥ 37 dB	(any)	99.99%	99.99%	99.99%

mal TV signal can be interrupted to allow near-continuous transmission of a particular test signal.

Television standards also consider the quality of the audio portion of the program. These can be stated more succinctly as the required *S/N* and the allowable amount of audio distortion in the received signal. Also, satellite networks that deliver multiple video channels from the same orbit position must also adopt a standard audio level to prevent contrast between video channels as the viewer tunes the home receiver across the transponders. This is mostly a concern in analog transmission systems where levels can drift over

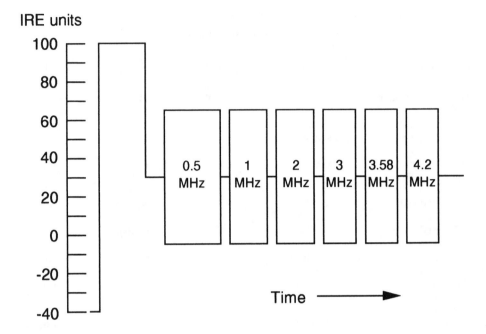

Figure 4.12 The multiburst signal used to measure baseband frequency response as part of EIA/TIA 250-C testing.

time. Standard 250-C specifies that the audio S/N must be greater than or equal to 66 dB, 58 dB, and 56 dB, for the short-haul, satellite, and end-to-end cases presented in Table 4.9. There is also a requirement that the time differential of the audio channel with respect to the video channel fall within the range of +25 ms to −40 ms.

The profile of 250-C testing is lengthy and complicated, using a test signal generator and a number of receiving measuring devices. Among the receiving test devices are the video analyzer and the vector scope. Recently, this equipment has been incorporated into an automated test system to both speed up and make consistent the entire procedure. This allows the full suite of 250-C tests to be performed at the touch of a button (or return key, as the case may be).

4.4.2.2 PAL and SECAM Transmission Requirements

Analog transmission systems that are designed for NTSC already meet many of the requirements for PAL and SECAM. This is not surprising because both

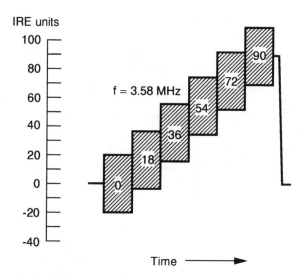

Figure 4.13 The modulated stairstep signal used to measure amplitude and phase nonlinearity as part of EIA/TIA 250-C testing.

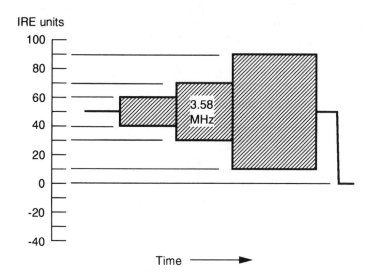

Figure 4.14 The three-level chrominance signal used to measure chrominance performance as part of EIA/TIA 250-C testing.

PAL and SECAM are derived from NTSC with respect to the use of interlaced scanning and the manner in which the luminance (black and white) information is carried. As shown in Table 4.8, PAL and SECAM differ from NTSC in that they use a basic frame rate of 25 per second as opposed to 30 for NTSC. In addition, with 625 lines per picture instead of 525, PAL and SECAM actually have better resolution. Because of these basic differences, PAL and SECAM require more baseband bandwidth, in the range of 5 MHz to 6 MHz, as compared to 4.2 MHz for NTSC. This has a direct impact on the transmission system, which must be designed to carry the greater bandwidth and to accept the 25 Hz frame rate and corresponding field frequency of 50 Hz. Since we do not increase RF power or bandwidth to compensate, the resulting S/N that can be achieved with PAL and SECAM is approximately 4 dB less than for NTSC.

Transmission standards for PAL and SECAM are generally covered under the ITU-BR and its series 400 recommendations. The most popular color TV system in the world, in terms of the number of countries and total population, is PAL. These countries have the 625-line, 50-Hz setup for the basic TV signal. The only country in the world with the combination of PAL and a 525-line/60-Hz standard is Brazil (all neighboring countries in South America have the same NTSC setup as the United States).

PAL is closest to NTSC in the manner in which the color information is transmitted, using the phase angle of the color subcarrier. Instead of requiring that the absolute value of phase be carried from camera to TV set, PAL improved upon the technique by doing it with a phase shift between alternating lines. This renders PAL less sensitive to several of the impairments found in the typical analog transmission system. For example, the 250-C specification for differential phase on a satellite link is 1.5 deg, which while significant for NTSC has effectively no meaning for PAL (or SECAM, for that matter). Both systems are equally susceptible to differential gain and noise.

The performance requirements for PAL can be determined with standard test equipment, similar in operation to that for NTSC. Table 4.10 summarizes

Table 4.10
Signal-to-Noise Requirements for PAL
Transmission Systems (CCIR
Recommendations 421-2 and J22)

Type of Link	Video	Audio
Short-haul microwave	62 dB	67 dB
Satellite link	52 dB	57 dB
End-to-end link	50 dB	54 dB

the most important requirements for a typical combination of a short-haul microwave and a point-to-point satellite link.

Other requirements apply to PAL and SECAM transmissions between studio and distribution point.

This chapter has taken us through common systems for TV distribution and many of the standards that apply. In Chapter 5, we review the developments in digital video, particularly compression systems that increase the channel capacity of the satellite by a substantial factor without reducing the visual quality as seen by the public. Chapter 6 covers how the foundation of satellite transmission, cable programming, and digital compression are producing broadcasting satellite systems with potential for success in DTH service.

References

[1] Adamson, S., et al., *Advanced Satellite Applications—Potential Markets*, Park Ridge, NJ: Booz, Allen, and Hamilton, Noyes Data Corporation, 1995.

[2] Vogel, H. L., *Entertainment Industry Economics*, Second Edition, Cambridge, UK: Cambridge University Press, 1990.

[3] Monash, B. (ed.), *International Television and Video Almanac*, 40th Edition, New York: Quigley Publishing Company, Inc., 1995.

[4] "Peoples Choice—Week 34, According to Nielson Ratings, May 8–14," *Broadcasting and Cable*, May 22, 1995, p. 33.

[5] "FCC to Mandate HDTV Capability in Receivers," *Broadcasting & Cable*, June 19, 1995, p. 37.

[6] Ciciora, W. S., "An Overview of Cable Television," *Engineering Handbook*, Eighth Edition, Washington, DC: National Association of Broadcasters, 1992, p. 731.

[7] Ciciora, W. S., "Inside the Set-Top Box," *IEEE Specturm*, APril 1995, p. 70.

[8] "Caught by the Web," *Cable and Satellite Europe*, February 1996, p. 46.

[9] Pelton, J. N., *Space 30—A Thirty-Year Overview of Space Applications and Exploration*, Society of Satellite Professionals International, Alexandria, VA, April 1989, p. 103.

[10] Williams, E. A., "Television Signal Transmission Standards," *Engineering Handbook*, Eighth Edition, Washington, DC: National Association of Broadcasters, 1992, p. 617.

Digital Video Compression Systems and Standards

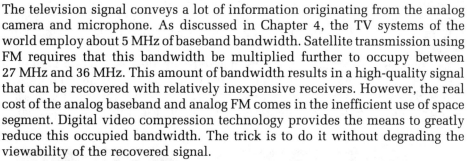

5

The television signal conveys a lot of information originating from the analog camera and microphone. As discussed in Chapter 4, the TV systems of the world employ about 5 MHz of baseband bandwidth. Satellite transmission using FM requires that this bandwidth be multiplied further to occupy between 27 MHz and 36 MHz. This amount of bandwidth results in a high-quality signal that can be recovered with relatively inexpensive receivers. However, the real cost of the analog baseband and analog FM comes in the inefficient use of space segment. Digital video compression technology provides the means to greatly reduce this occupied bandwidth. The trick is to do it without degrading the viewability of the recovered signal.

Digital compression plays a very important role in modern video transmission. Its principle benefits are:

- Reduced transmission bandwidth, which saves space segment costs and reduces the amount of power needed to transmit an acceptable signal;
- More channels available per satellite, which greatly increases the variety of programming available at a given orbit position, in turn promoting new services like impulse PPV and home education and making it feasible to expand a programming service through tailoring (e.g., packaging several different *feeds* of the same material with different advertising or cultural view) and multiplexing (i.e., sending the same channel at several different times);
- The potential of using a common format for satellite DTH, cable TV, and terrestrial broadcasting;
- That it provides a base for HDTV in the digital mode because the number of bits per second of a compressed HDTV signal is less than what was previously required for a broadcast-quality conventional TV signal.

Digital compression was first developed 25 years ago to save bandwidth by a factor of two on international satellite links. Some of the motivation also came from the sheer excitement of dealing with a challenge that was difficult given the technology available at the time. A system developed by COMSAT Laboratories in 1970 could multiplex two NTSC signals within the bandwidth of one transponder. They used a hybrid approach of analog filtering and digital processing. The cost of the equipment was projected to be so high that it never became popular, even though these researchers proved that a compressed signal could provide acceptable viewing.

Work in digital image processing continued for a very long time, yielding many innovations in the theory of digital video representation and electronic digital signal processing. The implementation of video compression has gone through a number of iterations, resulting finally in very affordable and usable consumer equipment. An example of this type of equipment is shown in Figure 5.1, produced under the Digital Satellite System proprietary standard for DIRECTV.

Compression systems that were marketed in the 1980s met a variety of needs, such as video teleconferencing, PC videophones, distance education, and early introductions of narrowband ISDN. Some examples of these early applications of digital video compression are listed in Table 5.1. The quality

(a)

Figure 5.1 An example of a digital DTH installation employing (a) the MPEG-based DSS integrated receiver-decoder and (b) a 45-cm dish.

(b)

Figure 5.1 (continued)

of the video portion is generally unacceptable for entertainment programming but probably adequate for a specific business purpose. For example, the P*64 systems are extensively used for point-to-point meetings to serve the needs of business and government users. The locations can be separated a few hundred kilometers (as in the case of communication between subsidiaries located in different cities of the same state, province, or region) to thousands of kilometers (when international service is needed). People who use P*64 conferencing systems find it convenient because no significant travel is required and more people may participate. Generally, these people already know each other and so can recognize who is doing the speaking and even pick up nonverbal clues

Table 5.1

Examples of Limited-Motion Digital Video Compression Applications, Carriage Return in Use on Satellites and Terrestrial Networks

Line Characteristic	Typical Line Speed	Applications
Digital carrier	3 Mbps	Distance education
T1, E1	1.544 Mbps	Meetings; distance education
P*64	384 Kbps	Teleconferencing; integrated networks
ISDN	128 Kbps	PC conferencing; videophones
Voice band dialup	19.2 Kbps	Technical communications, PC conferencing

from body language. Services that involve basic rate ISDN and analog dialup are not attractive in the meeting situation but would prove useful for desktop applications, as suggested in the table. This is also serving as the next step into multimedia communications, particularly on the Internet and World Wide Web.

A wide range of performance of compression systems results from the relationship between the data rate (which is proportional to the occupied bandwidth) and the quality of the picture. When quality can be sacrificed, then data rates below 1 Mbps are possible. On the other hand, if the intended application is in the field of education or entertainment, then significantly more than 1 Mbps is dictated. The first introduction of compression equipment with adequate quality for education applications was the Spectrum Saver system from CLI [1]. With a selectable data rate of either 3 Mbps or 6 Mbps, the user can determine the level of absolute quality against the cost of satellite transmission. Terrestrial transmission of the Spectrum Saver was not considered in the development of the facility.

Table 5.2 gives an indication of the relationship between bit rate and application in commercial broadcasting. A perfect video reproduction of analog TV standards (e.g., NTSC, PAL, and SECAM) is achieved with rates of 90 Mbps

Table 5.2

Typical Data Rate Requirement for Production and Distribution of Network TV Signals

Purpose	Data Rate (Mbps)
Acquisition (camera)	150
Production (studio)	150
Transmission (distribution)	30–45
Reception (direct-to-home)	3–6

or greater. Typical viewers cannot usually tell that anything is impaired when the signal is compressed to a rate of 45 Mbps. Below this value, it becomes subjective. For movies, a rate as low as 1.5 Mbps, the standard T1 in North America and Japan, is sufficient. However, for any live action as used in sports, at least 5 Mbps will be needed. In time, the performance of current compression systems and standards will allow improvements in motion performance so that 1.5 Mbps will meet the needs of all types of content.

Compression systems that operate at 45 Mbps or greater are designed to transfer the signal without permanent reduction of resolution and motion quality. They are said to be *lossless* in that the output of the decoder is identical to the input to the encoder. In contrast, operation below about 10 Mbps is *lossy* in that it introduces a change in the video information that cannot be recovered at the receiving end. Lossy compression can produce a picture of excellent quality from the viewer's point of view at data rates above about 3 Mbps. There is an intermediate position called quasi-lossless wherein a lossy compression service is augmented with the parallel transmission of an error signal that contains correction data to recreate a lossless image at a compatible receiver. The application of lossy transmission with reduced picture quality may be attractive since it can reduce transmission costs (or allow the user to employ an existing communications systems such as a VSAT network).

Our focus in the chapter is on modern compression technology and standards that are being applied to the consumer marketplace. While drawing on previous experience, the new approaches provide high-quality images and employ low-cost set-top equipment. This breakthrough in applying technology is revolutionizing the satellite TV industry and will have a profound effect on its extension over the coming decade.

5.1 COMPRESSION TECHNOLOGY

The analog waveform of standard NTSC or PAL is very effective in its ability to provide entertainment and business communications. Enjoyment has been further enhanced with the addition of stereophonic sound; and additional services like closed caption, second language, and teletext are available as well. These systems are also relatively simple in terms of generation and display. The transmission of the video signal is relatively straightforward, provided that the link has adequate bandwidth and linearity.

From a technical standpoint, video sequences scanned at the rate of either 30 or 25 frames per second with 525 or 625 lines each, respectively, contain a significant amount of redundancy both within and between frames. This provides the opportunity to compress the signal if the redundancies can be removed on the sending end and then replaced on the receiving end. To do this, the encoder at the source end examines the statistical and subjective properties of the frames and then encodes a *minimum set* of information that is ultimately placed on the link. The effectiveness of this compression depends on the amount of redundancy contained in the original image as well as on the compression technique (called the compression algorithm).

For TV distribution and broadcast applications over satellites, we wish to use data rates below 10 Mbps in order to save transponder bandwidth and RF power from the satellite and Earth station. This means that we must employ the lossy mode of compression, which will alter the quality in objective (numerical) and subjective (human perception) terms. A subjective measure of quality depends on exposing a large quality of the human subjects (viewers) to the TV display and allowing them to rate its acceptability. The TASO scale shown in Table 4.3 is an excellent example of such a subjective scale for measuring quality.

The ultimate performance of the compression system depends on the sophistication of the compression hardware and software and the complexity of the image or video scene. For example, simple textures in images and low video activity are easy to encode and no visible artifacts (defects) may result even with simple encoding schemes. The real test is for scenes with a great deal of detail, including varying textures, and fast-moving live action. Conventional movies that were filmed at 25 frames per second do not represent a challenge; however, TV coverage of live sporting events will severely test any compression system.

5.1.1 Digital Processing

Any analog signal can be digitized through the two-step process of sampling at discrete time intervals, followed by converting each sample (usually a voltage value) into a digital code. The latter process is also called quantization because it involves forcing the measurements to fit onto a scale with discrete steps. The example in Figure 5.2 shows a simple analog waveform on the top and its quantized version below. There are only eight quantization levels, which correspond to a digital representation of three bits per sample (because the numbers 0 through 7 are represented by the binary numbers 000 through 111). The number of bits determines the quality of reproduction, which can be specified in terms of the signal-to-quantization noise ratio (S/N_q). The more bits per sample, the better the reproduction, as evidenced by the equation

$$S/N_q = 3M^2 \tag{5.1}$$

where M is the number of bits per sample. Equivalently, in terms of decibels,

$$S/N_q = 4.8 + 20 \cdot \log(M) \tag{5.2}$$

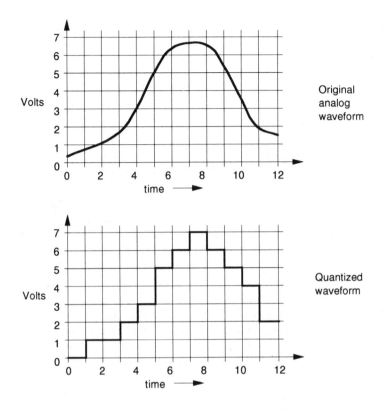

Figure 5.2 Comparison of the original analog waveform with the quantized version, based on eight levels (or three bits).

Typical values of M are in the range of 6 to 12, with 8 being the most common. At this level, the S/N_q is equal is 22.8 dB. This relation indicates that doubling the number of bits per sample reduces the quantization noise by 6 dB.

A further refinement is to compress this scale to emphasize the lower levels and de-emphasize higher levels, which are inherently easier to accept by the human watcher or listener. This is the familiar process of companding used in PCM applied in digital telephone networks. Companding provides a subjective improvement to the basic quality provided by the quantization process. If this companding advantage is 30 dB (a typical value), then the S/N_q of 22.8 dB in the previous example can be increased to 52.8 dB in terms of its subjective effect on humans. At this level, it would be rated comparably to a high-quality analog telephone line that has a measured value of S/N above 50 dB. A subjective evaluation of telephone communications is discussed further in Chapter 9 for mobile satellite service. In the following discussion, we consider how this performance will affect video images.

According to the Nyquist sampling theorem of communications engineering, a lossless sampling requires that samples be taken at a rate that is twice the baseband frequency. Therefore, for a typical video signal of 5-MHz bandwidth, the sampling rate would have to be at least 10 million times per second (e.g., 10 MHz). In the technique of subsampling, a reduced sampling rate below the Nyquist rate can still be lossless, provided that some additional conditions are met. This is because the subsampling produces a folding back, at a reduced level, of the low end of the spectrum onto the high end. The impairment that results is like another form of interference into the video baseband, yet this may still produce an acceptable picture.

Essentially all of the practical encoding and compression systems use subsampling and quantization prior to compression. Subsampling is applied by first reducing the horizontal and/or vertical dimension of the input video, which in turn reduces the number of picture elements (pels) that must be coded. The images at the receiving end are smoothed using the mathematical process called interpolation. This produces a more natural look to the image so that human observers cannot detect the potential impairment from subsampling. Interpolation is the first level of compression, ahead of the steps to be taken in the coding of each image and the compression from frame to frame.

The black-and-white (luminance) component of the picture has more information content and hence requires more bandwidth than the color component (chrominance). As a result, subsampling is not applied equally to the luminance and chrominance parts of the picture, particularly because the human eye is much less sensitive to compression applied to chrominance. In fact, the number of bits per sample for the color information is significantly less than for the luminance information. This is reflected in all of the video compression standards, particularly MPEG.

5.1.2 Spatial Compression (Transform Coding)

Transform coding is the most popular technique for reducing the number of bits required to represent a digital image. The basic idea is to replace the actual image data with a mathematically derived set of parameters that can uniquely specify the original information. The parameters, which are the coefficients of a mathematical transform of the data, require less transmitted bits to be stored or sent than the original image data itself because they can be compressed further. Over the years, the DCT has proven to be the most popular mathematical procedure and is now part of the JPEG and MPEG series of standards, which will be discussed later in this chapter. The mathematical formulation of the DCT in the forward direction is

$$F(u, v) = \frac{4c(u)c(v)}{N^2} \sum_{i=1}^{N-1} \sum_{j=0}^{N-1} f(i, j) \cos\frac{(2i + 1)u\pi}{2N} \cos\frac{(2j + 1)v\pi}{2N} \qquad (5.3)$$

and in the inverse direction is

$$f(i, j) = \sum_{i=1}^{N-1} \sum_{j=0}^{N-1} c(u) \, c(v) F(u, v) \cos\frac{(2i + 1)u\pi}{2N} \cos\frac{(2j + 1)v\pi}{2N} \qquad (5.4)$$

where

$$c(w) = \begin{cases} 1/52 & \text{for } w = 0 \\ 1 & \text{for } w = 1, 2, \ldots, N - 1 \end{cases} \qquad (5.5)$$

The DCT is similar to the FFT, which is a technique that allows a computer to generate the frequency spectrum for a time waveform and vice versa. Because of this, the DCT could be described as a way to convert from linear graphic display (which is in two dimensions) to an equivalent series of spatial frequency components. A spatial frequency is measured in cycles per unit of linear measure (cycles per inch, for example) instead of cycles per unit time (cycles per second, or hertz). A square wave in terms of a spatial frequency would look like an alternating sequence of black and white squares, like on a chessboard. In fact, a chessboard image can be transferred very efficiently using the DCT with a minimum number of bits. Any picture can then be viewed as the overlay of these frequencies, much the way an analog time signal can be viewed as the combination of frequency components. The difference with image data is that the time axis of the signal is replaced with a two-dimensional distance axis, as one scans across and down the particular picture element.

The basic concept of how the DCT is applied to two-dimensional compression of an image is shown in Figure 5.3. The image is divided into square or rectangular segments, and then the transform is applied to each individually. In this particular example, the image is split into blocks that are $N \times N$ pels on each side. If you examine one of these squares, you can see that a given horizontal string of pels can be represented by a combination of frequency components in the same way that a time waveform can be expressed by a combination of sine waves (a Fourier series). The first step taken by the DCT coder is to represent the block in the form of an $N \times N$ matrix of the pels and then apply the DCT algorithm to convert this into a matrix of coefficients that represent the equivalent spatial frequencies. More bits are removed by limiting the number of quantization steps and by removing some of the obvious redundancy. For example, coefficients that are zero are not transmitted.

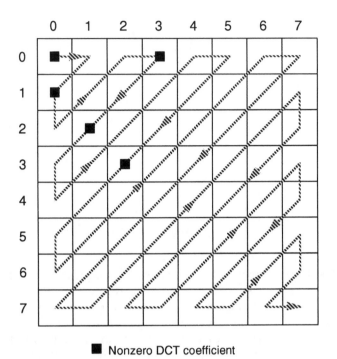

■ Nonzero DCT coefficient

Figure 5.3 The zig-zag scanning pattern used to collect DCT coefficients with an 8 × 8 block, selecting nonzero values at the locations of the dots.

Using frequencies to represent the individual blocks that comprise the image is very effective from the standpoint of compression and transmission. Information loss increases as the relative size of the block increases and the number of frequencies used to represent the content of the block decreases. What you see is that some elements of the image show up as small squares (like a cubist painting), and some shaded areas have odd patterns (like herringbones) running through them. In the worst case, the squares themselves are clearly apparent in the picture. Of course, when the system is properly optimized and the error rate is acceptable, the DCT performs very well and the image is natural. On close examination of the original as compared to the previously compressed image, the modifications are clearly discernible. This does not detract from the enjoyment or utility of the video service, unless of course one demands a perfect reproduction (which is impossible in the case of any of the standard analog TV systems).

5.1.3 Temporal Compression (Frame-to-Frame Compression)

The types of video sequences involved in NTSC, PAL, and SECAM are statistical in nature and, or course, contain a high degree of redundancy. Stable sequences, such as what happens when the camera is held in a fixed position during a scene, are highly correlated because only a few aspects change from frame to frame. Consider a video segment of the nightly news with a reporter at her chair behind a desk. During the entire time that she is speaking, the foreground and background never change; in fact, the only noticeable motion is of her head, mouth, and perhaps her upper body and arms. The result of this is that only the first frame needs to be encoded in its complete form; the remaining frames must only be replaced by information about the changes. This is possible because interframe correlation is high and, in fact, two or more consecutive intermediary frames can be predicted through interpolation. The formal way to state this is that an approximate predication of a pel can be made from the previously coded information that has already been transmitted. For greater resolution, the error between the predicted value and the previous one can be sent separately, which is a technique called *differential pulse code modulation* (DPCM). Both DCT and DPCM can be combined to provide a highly compressed but very agreeable picture at the receiving end.

5.1.4 Motion Compensation

The technique used to reduce the redundant information between frames in a sequence is called *motion compensation*. It is based on estimating the motion between video frames by observing that individual elements can be traced by their displacement from point to point during the duration of the sequence. Shown in Figure 5.4, this motion can be described by a limited number of motion parameters that are defined by vectors. For example, the best estimate of the motion of a given pel is provided by the motion-compensated prediction pel from a previously coded frame. To minimize error, both the motion vector and the prediction error are transmitted to the receiving end. Aggregating nearby pels and sending a single vector and error for them as a group is possible because there is a high degree of correlation between adjacent pels as well.

Motion compensation by its nature is a very complicated and computation-intensive process. Therefore, it only became practical for commercial video compression systems in the 1990s. As techniques become more refined and as hardware implementations in VLSI proliferate, we will see substantial improvements in the observed quality of digital DTH and cable TV in coming years. For example, the MPEG 2 standard, as well as the upcoming U.S. standard for HDTV, exploit these improvements and it is only a matter of time before the majority of the motion impairments are removed through this process.

5.1.5 Hybrid Coding—Other Advanced Techniques

Two or more coding techniques can be combined to gain more advantage from compression without sacrificing much in the way of quality. A technique com-

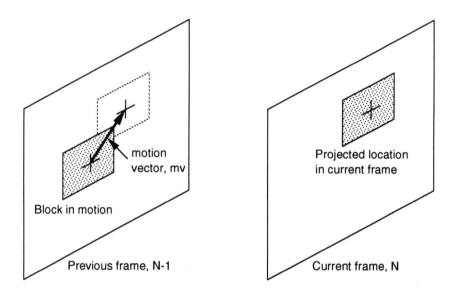

Figure 5.4 A technique for frame-to-frame projection using motion compensation. The motion vector, *mv*, indicated in frame $N - 1$ points toward the new location to be used in subsequent frame N.

monly applied is to combine frame-to-frame (temporal) DPCM with the spatial DCT method. Temporal correlation is first reduced through prediction, and then the DCT is applied to the prediction. The DCT coefficients in the $N \times N$ matrix are quantized and compressed further. This method is central to the MPEG standard to be discussed in the next sections.

Hybrid coding can be applied in levels as a way to enhance the quality of a service that is tailored to a particular need or income level. The basic service using DCT may be transmitted with maximum compression for minimum cost of transmission. Enhancement through a second channel that adds the hybrid coding feature would require more capacity and processing on the sending and receiving ends. This opens up the possibility of offering a higher quality service with better resolution and motion performance at a higher price. It is a way of making a service backward-compatible where the first service is based on the current standard for MPEG 2 and the next generation is implemented through hybrid coding.

The other area of significant improvement in compression performance is that of fractal coding. Fractal compression, which is still under investigation, holds promise for another order of magnitude reduction in the number of bits per second as compared to the DCT. It is recognized that fractal compression is lossy because of the type of decomposition of the image that is applied, but it will likely find some significant applications in telecommunications and information storage and retrieval. A hybrid combination of fractal and statistical compression is another potential direction for the highest possible degree of data reduction.

5.2 JOINT PHOTOGRAPHIC EXPERTS GROUP (JPEG)

The *Joint Photographic Experts Group* (JPEG) was created in 1986 by members of the ISO and the CCITT (now ITU-T) for the purpose of creating an open standard for coding of continuous tone still images. The JPEG standard was issued in 1991 as ISO 10918. Initially developed on black and white, the standards have been extended to high-resolution computer color images to take advantage of high-definition video displays and output devices. By 1992, JPEG had become one of the most popular digital image compression standards for PC applications, particularly hard disk storage of photographs and CD-ROM media. The standard is implemented both in software and as hardware available in the form of plug-in boards.

The basis of JPEG is a spatial DCT compression system, providing a variety of modes and degrees of lossy compression. In this way, the user can determine the degree of compression and therefore the amount of information loss, if any. JPEG is really a set of standards that allow the user to make his or her own tradeoff of compression versus quality. It offers a lossless mode using the DPCM algorithm. For example, an uncompressed 480×640 pixel image (307,200 total pixels) suitable for display on a VGA monitor might require 5 MB of hard disk storage space. With lossless compression using DPCM, the same image file could be stored using only 1 MB. A still-usable high-resolution image that has been compressed on a lossy basis with DCT might require 20 kB. An uninformed observer probably could not tell the difference among these images, but the alterations produced by DCT would be apparent upon close examination using a magnifying glass.

The JPEG standard in the lossy mode processes blocks of 8×8 pels at a time, the first step being to transform the digitized image data using the DCT. Each set of 64 DCT coefficients is uniformly quantized to provide some degree of weighting according to human perception. The DC component of the output is treated differently from the rest by encoding it using a differential DC prediction method. The remaining coefficients, corresponding to various spatial frequencies, are encoded using a variable-length type of code. Also referred to as entropy coding, the idea is to encode a given coefficient so that the greater the information content, the longer the code. The limiting case is for coefficients that are zero and therefore are not transmitted at all. As shown in Figure 5.3, the coefficients are actually zig-zag scanned across the block of 8×8 pels.

The power of the JPEG standard was clearly shown to this author in 1990 by a VHS videotape produced by C-cube Corporation of Mountain View, CA. In this simple demonstration, spatial compression to each frame using the JPEG algorithm was implemented in hardware. The action scene was of a snow skier traveling down a hill. Three versions of the scene were included, namely, the original analog material, a digitized version with no compression, and a compressed version using JPEG at 1.9 Mbps. All three were presented in analog form on the video monitor. To this author's eyes, it was impossible to tell one version from the other, clearly showing the commercial potential of the JPEG standard, which used the DCT algorithm without any frame-to-frame compression. No audio was provided since JPEG relates only to images and neither supports motion sequences nor the associated audio. An exception is *Motion JPEG* (MJPEG), which several companies support as a less-complex substitute for the now-adopted MPEG standard for motion sequences [2]. MJPEG compresses each frame without using temporal compression and can be decompressed directly with the JPEG algorithm. As we will see in the next discussion, MPEG adds these features and more, to provide a complete package for the TV broadcaster, DTH service provider, and cable network.

5.3 MOTION PICTURE EXPERTS GROUP (MPEG)

The MPEG, also affiliated with ITU-T and ISO, has completed work on the compression standard for full motion pictures, making use of frame-to-frame compression. MPEG can eventually permit the transmission of full-color, full TV motion images at a rate as low as 1.5 Mbps. A key point is that the compression is done in real time, without the need for digital processing by large mainframes or supercomputers.

The MPEG series of standards for motion pictures and video provides many desirable features.

- The MPEG series of standards supports a wide variety of picture formats with a very flexible encoding and transmission structure.
- It allows the application to use a range of data rates to handle multiple video channels on the same transmission stream and to allow this multiplexing to be adaptive to the source content.
- The algorithms can be implemented in hardware to minimize coding and decoding delay (typically less than 150 ms at the time of this writing).
- Developers can include encryption and decryption to comply with content restrictions and the needs for business integrity.
- Similarly, provisions can be made for an effective system of error protection to allow operation on a variety of transmission media such as satellite and local microwave links.

- The compression is adaptable to various storage and transport methods, an excellent example of which is the DVB standard, which will be discussed at the end of this chapter.
- The frame-to-frame compression approach with the use of *intra* (I) pictures (discussed later in this chapter) permits fast forward and reverse play for CD-ROM applications, impacting the degree of compression since frames cannot be interpolated if used for these features.
- Transcoding will permit conversion between compression formats since MPEG is widely popular.
- MPEG-processed videos can be edited by systems that support the standard.
- Random access can be allowed using the I pictures.
- The standard will most probably have a long lifetime since it can adapt to improvements in compression algorithms, VLSI technology, motion compensation, and the like.

These properties relate to the evolving MPEG family of standards. As of the time of this writing, two of these standards were complete and already available on commercial markets. These include the MPEG 1 standard, which provides for encoding video sequences intended for CD-ROM and other multimedia applications, and MPEG 2, which is the standard for commercial digital television. The obvious focus of this book is MPEG 2, but MPEG 1 is important because it is the predecessor and provides an important technical foundation. We review each of these standards in the following subsections.

5.3.1 MPEG 1

The first full-motion compression standard to be produced was MPEG 1, which is aimed at nonbroadcast applications like computer CD-ROM. It draws from JPEG in the area of image compression using DCT and is also intended to provide a broad range of options to fit the particular application. For example, there are various profiles to support differing picture sizes and frame rates and it can encode and decode any picture size up to the normal TV with a minimum number of 720 pixels per line and 576 lines per picture. The minimum frame rate is 30 (noninterlaced) and the corresponding bit rate is 1.86 Mbps. Principle technical parameters for MPEG are listed in Table 5.3.

Because MPEG 1 is aimed at multimedia applications, there was a need to allow convenient fast-forward and fast-backward capability. This means that complete frames are needed at relatively frequent intervals to permit scanning by the user. Otherwise, the various forms of frame-to-frame compression and motion compensation would have greatly reduced the possibility to cue material

Table 5.3
Summary of MPEG 1 Technical Characteristics

Characteristic	Value or Description
Type of coder	Hybrid
Spatial transform	DCT, 8 × 8 block
Quantization	Separate luminance and chrominance matrices
	Can be user-supplied
	User-supplied scaler used to adjust absolute level
	Can be varied macroblock by macroblock
Variable length code	Default: two-dimensional Hoffman
	Can be user-supplied
Temporal compression	Motion compensation by motion vectors
	±15 pixel search in both axes
	Difference image coding
Active pixels	Up to 4096 by 4096
Rate	8 frame rates, up to 72 frames per second
Raster	Progressive scan

at intermediate stages in a video sequence. The special multimedia features of MPEG 1 include:

- Fast forward and fast reverse (FF/FR);
- Reverse playback;
- Ability to edit a compressed bit stream.

These features are provided by encoding two types of pictures: *intra* (I) pictures and *interpolated* (B) pictures. I pictures are encoded individually, without considering other pictures in a sequence. This is analogous to taking an image and compressing it by itself with JPEG. These pictures can therefore be decompressed individually in the decoder and displayed one-by-one, providing random access points throughout a CD-ROM or video stream. *Predicted* (P) pictures, on the other hand, are predicted from the I pictures and incorporate motion-compensation as well. They are therefore not usable for reference because they require other pictures to be decoded properly. Another aspect is that it takes a sequence of I and P pictures to reproduce a video sequence in its entirety, which introduces a delay at the receiver. If the sequence had been encoded exclusively with I pictures, then the delay is minimal, amounting only to the time needed to convert back from the DCT representation to the equivalent uncompressed sequence. Like the MJPEG standard discussed in the last section, this greatly increases the amount of data needed to record and/or transmit the sequence.

An example of an MPEG frame sequence is shown in Figure 5.5, displaying the relationship between three classifications of pictures (I, B, and P). As stated previously, the I pictures are stand-alone DCT images that can be decompressed and used as a reference. The B pictures are interpolated between I pictures and are therefore dependent on them. P pictures are computed from the nearest previously coded frame, whether I or P, and typically incorporate motion compensation. Interpolated B pictures require both past and future P or I pictures and cannot be used as reference points in a video sequence.

The temporal compression of MPEG 1, based on interpolation, means that one can no longer transmit a true time-sequential stream. This is shown in Figure 5.6 for one complete intraperiod of a digital video sequence. Every transmitter and receiver requires adequate video frame memory to hold the forward I pictures and computational power to calculate P and B pictures from them. With high-speed VSLI and *application-specific integrated circuits* (ASICs), the cost and complexity of accomplishing this have been reduced to a very accessible level. MPEG 1 decoders are available on chips that can be introduced right into the CD-ROM drive itself or the display device. MPEG 1 encoders and decoders are available at low cost as computer plug-in boards for low-cost multimedia applications in education and business.

Many specific tradeoffs are possible because the three types of pictures can be introduced and arranged to suite the needs of the application. The most

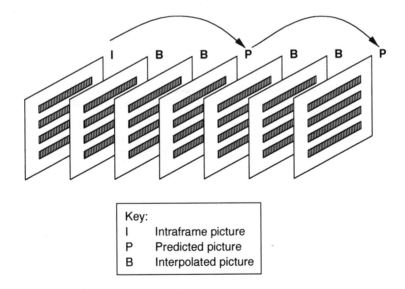

Key:
I   Intraframe picture
P   Predicted picture
B   Interpolated picture

Figure 5.5 The use of I, P, and B pictures for frame-to-frame compression. The coding sequence and order are selected based on the content of the video sequence.

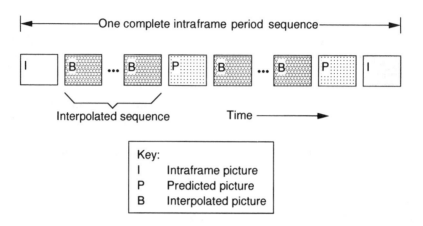

Figure 5.6 An example of an actual sequence of pictures used in an MPEG sequence.

flexible approach is to employ the I picture format exclusively because it will provide random access to any frame in the sequence. However, it is also the most expensive in terms of storage or bandwidth. A sequence with a moderate quantity of P pictures provides a degree of random access and FF/FR functionality. If storage and access are not contemplated, then B pictures may be used along with the other types as this provides the greatest opportunity for bandwidth reduction.

The MPEG 1 standard was clearly a pioneering effort in the journey to a highly efficient digital TV system. It grew out of the work on computer images, which is clearly a mass market of the type to capture the attention of the leading electronics and media companies. The next step was to recognize the special needs of the broadcasting industry. However, the participants in conventional TV have long resisted change. In comparison, we in satellite communications have experimented with digital technology for decades in the hopes of finding that *killer application* that would allow us to develop new markets for our technology. To enter into this particular field, we first discuss a video standard that predates MPEG, namely, CCIR Recommendation 601. The significance of 601 is that it was adopted by the international TV community and therefore provides an agreed baseline for the encoding and compression process.

5.3.2 CCIR Recommendation 601

Recommendation 601 is the result of an effort to standardize digital TV that began around 1976. MPEG 1 operates on noninterlaced video inputs but has been adapted to normal TV for 525- and 626-line systems. The intersection of

MPEG and CCIR Recommendation 601 comes in the form of a transcoder between the two systems. The focus of Recommendation 601 is on studio applications where the bandwidth of communication is less of a problem. The exception to this, of course, is with respect to storage of video programs for editing and archival purposes. A basic agreement on this recommendation was reached in 1981 between the U.S. *Society of Motion Picture and Television Engineers* (SEMPTE) and the EBU through the efforts of CCIR Study Group XI.

CCIR 601 is a standard digital video format that can accommodate the three analog TV systems in use in the world [3]. The application as a primary studio format for program acquisition is based on the principle of component coding and its extensibility to various analog formats. It is a family of standards rather than a single unified format. Sampling is accomplished with a common sampling rate of 13.5 MHz, which is 858 times the 525 horizontal frequency and 864 times the 625 horizontal frequency. Color-difference signals are encoded at half this rate.

The specific parameters are summarized in Table 5.4. The amount of information contained in the luminance signal is at least twice that of each of the two color-difference signals. Frame repeat rates differ for the 525- and 626-line systems, being 29.97 Hz and 25 Hz, respectively. These properties are reflected in Recommendation 601 by the relative sampling rates and the corresponding effective bit rates; the ratios that apply to the luminance channel and the U and V color-difference channels are 4 to 2 to 2, respectively.

Table 5.4
Main Parameters of CCIR Recommendation 601 (4:2:2)

Analog Standard Input	*525 Line/60 Hz*	*625 Line/50 Hz*
Samples per line		
Luminance component	858	864
Color component (each)	429	432
Sampling frequency		
Luminance component	13.5 MHz	13.5 MHz
Color component (each)	6.75 MHz	6.75 MHz
Samples per active digital line		
Luminance component	720	720
Color component (each)	360	360
Correspondence between number of quantizing bits and signal level	on a scale of 0–255	on a scale of 0–255
Luminance component	16 (black) to 255 (white)	16 (black) to 255 (white)
Luminance component	128 (no color) ± 112 (16 to 240 full saturation)	128 (no color) ± 112 (16 to 240 full saturation)

The standard manner by which to transfer 601-encoded video is via the serial interface, which is lossless and has a bit rate of 216 Mbps. The engineering community has determined that a gross rate of 243 Mbps is to be employed to allow other forms of information to be added, particularly stereo audio. The advantage of using this format is that it is an international standard to which manufacturers and operators can comply.

Recommendation 601 is important because of its pioneering status as a worldwide digital video standard. With a serial information transfer rate of almost 250 Mbps, it is unlikely that 601 signals will find their way into the home. Rather, it is an interface standard or baseline upon which the practical digital compression systems will be evaluated. This will be apparent in the discussion of MPEG 2 that follows.

5.3.3 MPEG 2

The second phase of consumer digital video standards activity took the innovations of MPEG 1 and added features and options to yield an even more versatile system for broadcast TV. From standard-setting activity begun in 1992, MPEG 2 has quickly become the vehicle for bringing digital video to the mass market. The purpose of this standard is to provide lossy video quality equal to or better than NTSC, PAL, and SECAM, along with the facility to support the lossless performance of CCIR Recommendation 601. The developers of MPEG 2 had in mind the most popular applications in cable TV, satellite DTH, digital VCRs, and future terrestrial networks that would employ ATM to deliver VOD. In 1994, the draft specification was produced; yet, an operational system based on this standard was already being introduced in the United States by DIRECTV, Inc., a subsidiary of Hughes Electronics Corp.

MPEG 2's important contribution is not in compression (that was established by JPEG and MPEG 1) but rather as an integrated transport mechanism for multplexing the video, audio, and other data through packet generation and time division multiplexing. It is an extended version (or superset) of MPEG 1 and is designed to be backward compatible with it. The definition of the bit structure, called the syntax, includes a bit stream, a set of coding algorithms, and a multiplexing format to combine video, audio, and data. New coding features were added to improve functionality and enhanced quality in the conventional video environment of interlaced scanning and constrained bandwidth. The system is scalable in that a variety of forms of lossless and lossy transmission is possible, along with the ability to support a future HDTV standard. Robust coding and error correction are available to facilitate a variety of delivery systems including satellite DTH, local microwave distribution (e.g., MMDS), and over-the-air VHF and UHF broadcasting. This would assure effective reception by the public in the face of link fades that produce what would otherwise be unacceptable transmission errors.

Because delivery systems and applications differ widely, MPEG 2 provides a variety of formats and services within the syntax and structure. This is the concept of the profile, which defines a set of algorithms, and the level, which specifies the range of service parameters that are supported by the implementation (e.g., image size, frame rate, and bit rate). The profiles and levels are defined in Tables 5.5 and 5.6, respectively. Table 5.5 begins at the lowest profile called SIMPLE, which corresponds to the minimum set of tools. Going down the table adds functionality.

The Main profile is the current baseline for MPEG 2 applications and has been implemented in a number of DTH systems. As suggested in Table 5.5, this profile does not include scalability tools and therefore is a point of downward-compatibility from the higher levels that provide scalability. The scalability tools for SNR and Spatial profiles add power to the standard for applications that are not available at the time of this writing. One can imagine how this might make a mobile video service or one based on less dependable transmission

Table 5.5
Profiles and the Associated Algorithms for the MPEG 2 Standard

Profile	*Algorithms*
Simple	Provides the fewest tools but supports the 4:2:0 YUV representation of the video signal.
Main	Starts with Simple and adds bidirectional prediction to give better quality for the same bit rate. It is backward compatible with Simple as well.
Spatial Scalable	This and the profile that follows include tools to add signal quality enhancements. By Spatial it is meant that the added signal complexity allows the receiver to improve resolution for the same bit rate. There would be an impact on the receiver in terms of complexity and hence cost. It can also be a means to add HDTV service on top of conventional resolution (i.e., only the appropriately designed receivers can interpret and display the added HDTV information).
SNR Scalable	The added signal information and receiver complexity improve viewable *signal-to-noise ratio* (SNR). It provides graceful degradation of the video quality when the error rate increases.
HIGH	This includes the previous profiles plus the ability to code line-simultaneous color-difference signals. It is intended for applications where quality is of the utmost importance and where there is no constraint on bit rate (such as within a studio or over a dedicated fiber optic link).

Table 5.6
Levels and the Associated Parameters for the MPEG 2 Standard

Level	Samples/Line	Lines/Frame	Frames/s	Mbps	BW (MHz)*
High	1920	1152	60	80	40
High 1440	1440	1152	60	60	30
Main	720	576	30	15	7.5
Low	352	288	30	4	2.0

*The bandwidth (BW) indicated in the last column is based on direct QPSK modulation of the bit stream and does not include forward error correction coding

at higher frequencies in the Ka-band portion of the spectrum practical. A possible implementation of a scalable MPEG 2 system is diagrammed in Figure 5.7.

The other dimension of MPEG 2 takes us through the Levels, which provide a range of potential qualities from low definition to HDTV. There is an obvious impact on the bit rate and bandwidth. These levels are reviewed in Table 5.6.

These levels are associated with the format of the originating source video signal and provide a variety of potential qualities for the application. It ranges

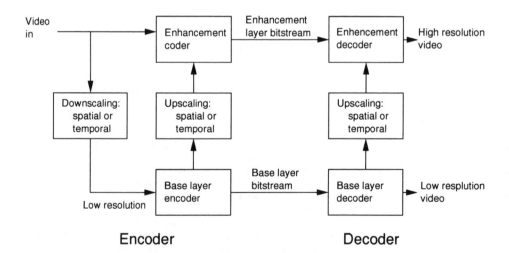

Encoder   **Decoder**

Figure 5.7 Conceptual configuration of scalable video encoding, to allow both a high-resolution and a low-resolution mode of transmission and decoding. For low-resolution, only the base layer bit stream is required at the receiving end. Enhancement layer bits are transmitted to permit compatible receivers to display full quality video at high resolution.

from limited definition and the associated low data rate all the way up to the full capability of HDTV. Another feature of the standard is that it permits the normal TV aspect ratio (width to length) of 4:3 as well as the more appealing movie screen or HDTV aspect ratio of 16:9. This particular part of MPEG 2 covers the base input and does not consider the degree of compression afforded by the profiles covered in Table 5.5. The basis of each of the levels is as follows.

- The Low level is an input format that is only one-quarter of the picture defined in Recommendation 601.
- The Main level has the full 601 input frame format.
- The High-1440 level is the HDTV format with 1440 samples per line.
- The High level is an even better HDTV format with 1920 samples per line.

The tools and formats of MPEG 2 allow as many as 20 different combinations, which the standard calls convergence points. As with any such standard, not every combination is either useful or viable. At the time of this writing, the Main profile and Main level represent the convergence point of all practical implementations. This is the case in North America with the various systems already in use as well as with the European DVB standard, which will be discussed in the next section.

To summarize, MPEG 2 is an attractive digital video standard that was developed for wide consumer application. Its core algorithm at the Main profile features nonscalable coding for both progressive and interlaced video information sources (e.g., broadcasting and specialized applications like video teleconferencing and multimedia). The Main level further specifies an input, which meets the needs of commercial television at 25 or 30 per second with an input rate of 15 Mbps. This, or course, is compressed down to as little as 1.5 Mbps, based on the tradeoff between transmission cost and application quality.

The other important element of MPEG 2 is the provision of stereo audio and data. The audio compression system employed in MPEG 2 is based on the European MUSICAM standard as modified by other algorithms [2]. It is a lossy compression scheme that draws from techniques already within MPEG. Like differential PCM, it transmits only changes and throws away data that the human ear cannot hear. This information is processed and time division multiplexed with the encoded video to produce a combined bit stream that complies with the standard syntax. This is important because it allows receivers designed and made by different manufacturers to be able to properly interpret the information. However, what the receiver actually does with the information depends on the features of the particular unit. All of these capabilities have been provided in the DVB standard, which is covered in the following section.

5.4 DIGITAL VIDEO BROADCASTING (DVB) STANDARD

The MPEG series of standards was the result of international cooperation among world-class organizations from several continents. From this, engineers and manufacturers must create the specific implementations and products that allow the public to enjoy the versatility of digital video. The first MPEG-based products were created for the U.S. market, but, as has been the usual case, these systems are incompatible with each other. At the same time that pioneers were addressing the most attractive single consumer market in the world, their counterparts in Europe set about to build a better mousetrap and to do it in a way that a unified approach might be produced. This is the effort that has resulted in DVB.

5.4.1 DVB Requirements and Organization

We cover DVB in some detail because of its technical relevance, openness, and success in the satellite communication field. The DVB system is intended as a complete package for digital television and data broadcasting [4]. It is built on the foundation of the MPEG 2 standard, providing full support for encoded and compressed video and audio, along with data channels for a variety of associated information services. The MPEG standard provides for a data stream syntax, discussed in the previous section, to multiplex the required functions together. On top of this, the standard considers the modulation and RF transmission format needed to support a variety of satellite and terrestrial networking systems.

The overall philosophy behind DVB is to implement a general technical solution to the demands of applications like cable TV, as discussed in Chapter 4, and DTH, to be discussed in Chapter 6. It includes the following systems and services:

- Information containers to carry flexible combinations of MPEG 2 video, audio and data;
- A multiplexing system to implement a common MPEG 2 *transport stream* (TS);
- A common *service information* (SI) system giving details of the programs being broadcast (this is the information for the on-screen program guide);
- A common first-level *Reed-Solomon* (RS) forward error correction system (this improves the reception by providing a low error rate to the decoded data, even in the presence of link fades);
- Modulation and additional channel coding systems, as required, to meet the requirements of different transmission media (including FSS and BSS

satellite delivery systems, terrestrial microwave distribution, conventional broadcasting and cable TV);
- A common scrambling system;
- A common conditional access interface (to control the operation of the receiver and assure satisfactory operation of the delivery system as a business).

The origin of DVB is a pan-European program of industrial and government cooperation that began in 1990. Over the course of a year, the group expanded to include consumer electronics manufacturers and common carriers. A *Memorandum of Understanding* (MOU) was signed in 1993 that established DVB as a set of standards, with digital satellite and cable TV drawing the most immediate attention. As of May 1995, almost 200 companies and agencies had signed the MOU, many from the United States (including AT&T, CLI, DEC, General Instruments, Hewlett Packard, Hughes Electronics, Motorola, and Texas Instruments), Japan (including NEC, Mitsubishi Electric, Pioneer, and Sony), and of course Europe (including ALCATEL, the BBC, EUTELSAT, France Telecom, News Corp., Nokia, RTL, Thomson, and ZDF).

5.4.2 Relationship Between DVB and MPEG 2

From the outset, DVB was made to follow the spirit and the letter of MPEG 2. This meant that a close tie was needed, which was possible due to the association among the organizations that signed the MOU and those that were part of MPEG. The group specified a family of DVB standards, including the following.

- *DVB-S:* The satellite DTH system for use in the 11/12-GHz band, configurable to suit a wide range of transponder bandwidths and EIRPs;
- *DVB-C:* The cable delivery system, compatible with DVB-S and normally to be used with 8-MHz channels (e.g., consistent with the 625-line systems common in Europe, Africa, and Asia);
- *DVB-CS:* The satellite master antenna TV (SMATV—pronounced "smat-vee") system, adapted from the above standards to serve private cable and communities;
- *DVB-T:* The digital terrestrial TV system designed for 7- to 8-MHz channels;
- *DVB-SI:* The service information system for use by the DVB decoder to configure itself and to help the user navigate the DVB bitstreams;
- *DVB-TXT:* The DVB fixed-format teletext transport specification;
- *DVB-CI:* The DVB common interface for use in conditional access and other applications.

5.4.3 The Satellite Standard—DVB-S

Since the topic of this book is satellite communication applications, the basic standard of most interest is DVB-S. (The requirements for DVB-SI are covered later in this section.) It provides a range of solutions that are suitable for transponder bandwidths between 26 MHz and 72 MHz, which includes all of the BSS and FSS satellite systems in existence or under development. The basis of transmission is a single carrier that has multiple digital video and audio channels multiplexed onto it. We introduce this concept here because it is an important part of DVB, which is more than a compression system. As will be discussed in the next chapter, this is the principle technique in use for coming generations of dedicated DTH systems like the DSS in the United States and Astra in Europe.

DVB-S is a layered transmission architecture [5]. At the highest layer we find the payload, which contains the useful bit stream. As we move down the layers, additional supporting and redundancy bits are added to make the signal less sensitive to errors and to arrange the payload in a form suitable for broadcasting to individually owned IRDs. The system uses QPSK modulation and concatenated error protection based on a convolutional code and a shortened RS code. Compatibility with the MPEG 2–coded TV services, with a transmission structure synchronous with the packet multiplex, is provided. All service components are time division multiplexed on a single digital carrier. Bit rates and bandwidths can be adjusted to match the needs of the satellite link and transponder bandwidth and can be changed during operation.

The video, audio, and other data are inserted into payload packets of fixed length according to the MPEG Transport System packet specification. This top level packet is then processed as follows (adding additional bits).

The payload is converted into the DVB-S structure by inverting synchronization bytes in every eighth packet header (the header is at the front end of the payload). There are exactly 188 bytes in each payload packet, which includes program-specific information so that the standard MPEG 2 decoder can capture and decode the payload. These data contain picture and sound along with synchronization data for the decoder to be able to recreate the source material.

The contents are then randomized.

The first stage of *forward error correction* (FEC) is introduced with the *outer code*, which is in the form of the RS fixed-length code. This adds 12% of overhead bits. RS is the most common outer code in use in this type of application.

A process called *convolutional interleaving* is next applied, wherein the bits are rearranged in a manner to reduce the impact of a block of errors on the satellite link (due to short interruptions from interference or fading). Convolutional interleaving should not be confused with convolutional coding, which is added in the next step. Experience has shown that the combination of RS coding and convolutional interleaving produces a very hearty signal in the typical Ku-link environment. The process is illustrated in Figure 5.8.

The inner FEC is then introduced in the form of punctured convolutional code. The amount of extra bits for the inner code is a design or operating variable so that the amount of error correction can be traded off against the increased bandwidth. This permits the coding rate, R, to be varied between 1/2 and 7/8 to meet the particular needs of the service provider. Specified values of R and the typical values of required Eb/No are indicated in Table 5.7.

The final step is at the physical layer where the bits are modulated on a carrier using QPSK.

The variables available to the service provider under DVB cover the multiplexing of individual video and audio channels along with key aspects of the

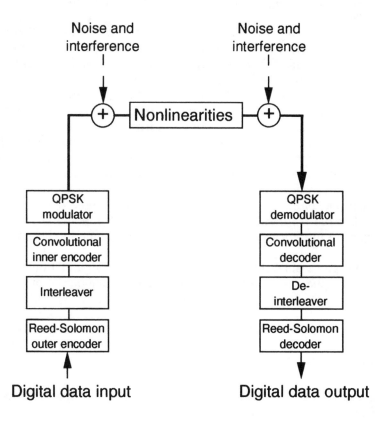

Figure 5.8 A typical concatenated coding system, using a Reed-Solomon outer code and a convolutional inner code.

Table 5.7
Coding Rate, R, and Typical Values of Minimum Eb/No for Use in the DVB-S Standard
IRD (the Link is Assumed to Operate at a 2×10^{-4} Error Rate)

Inner code rate, R	1/2	2/3	3/4	5/6	7/8
Eb/No (dB)	3.3	3.8	4.3	4.8	5.2

link (i.e., coding and interleaving). Burst errors are compensated for through the randomization process, and the amount of FEC is adjusted to suite the frequency, satellite EIRP and receiving dish size, transmission rate, and rainfall statistics for the service area. The system therefore can be tailored to the specific link environment, which was discussed in Chapter 2. The standard DTH operator would consider a range of availability requirements, including 99.7, 99.9, and 99.99%.

Block diagrams of basic DVB *transmitting and receiving* (IRD) systems are provided in Figure 5.9. An example of a typical transmission design, indicating all of the features needed to encode, process, filter, and modulate the composite MPEG 2 signal, is outlined in Table 5.8.

Further details concerning the options for bit rates and transponder bandwidths are provided in Table 5.9. Figure 5.10 provides a typical example of the coding performance on the satellite link from end to end.

There is an obvious relationship in a business sense between DVB-S and the DVB-C standard that applies to cable TV delivery. This is because both services rely on much the same programming and address very similar markets. In DVB-C, the physical layer is different because of the nature of the cable environment as compared to satellite transmission. Here, there is essentially no fading but the bandwidth available per channel is potentially less. Cable also can pass amplitude variations and therefore can support hybrid modulation systems.

The designers of DVB-C therefore chose a hybrid modulation method that combines both phase and amplitude modulation. The modulation is based on *quadrature amplitude modulation* (QAM) and no inner FEC is applied (because the error rate on cable would be lower and more stable as well). The system can support varying levels of QAM, including 16-, 32-, and 64-QAM. A typical cable system with 8-MHz video channels can accommodate a 38.5-Mbps information rate with 64 QAM.

5.4.4 Supporting DVB Services—Sound, Service Information, and Conditional Access

There are three supporting areas of DVB that are particularly important to the operation and success of a DTH system. These are the arrangements for high-

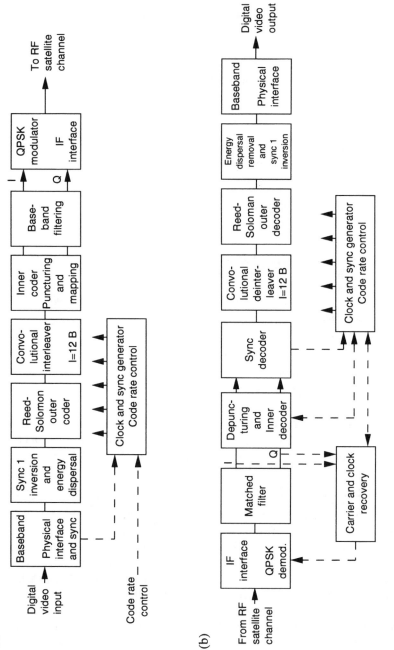

Figure 5.9 The DVB-S baseline system: (a) the transmitter and (b) the receiver.

Table 5.8
The DVB-S Baseline System

a)	Synchronization Inversion and Energy Dispersal
	– Synchronous scrambling
	– Every 8th synchronization byte is inverted to synchronize the deinterleaver and RS-decoder
	– No additional synchronization byte
	– Phase ambiguity recovering
b)	Outer Coding
	– Shortened Reed-Solomon code: RS (204, 188)
c)	Interleaving
	– Forney interleaver
	– Interleaving depth: I = 12
d)	Inner Coding
	– Mother convolutional code: 1/2
	– Punctured convolutional code: 2/3; 3/4; 5/6; 7/8
	– Constraint length K: 7
e)	Baseband Filtering
	– Square-root-shaped cosine filter, 35% roll-off.
f)	Modulation
	– QPSK Gray coded

quality stereo audio, service information (DVB-SI), and the common interface for conditional access (DVB-CI). The following paragraphs review each of these services and relate them to the overall capability of the system.

5.4.4.1 MPEG-2 Sound Coding

The sound coding within DVB provides near audio CD quality at a reduced bit rate. It is based on the MPEG Layer II MUSICAM standard, which is being applied to a variety of digital audio products from manufacturers in the United States, Asia, and Europe. The digital compression of audio takes advantage of the fact that a sound element will have a masking effect on other nearby sounds that are at a lower level of volume. White noise has the same effect. This is used to increase the compression by not sending this unheard information. MUSICAM can provide sound quality that is very close to that of the familiar audio CD and can be used for digital channels that provide stereo, mono, multilingual sound, and surround sound.

Table 5.9
Examples of Bit Rates Versus Transponder Bandwidth

BW (at −3 dB) (MHz)	BW* (at −1 dB) (MHz)	R_s (for BW/R_s = 1.28) (Mbaud)	R_u (for QPSK+ 1/2 convol) (Mbps)	R_u (for QPSK+ 2/3 convol) (Mbps)	R_u (for QPSK+ 3/4 convol) (Mbps)	R_u (for QPSK+ 5/6 convol) (Mbps)	R_u (for QPSK+ 7/8 convol) (Mbps)
54	48,6	42,2	38,9	51,8	58,3	64,8	68,0
46	41,4	35,9	33,1	44,2	49,7	55,2	58,0
40	36,0	31,2	28,8	38,4	43,2	48,0	50,4
36	32,4	28,1	25,9	34,6	38,9	43,2	45,4
33	29,7	25,8	23,8	31,7	35,6	39,6	41,6
30	27,0	23,4	21,6	28,8	32,4	36,0	37,8
27	24,3	21,1	19,4	25,9	29,2	32,4	34,0
26	23,4	20,3	18,7	25,0	28,1	31,2	32,8

Note 1: R_u stands for the useful bit rate after MPEG-2 MUX. R_s (symbol rate) corresponds to the −3dB bandwidth of the modulated signal.
Note 2: The figures of table C.1 correspond to an E_b/N_0 degradation of 1.0 dB (with respect to AWGN channel) for the case of 0,35 roll-off and 2/3 code rate, including the effects of IMUX, OMUX and TWTA.

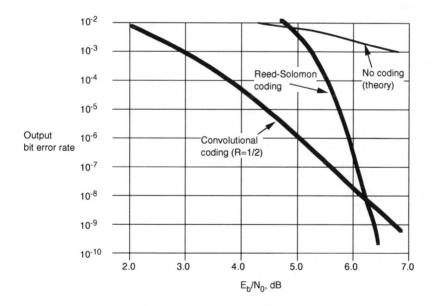

Figure 5.10 Performance of a typical DVB-S system under a certain set of assumptions as to coding rates and types.

5.4.4.2 DVB-SI Service Information

The DVB-SI portion of the transmission adds information that groups the individual video/audio services into categories and allows the IRD to tune to a particular service. This facilitates the creation of the now-familiar on-screen menu of programming, also called the *electronic program guide* (EPG). Relevant schedule information and descriptions of the programs are broadcast over the same link with the video and thereby provide the EPG directly to viewers. The typical DVB environment will support hundreds of video channels and other options, so the DVB-SI standard and resulting EPG are vital to delivering a service that subscribers will find both entertaining and usable.

The DVB-SI standard implements the service delivery model shown in Figure 5.11 [6]. It provides a lot of information (see Table 5.10), some of which is important to the satisfactory operation of the service provider's business. Technical attributes of each service are of interest to individual broadcasters, such as program start times, name of service provider, and classification of events. To support these needs as well as the EPG, the SI includes the service categories shown in Table 5.10, which are themselves arranged in tables.

These tables are the tools that allow DVB-SI to cover the expected range of practical scenarios that the network and user are likely to experience. It also assures a good interface between a satellite DTH service and its counterpart over terrestrial cable TV.

5.4.4.3 Conditional Access

The foundation of any DTH system that achieves its goals in a business sense is the manner in which set-top boxes are controlled so that the user gets only those programming and information services that he or she is authorized to receive. Conditional access starts with the initial granting of access when the user first subscribes but must be extended to cover the particular set of services. These can change from time to time as the user adds and delete services, including PPV movies and events and teletext services that will become a standard in DTH in years to come. The other element to this is that delivery of programming is subject to a variety of controls and restrictions that depend on aspects such as intellectual property rights (copyrights in particular), local government regulatory controls, and limits that are self-imposed by the subscriber (such as blocking adult material in a family environment). A particular threat to the financial integrity of an otherwise viable DTH service is piracy. These aspects are covered in the following paragraphs as they have been addressed in the DVB Conditional Access Package.

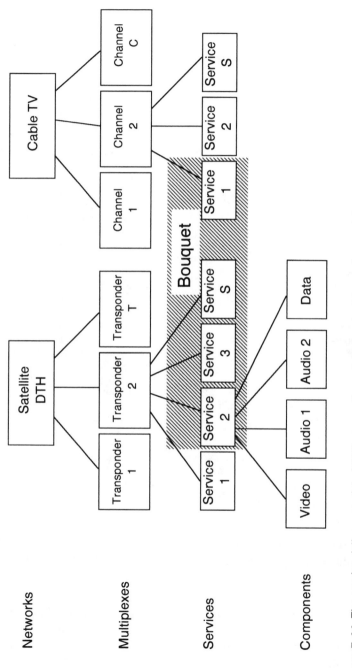

Figure 5.11 The service delivery model of DVB, supporting satellite and cable service provision.

Table 5.10
The Required and Optional Tables of Information that Make Up the SI Element of DVB

Table Name	Abbreviation	Description
Network information	NIT	Groups together services belonging to a particular network provider; contains all the tuning information that might be used during the set-up of an IRD; also used to signal a change in the tuning information.
Service description	SDT	Lists the names and other parameters associated with each service in a particular MPEG multiplex.
Event information	EIT	Used to transmit information relating to all events that occur or will occur in the MPEG multiplex; contains information about the current transport and optionally covers other transport streams that the IRD can receive.
Time and date	TDT	Used to update the IRD's internal clock so that the EPG and other functions are properly synchronized with the network and local times.
Optional tables Bouquet association	BAT	Provides an effective means of presenting services to the user. A bouquet is a grouping of services, such as sports programming, so that the IRD can present them to the subscriber. A particular service can belong to one or more bouquets.
Running status	RST	Used for the rapid updating of the running status of one or more events. The running status is sent out only once, at the time when the status of an event changes, unlike other tables that are transmitted repetitively.
Stuffing	ST	Used to replace or invalidate either subtables or complete tables.

The committee that defined DVB came to a consensus on the following seven points which the conditional access package was to address.

1. Two basic options are available, namely, a single conditional access system (the "Simulcrypt" route) and a common interface that allows for the use of multiple conditional access systems (the "Multicrypt" route). A key difference between the two schemes is that in Multicrypt the entire video stream goes through a removable *smart* card, whereas in Simulcrypt

the video stream is decoded in a possibly piratable (and more expensive) set-top box. An IRD produced with Simulcrypt would only work on a network that is set up for this conditional access arrangement, whereas the Multicrypt route permits an IRD to be able to work with several different networks and their associated conditional access arrangements.

2. The definition of a common scrambling algorithm and its inclusion, in Europe, in all receivers, which enables the concept of a single receiver per home even if different services are to be subscribed to.

3. The drafting of a code of conduct for access to digital decoders, applying to all conditional access providers.

4. The development of a common interface specification.

5. The drafting of antipiracy recommendations. Having this done up front will go a long way to alleviating the difficulties associated with tracking down and prosecuting pirates.

6. The licensing of conditional access systems to manufacturers should be on a fair and reasonable terms and should not prevent the inclusion of the common interface.

7. The conditional access systems should allow for simple transfer of control, for example, at cable headends, where cable operators should have the possibility to replace the conditional access data with their own data.

The ability of the conditional access system to control and regulate usage is highly dependent on the security offered by the DVB common scrambling system. At the IRD, this consists of two parts: decryption and descrambling. The purpose of decryption is to translate the scrambling key that is transmitted over the satellite link along with the programming and other data. This is the most efficient and effective means of delivering the scrambling key. However, using the same broadcast link for this purpose greatly simplifies the pirate's job, since they have continuous access to this critical information. The encryption/ decryption process was designed to make its compromise as difficult as possible. The ultimate success depends on the strength of the system coupled with the ability of the operator to increase strength or modify the technique in response to a compromise of security. Because of the sensitive nature of this information, the technical details of decryption are tightly controlled and are only available through a rigorously enforced system of confidentiality agreements and custodians.

Simulcrypt allows the delivery of one program to a number of different decoder groupings that contain different CA systems. It also provides for the transmission between different CA systems in any grouping, in particular for the recovery after compromise of the particular implementation by pirates. Multicrypt is the approach that puts the intelligence of the encryption system on a separate module such as a smart card. It was proposed for standardization by the European Committee for Electrotechnical Standardization (CENELEC) and is provided through DVB-CI. The type of connector has been specified as the *Personal Computer Memory Card International Association* (PCMCIA) connector, used heavily in personal computers for functions like modems and flash RAM.

The conditional access data is broadcast in the MPEG 2 syntax structure through the common interface (DVB-CI). This interface physically lies between the DVB decoder that recovers the payload and a conditional access module within the IRD that provides the decryption and descrambling.

When we examine the approach taken by the developers of the DVB family of standards, we come to realize that an excellent structure has been created. In many ways, it mirrors the popularity of the GSM digital cellular standard, which resulted from a lot of hard work by European organizations. DVB facilitates the early introduction of digital video throughout the satellite communications industry. Already, DVB-based DTH networks are operating in Asia and Europe and more are likely throughout the world. The prospects in the United States are also excellent, as sources of IRDs and uplink equipment expand.

References

[1] Howes, K. J. P., "Digital Video Via Satellite," *Via Satellite*, March 1995, p. 34.

[2] Murray, J. D., and W. van Ryper, *Encyclopedia of Graphics File Formats*, O'Reilly & Associates, 1994.

[3] Kretz, F., and D. Nasse, "Digital Television: Transmission and Coding," Proc. IEEE, Vol. 73, No. 4, April 1985, p. 575.

[4] *Digital Video Broadcasting—Television for the Third Millennium*, Geneva, Switzerland: DVB Project Office, European Broadcasting Union, May 1995.

[5] *Digital Broadcasting Systems for Television, Sound and Data Services; Framing Structure, Channel Coding and Modulation for 11/12 GHz Satellite Services*, European Telecommunication Standard ETS 300 421, ETSI Secretariat, Sophia Antipolis—Valbonne, France: European Telecommunications Standards Institute.

[6] Elbert, B. R., *Introduction to Satellite Communication*, Norwood, MA: Artech House, 1987.

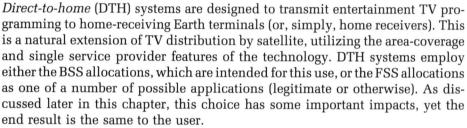

Direct-to-Home TV Broadcasting by Satellite

6

Direct-to-home (DTH) systems are designed to transmit entertainment TV programming to home-receiving Earth terminals (or, simply, home receivers). This is a natural extension of TV distribution by satellite, utilizing the area-coverage and single service provider features of the technology. DTH systems employ either the BSS allocations, which are intended for this use, or the FSS allocations as one of a number of possible applications (legitimate or otherwise). As discussed later in this chapter, this choice has some important impacts, yet the end result is the same to the user.

To make this type of service effective, however, there a number of factors that must be addressed properly, such as:

- Incompatibilities with the other DTH or cable TV systems, which are dependent on the nature of the business plan;
- The programming mix, for example, the quantity, variety, language options, and degree of interactivity, which must compete with other DTH systems and delivery mechanisms (e.g., cable TV, multichannel microwave distribution service, cassette rental);
- Receiving equipment, that is, its affordability, convenience of installation and use, and aesthetics;
- Acceptability of the service price and an effective means to collect payment;
- Conditional access and scrambling in order to deal with copyrights, privacy, collection, regulations, and content rules (which may exist in the country markets of interest);
- Uplinking system, including the redundancy, strength, and program acquisition facilities.

Experience with DTH systems has shown that the service must be attractive as compared to other forms of video distribution and use by the consumer must be properly controlled. This competition between delivery media is highly variable between countries. The U.S. market is the most competitive in the world, with high cable TV penetration, three DTH systems in operation as of the time of this writing, and newer broadband options coming in line in the near term. Other environments like the United Kingdom or Japan are less dependent on cable or MMDS, so DTH has greatly improved prospects. In developing countries, the limitations have less to do with competition and relate mainly to the ability of consumers to afford the price of the equipment and the service. Each of these factors, along with the historical background, are addressed in the following sections.

6.1 DIFFERENCES AMONG DTH SYSTEMS

The basic architecture of a DTH system is probably the simplest of all satellite application systems because it relies on a single uplink to gather and transmit the programming and literally millions of home receivers to act as individual downlinks. An example of the type of receiver is shown in Figure 5.1. Later in this chapter, we explore the design of the uplink (broadcast center) and the receiver.

DTH systems have multiplied in number and capability over the years, resulting in a rather substantial receiving public who own their own antennas. As a result, developers of these systems and services need to pay attention to the differences among systems, resulting in potential incompatibilities. The degree to which these impact the viability of a DTH service depend on the business strategy. For example, a totally stand-alone system that does not wish to attract viewers from other operators can implement an incompatible system. The advantage of doing this is that the technical or business parameters can be optimized. One such approach is to use very high satellite EIRP and corre-spondingly small receiving dishes that cannot also receive from existing medium-power DTH services transmitted from FSS satellites. On the other hand, an existing base of DTH viewers must adhere to an appropriate number of these factors in order to succeed. A given incompatibility can, of course, be resolved by modifying or adding equipment on the ground—one DTH receiver at a time.

6.1.1 Downlink Frequency

The choice of downlink frequency is probably the most strategic in a technical sense. As of the time of this writing, the bands either in use or being introduced within a two-year time frame include (frequencies indicated are for the downlink):

- S-band BSS (2100 to 2128 MHz);
- C-band FSS (3.4 to 4.2 GHz);
- Ku-band FSS (10.75 to 12.2 GHz);
- Ku-band BSS (12.2 to 12.9 GHz).

Most DTH receivers are designed for a specific downlink frequency and total bandwidth. Some offer multiple bands, using dishes with dual frequency feeds that employ separate LNBs. An example of such a system at C band is shown in Figure 6.1. This approach permits reception from different satellites and service providers since the dish may be repointed using a motor drive. It is more for the enthusiast or hobbyist and not the typical consumer. For a consumer installation, the dish is fixed in position and the particular band is permanently part of the design, so access to alternative programming and satellites is restricted.

From an uplink standpoint, there are nearly as many options from which to choose. Developers of satellites and DTH systems usually employ the paired uplink band for the particular downlink, which is required in the case of the BSS allocations where the 12-GHz segment of Ku is paired with the 18 GHz of the Ka band for uplink access to the satellite. Use of the paired band is usually a good idea because the spectrum is already allocated and probably available at the assigned orbit position. On the other hand, the 18-GHz uplink frequency is subject to increased rain attenuation, particularly in tropical regions with a

Figure 6.1 A typical C-band DTH installation.

lot of thunderstorm activity. The objective of the uplink design is to maintain the signal for a high percentage of the time, typically 99.95%. This is relatively easy to do at frequencies below about 10 GHz in tropical regions and below about 20 GHz in more temperate climates with less thunder storm activity.

6.1.2 Significant Differences in Satellite EIRP

Satellite EIRP has the greatest single impact on DTH dish diameter. As discussed in Chapter 2, the relationship is that every doubling of EIRP (in power, not decibels) allows the receiving dish diameter to be reduced by a factor of the square root of two, or 1.414. Thus, if we increase the satellite EIRP by 3 dB (a factor of 2 in power), a 1m dish can be squeezed down to 1/1.414m or 71 cm. Another 3 dB (or a total of 6 dB of increase) brings it exactly to 50 cm. A DTH system designed with a high EIRP produces the smallest sized receiving dish. This minimizes installation cost and improves esthetics; however, a smaller dish may be incapable of adequate reception from a medium-power FSS satellite and could receive unacceptable levels of adjacent satellite interference.

Dish size is one of the most easily seen differentiators among services. The smaller the dish, the less obtrusive and potentially objectionable is the installation at the home. The DIRECTV service in the United States has benefited from a small dish size because of the lack of a perceived problem with appearance. In contrast, there is a general feeling that 3m C-band dishes are only acceptable in rural areas where normal TV reception is limited. This rule is perhaps violated in some developing countries where a rather obvious C-band receive antenna is a sign of prosperity. Developing the power to produce a small receiving dish is no longer the problem that it was once considered to be. Space power amplifiers, particularly the high-power *traveling wave tube* (TWT), are available up to 200W/channel. These deliver very high EIRP over a wide coverage footprint, affording consumers the opportunity to use small receiving dishes in the broadest possible market.

6.1.3 Polarization Selection (LP or CP)

Designers of DTH systems can select either *linear polarization* (LP) or *circular polarization* (CP) for the downlink. The standard BSS assignments at Ku band all employ CP to simplify the installation of home dishes because the feed does not need to be rotated after the antenna is aligned in azimuth and elevation. This is a factor in large DTH networks, where antennas are supposed to number in the tens of millions. FSS systems typically employ LP since this tends to simplify the design of the Earth station transmit and receive feed system. In addition, depolarization caused during heavy rain is less of a factor with LP.

The INTELSAT system standardized on CP for its C-band satellites long ago, and DTH systems like DIRECTV have retained CP at Ku band to simplify DTH dish installation. A single polarization system will experience a 3 dB loss when a LP-to-CP or CP-to-LP connection is attempted. Even worse, the polarization isolation between CP and LP transmissions is 3 dB in the ideal case, independent of the sense of polarization of either link. Because of an electrical property of typical offset-fed dishes, CP experiences a slight squint of the beam, left or right of center, depending on which sense (left hand or right hand) the signal is following.

6.1.4 Frequency Plan Differences (Channel Spacing)

The arrangement of the frequency plan for the downlink impacts the design of the tuner in the home receiver. Examples of the most popular channel spacings for C and Ku band satellites is provided in Table 6.1. Channelization of the BSS frequencies is according to an international plan and so standardizes receiver design and use. These were selected to permit a single FM TV carrier to occupy the bandwidth. With the advent of digital compressed video, multiple video channels are time division multiplexed onto a carrier that can pass within the same bandwidth and channel characteristics. FSS assignments at C band follow a defacto standard (e.g., 40 MHz, starting at a center frequency of 3,720 MHz), while those at Ku band generally do not. The potential use of Ka band as a DTH medium is likewise not standardized.

Most receivers have standardized on certain transponder frequencies, bandwidths, and spacings. The most flexible design allows the user to select literally any frequency in the downlink range using a digital frequency synthesizer. However, this type of device must be properly programmed, which can be a burden for a nontechnical user (i.e., having to tune to a specific frequency, like 11.7255 GHz, as opposed to a simple channel assignment, like channel

Table 6.1
Typical Transponder Channel Spacings in the C and Ku Bands

Band	Lowest Center Frequency (MHz)	Channel Spacing (MHz)	Usable Bandwidth (MHz)
S	2114	28	20
C	3720	40	36
C (extended)	3400	40	36
Ku (FSS)	10750	60	54
Ku (BSS)	11715	30	27

10). Alternatively, the tuning is managed over the satellite link, making the process transparent to the user. Instead, the user selects programs using a remote control unit and on-screen channel guide. In a true DTH system with a uniform receiver design, there is only one channel bandwidth allowed. This assures that all receivers behave in exactly the same manner, with a minimum of user involvement.

6.1.5 Transmission Format—Analog (FM) or Digital (QPSK)

With twenty years of history, FM-TV is essentially a unified standard. The specific transmission parameters for this format are presented in Chapter 4. Analog receivers bear little resemblance to the digital variety and so it is unlikely that there will be a serious attempt to make dual-mode models available to the general public. Rather, the move to digital formats means that older analog units eventually will be retired. This is not a serious problem because a DTH "megamarket" can only develop in the digital format. From that point, the specifics of the digital modulation and coding scheme must be selected. It is possible to implement a programmable unit, as discussed in Chapter 5 for the DVB standard.

6.1.6 Video Signal Format

In Chapter 4, we identified the three standard analog TV signal formats, namely, NTSC, PAL, and SECAM. These represent an important differentiator among DTH systems. A DTH system for North America can conveniently employ NTSC, while in China the PAL system is the appropriate vehicle to reach the mass market. Southeast Asia and South America are more of a challenge because both employ NTSC and PAL within the respective regions. Africa and Europe are each split between PAL and SECAM.

While multiple standard TV receivers exist in markets like western Europe and east Asia, they cannot be assumed as a given in DTH service. Therefore, it may be necessary to transmit programming in more than one format. Conversion between formats is certainly possible but is not currently available within set-top boxes. This is because these are unique facets of, say, NTSC that the digital format must represent and transfer along with the images. One can imagine that a time will come when TV sets will be more like PCs, and so the video signal format will cease to be a differentiator.

In the case of digitally encoded TV, the clear trend is toward the MPEG standards. MPEG 2 is the leader in the market at the time of this writing. We reviewed in Chapter 5 some of the specifics of the current implementation of MPEG 2 with the Main profile and level. This provides a good compromise among the factors that designers can currently control. However, in time, digital processing technology will advance to the point that higher levels and more advanced profiles can be adopted into low-cost equipment. Add to this the prospect of improved or high-definition TV, which is something that satellites can deliver ahead of existing terrestrial cable and over-the-air resources.

6.1.7 Scrambling

The purpose of the scrambling technique is to render the picture unwatchable unless and until the subscriber has been authorized by the service provider. The typical reasons why this is necessary or desirable are to:

- Prevent piracy of the programming by a functioning but unauthorized receiver;
- Control the customer base to assure payment for services obtained;
- Protect intellectual property rights such as copyrights that might not have been granted within a particular service area;
- Implement PPV services to the end user so that specific programming can be turned on and off either by a central authorization center or directly by the user;
- Provide levels of service and privacy to satisfy domestic regulations and content rules.

   Because scrambling systems fall into a number of categories, they can be difficult to implement as an open system. One approach is to scramble the picture while it is in analog form, which is comparable to what is done on current cable TV systems (see Chapter 4). For a digital format, the information in the picture has been compressed using the DCT, which requires a properly designed receiver to recover it. Scrambling amounts to rearranging the bits in the sequence according to a secure process driven by a key. The strongest and most secure way to do this is through encryption using algorithms like the *digital encryption standard* (DES). As is usually the case, the process itself is typically not keep secret since it is likely part of a standard like DVB. The key, on the other hand, must be handled properly because with it anyone can recover the picture, sound, and data. Part of the key can be delivered over an encrypted satellite link as long as the second key used to recover it is provided by an independent path (e.g., a password or smart card). All of these aspects of scrambling represent areas of possible incompatibility. This is why scrambling systems are usually supplied either as part of a validated standard (where an agency performs type acceptance of products from various vendors) or as a proprietary package from a single vendor.

6.1.8 Conditional Access System

The last step of compatibility has to do with the way the subscriber's set-top box is controlled as part of service delivery. Since a geostationary satellite can broadcast its signal over a wide area to a potentially very large subscriber base, DTH operators and programmers must have an efficient and effective means to control the ultimate distribution of the programming product. The VideoCipher II system, discussed later in this chapter, was one of the first to introduce conditional access on a wide scale. Also, DVB conditional access features reviewed at the end of Chapter 5 provide a workable baseline for modern digital DTH applications.

The satellite link offers an attractive medium for downloading control information to remote equipment, hence the preference to use it for conditional access. However, a physical connection is more secure and flexible since it provides more direct control to the DTH programmer as well as a mechanism for return data. The smart card performs many useful tasks under conditional access, such as holding the subscriber's key and securing the process of unlocking the decoder. The card also can store the viewing history for the particular installation. An alternative is to have the subscriber enter the authorization code manually (like a password) into the decoder box. A connection to the telephone network is another way to isolate the key from the satellite link.

DTH is being considered for other one-way communications applications such as data broadcasting to small dishes. CD-quality audio broadcasting is being advanced as a viable application. The EBU has developed standards, and some U.S. companies are already offering CD-quality audio to cable systems. For return data, it is possible to use fixed and mobile telephone networks. There is the prospect of using the satellite for both directions of data transmission to serve consumers, but this will have to await the introduction of low-cost VSAT technology.

6.2 HISTORY OF DTH SYSTEMS

The history of DTH as a legitimate satellite offering provides a good introduction into the development of a modern system and service. The old adage, "those who fail to heed the lessons of history may be doomed to repeat them" would seem to apply. Therefore, we want to make sure that anyone who considers making the kind of investment that DTH requires has the benefit of this background. Readers are encouraged to conduct their own research into the fine points of whichever of these systems represent the closest parallel to what they plan to do.

6.2.1 The First BSS Plan

At the *World Administrative Radio Conference* in Geneva, Switzerland, in January 1977 (WARC-BS), representatives of most of the nations of the world and international regulators working under the auspices of the ITU provided the first DTH impetus by assigning BSS orbit positions and channels to every member. This meant that big countries like Russia and China as well as small countries like Kuwait and Mauritius each received their dedicated assignments for domestic BSS satellites. To be consistent with expected demand, larger countries with larger populations were assigned multiple orbit positions. Characteristics of the assumptions used in the Plan for ITU Regions 1 and 3 are provided in Table 6.2. The details of the plan for Region 2 were put forth at a subsequent Regional Administrative Radio Conference. The proceedings of both conferences can be found in Appendix 30 and 30 B to the ITU Radio Regulations.

This was a classical planning effort by regulatory experts and technologists who worked very hard but lacked a practical feel. They made several assumptions that may have seemed sensible at the time but failed to take account of future improvements. For example, the WARC-BS plan assumed that the maximum spacecraft antenna size would be 3.5m, the typical low-cost home receiver would have a system noise temperature of 1000K, and the maximum

Table 6.2
Assumed Characteristics of BSS Satellites in the WARC-BS Plan

Feature	Specifications
EIRP	Typically 63 dBW
Minimum satellites per country	1
Channels per satellite	5 (or multiples thereof)
Frequency plan	23-MHz transponders
Polarization	Circular polarization (left and right hand)
Orbit spacing between satellites	Fixed orbital spacing between satellites (3 deg minimum separation) 9 deg for cofrequency assignments Satellites are colocated where possible
Maximum spacecraft reflector size	2m
Beam shape	Elliptical
Sidelobe radiation pattern	Tightly specified
Assumed home receiver characteristics	
Dish size	No smaller than .9 meters
Receiver noise figure	No lower than 9 dB
Cross-polarization isolation	Relatively poor
Angular discrimination	Relatively poor

transmit power would be 1 kW. Over time, it would be established that LNAs could be produced cheaply with less than 100K, and therefore transmit power could be held to 200W. Another technical assumption that has turned out to be too pessimistic was that satellites could not carry sufficient reserve battery power to sustain full capacity during an eclipse. As a result, satellites were all placed to the west of the service area so that the eclipse would occur after local midnight.

Consequently, WARC-BS was little more than an exercise in logic because only one country, Japan, used its assigned slot during the first ten years that the plan existed. This first BSS system employed a Ku-band satellite specifically designed for the purpose. Japan has worked hard and invested a great deal of money to become the country with one of the most successful BSS networks in the world.

6.2.2 Early U.S. DTH Experience

The early 1980s saw a number of start-up ventures that intended to provide DTH service to the general public. The first two attempts failed in business terms, and it was primarily the C-band "backyard dish" approach that survived the period of the 1980s. This is the backdrop for the successful introduction of digital DTH services in the 1990s.

6.2.2.1 Satellite Television Corporation (STC)

The concept at the time was to use the new WARC-BS orbit and frequency allocations for BSS systems. COMSAT, in particular, established a subsidiary called *Satellite Television Corporation* (STC) and set about to implement a five-channel analog service. Their goal was to have the service up and running before cable TV reached critical mass in the U.S. market. They viewed their competition correctly because the rate of increase in U.S. cable penetration during the 1980s was extremely fast, ending the decade at around 65% of U.S. households. Meanwhile, STC could not find the right programming partner nor could it gather the financial resources to initiate the operational part of the program. After this series of misfortunes, COMSAT closed down STC and sold the developmental spacecraft to NHK of Japan.

6.2.2.2 USCI

Another major effort was engaged at about the same time by a start-up company called USCI. Their technical strategy, being based on the use of Ku-band FSS capacity, was less aggressive than that of STC. By renting transponders on an

already operating US Ku-band satellite, they avoided the large startup investment that partially hampered STC. This resulted in a cheaper entry price for
the business and a dependence on a somewhat larger receiving dish. USCI
introduced the service to customers in the mid-Atlantic region in the form of
five channels of movies and cable programming. Signals were transmitted in
analog FM TV format with scrambling.

Unlike STC, USCI actually entered service and achieved a customer base
of about 100,000 subscribers. The service lasted only about a year because the
backers could not sustain the operating costs throughout the time it was taking
to reach a break-even point. USCI also suffered from the problem of trying to
reach critical mass at a time when cable TV was growing very rapidly along
with the new popularity of C-band DTH, which offered over 100 channels of
cable and network TV programming.

6.2.2.3 C-Band Backyard Dish

DTH was finally introduced in the United States as an ancillary service to the
C-band cable TV distribution system. The backyard dish market, as it is called,
was built up by entrepreneurs who manufactured and distributed relatively
low-cost installations with antennas around 3m in diameter. These systems
typically cost $3,000 and were sold to rural customers who did not have access
either to cable or good over-the-air TV reception. Another market that opened
up consisted of hobbyists and higher income people who wanted to experience
the joy of obtaining a very wide array of programming directly from the source.
This included the primary TV networks who transmitted their source material
in the clear on satellites like Telstar 3 and Satcom 2. Because signals were not
yet scrambled, the programming was available for free once the user paid for
the dish, receiver, and installation service.

The leading cable networks, including HBO, CNN, ESPN, MTV, and Disney, did not pay attention to this market due to its perceived small size and
the complexity (impossibility) of collecting money. Subsequently, HBO set
about to protect its signals through the General Instruments (then a division
of M/A-Com) *Videocipher II* (VC-II) scrambling system, which forced many
backyard dish people to buy decoders and, for the first time, pay for the service.
When HBO and other leading cable networks began VC-II operations, new sales
of backyard dishes nearly dried up.

The U.S. experience with Videocipher II, the first widely used scrambling
system, was both good and bad. On the positive side, VC II achieved critical
mass quickly, giving cable programmers the control they needed over their
principal customer base, that is, the cable TV systems. Piracy had been widespread among the smaller "mom and pop" cable systems (e.g., systems owned
by individuals and serving a relatively small customer base, around 200 to

1,000 subscribers); private individuals were buying dishes by the hundreds of thousands; entrepreneurs were stealing the signals and reselling them through private cable systems (SMATV) in trailer parks and apartments buildings; and commercial establishments like bars and motels enticed groups to partake without paying the appropriate royalties. Revenues were lost by programmers and producers alike. HBO, the leader in satellite pay programming and an innovator in delivery system technology, chose to proceed with the VC II system in 1984.

6.2.2.4 VideoCipher II

The VC II system developed by M/A-Com relies on encrypting video, digitizing and embedding the audio in the video, and protecting the entire information package with the DES algorithm. DES was developed jointly by the U.S. government and IBM to improve commercial security for the transmission of various forms of data over the airwaves and public networks. It is generally considered to be very strong in a cryptographic sense and has not, to this author's knowledge, been broken directly. The U.S. government maintains control over the export of DES hardware, which has been a problem for non-U.S. operators who wish to employ VC II.

Scrambling of the picture is done through removal of the sync and introducing some switching of the frame. The VC II system performs the following functions.

- The *horizontal blanking interval* (HBI) signal is removed and the video information is digitized.
- The incoming audio is companded to improve quality, digitized, and encrypted.
- The digitized audio is inserted into the time formerly occupied by the HBI.
- Part of the audio (approximately 1/480th of a second's worth) is fed with one of the video lines, based on the encryption algorithm. This introduces a "random" scrambling sequence into the picture.

The resulting signal is a hybrid digital/analog waveform that can be handled by a conventional TV and cable plant and transmitted over a single satellite transponder with little trouble. Because of the digital components, the signal is dispersed over the transponder bandwidth and hence tends to cause less interference to adjacent satellites than a conventional FM TV signal.

A powerful feature of VC II is its ability to control the delivery of the signal to each subscriber (e.g., conditional access). The system, indicated in Figure 6.2, allows the programmer and network operator to activate individual decoders over the satellite link itself. This is done by encrypting the appropriate

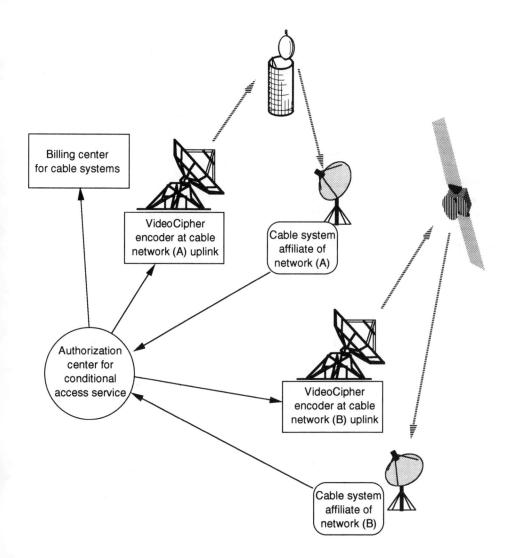

Figure 6.2 Typical VideoCipher II network arrangement for conditional access and billing.

key that the decoder needs to unlock the given channel and decrypt the signal. For the United States, GI maintains an administrative hub in San Diego, CA, for the purpose of processing authorization requests from programmers and subsequently unlocking remote VC II descramblers. The programmers must own and operate the uplink scrambling systems themselves. These are very costly, which is only justified for the large networks that programmers operate.

The unfortunate part of the history, one that taught us all a lesson, is that VC II was broken. HBO and GI claimed that the system was completely safe from pirating and that they had enlisted several university professors and encryption experts to try to break the system. The fact that the basic encryption is done with DES would seem to support these claims. However, it was vulnerable from the start, and this vulnerability had nothing to do with the security of video scrambling or audio encoding. Rather, it was the fact that the unique address of each box was in the format of a plug-in ROM chip, which easily could be duplicated. The address of an authorized receiver can be copied onto as many blank chips as desired (called *clones*), and these can be used to activate a like quantity of legitimate VC II decoders. As long as this account is maintained (by paying the appropriate subscription fees), the cloned units will receive the satellite transmissions correctly. There is no system for verifying which units are legal and which have illegal clone chips in them. Ironically, GI was able to sell every VC II unit they could make (it was rumored that as many as half of all units were subsequently unlocked fraudulently through the address cloning process).

General Instruments introduced VC II Plus as its response to the security failure of its original system. The decoder unit is totally modified from the original version to place the address and other functions within the basic processing chip. This makes it much more difficult to clone an address since it would require the reengineering of a major part of the device. Security was also enhanced by increasing the address size and providing a smart card feature to simplify future upgrades. As of the date of this writing, VC II Plus has remained secure.

General Instruments has not given up with their technology. Its principal advantage is its simplicity and low cost. IRDs now sell for under $500. The technology has been moved to Europe and is the core of the EuroCipher system. This permits PAL to be used as the transmission and video format. Furthermore, GI has started its rollout of their digital product called DigiCipher. Initially provided to HBO and other front-line cable TV networks, it is being upgraded for MPEG II compliance. The *DigiCipher II* (DC II) product, in particular, offers three modes of operation, namely, DC II digital mode, VC II analog mode, and straight-through of analog video.

6.2.3 Digital DTH in the United States

The U.S. home TV market poses a special challenge to developers of DTH systems as witnessed by the demise of STC and USCI. Unlike the experience of Sky in the United Kingdom, systems in the United States that offered 5 to 10 channels of programming did not have a good experience. This happened principally because of competition from cable TV, with typically 50 channels, and C-band backyard systems that can access more than 100 channels. A theory evolved that a successful DTH business would also require as many as 100 channels or more and be as convenient as cable to use. Also, the equipment must be unobtrusive and have a cost comparable to that of, say, a 30-in TV or high-quality VCR. To achieve this number of channels, the advances in the area of digital compression had to be harnessed. Thanks to the work underway by MPEG (discussed in Chapter 5), the technical base became available in 1991.

The first company to conceive such a system and bring it to market was Hughes Electronics Corp., a subsidiary of General Motors. Their DIRECTV service in the United States expanded in the digital field, having reached two million subscribers in 1996 after two years of operation. Digital DTH was subsequently introduced in Latin America, Asia, and Europe in much less time than it took in the United States. As reviewed at the end of Chapter 5, the DVB standard is allowing MPEG 2-based systems to proliferate. Satellite capacity needed for DTH service must have the power and coverage to deliver enough programming to small antennas over a sufficient market to create a profitable business. Extensions of satellite video broadcasting to permit a degree of interactivity could make DTH satellites into the primary vehicle for multimedia.

Studies conducted by Hughes in 1991 showed that a subscriber base of around 10 million U.S. households was achievable provided that the service meet the following goals [1]:

- Receiving dish size of approximately 45 cm to aid in installation and for aesthetics;
- A diverse programming mix offering approximately 150 channels;
- An initial cost of less than US$1,000, preferably in the neighborhood of $700;
- An average monthly service charge of approximately $25.

The original target of DIRECTV, Inc. (set in 1991), was to begin marketing to consumers in late 1994, during the peak Christmas selling season. The *Digital Satellite System* (DSS), as the DIRECTV consumer equipment is known, was introduced on schedule and subsequently recorded as the most successful new consumer electronics product ever. These first units were produced by Thomson Consumer Electronics under the RCA brand, a very familiar name in U.S. home electronics. The on-screen display from the Thomson set-top box is provided in Figure 6.3.

It was critical that the DIRECTV service quality be as close to perfect as possible at the time of business launch. A delay in the roll-out would be preferable to failing to meet a high standard. This goal was also met, and consumer reaction to video and audio quality has been excellent. A second

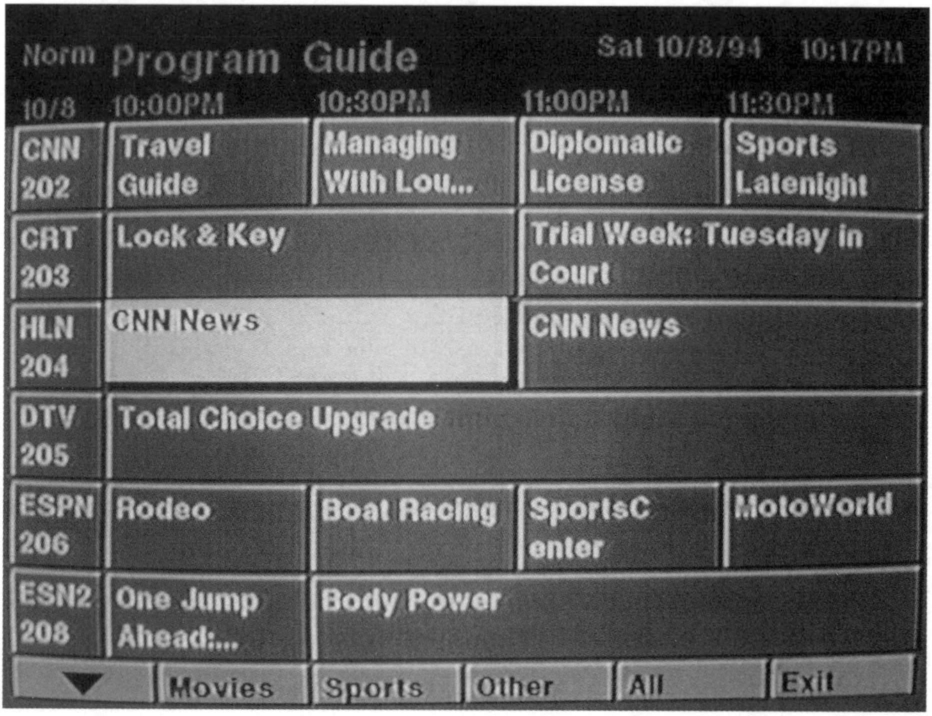

Figure 6.3 An example of the actual on-screen display for the Thomson DSS set-top box (photography courtesy of DIRECTV, Inc.).

supplier, Sony, introduced their DSS system to the U.S. market in 1995. An example of their antenna and set-top box is shown in Figure 6.4.

The high-power Ku-band satellites for DIRECTV, Inc., built by *Hughes Space and Communications* (HSC), employ the HS-601 three-axis spacecraft design. The coverage is confined to the continental United States and southern Canada, delivering EIRP in the range of 48 dBW to 58 dBW. Initial service was provided by DBS-1, which is shared by DIRECTV, Inc., and *U.S. Satellite Broadcasting* (USSB), a part of the Hubbard Broadcasting family of companies. The DIRECTV service contains a mix of more than 150 channels, which is even more extensive than what is available over a larger cable TV systems such as one would find in the New York City or Los Angeles markets. This includes the principle cable TV networks (e.g., CNN, Disney, A&E, The Discovery Channel) and a unique array of PPV movies and sports events. USSB, on the other hand, is more focused, offering five channels of HBO and Cinemax, Viacom's networks (Showtime, MTV, VH-1, Nickelodeon, and Nick at Night), along with

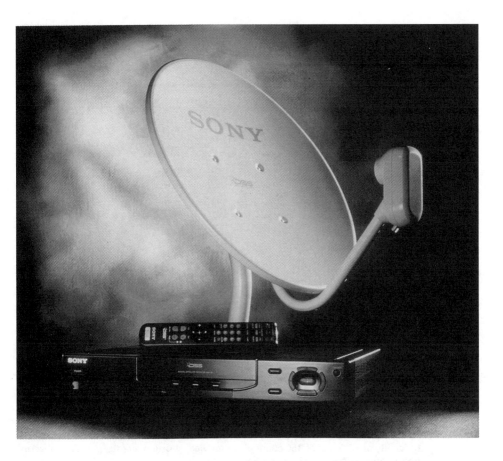

Figure 6.4 The DSS equipment offered by Sony Electronics Corporation, consisting of the 18-in satellite antenna, s decoder box, and a remote control (photograph courtesy of DIRECTV, Inc.).

Hubbard's own news channel. The two firms agreed in 1991 to share a common technical infrastructure but have separate billing and customer service organizations. The DIRECTV uplink center in Castle Rock, CO, is shown in the photograph in Figure 6.5. Each of the 16 transponders on DTH-1 has 24 MHz of bandwidth, with five belonging to USSB and the remaining 11 belonging to DIRECTV, Inc. Two additional satellites have been launched, and their combined capacity of 32 transponders is dedicated to entertainment service.

6.2.4 European DTH Experience

The European experience with satellite TV in general and DTH in particular is almost exclusively at Ku band. A difference from the United States is that

Figure 6.5 The DIRECTV Broadcast Center in Castle Rock, CO, is recognized as the most sophisticated television transmission facility ever built. As the heart of DIRECTV, the 55,000-ft^2 facility provides over 150 channels of movies, sporting events, popular subscription networks, and special attractions.

both cable TV and DTH grew up at the same time. This is because cable TV is strong in only a few countries like Germany and Belgium, while in others like France and the United Kingdom DTH has always been the primary market for reaching the consumer. There has also been an explosion in the use of DTH in Eastern Europe, where the existing broadcasting and cable infrastructure tends to be extremely weak.

The points already made about simplicity, cost, and programming have been validated over and over again. Sky TV, the first service to be offered directly to the public, is staying ahead precisely because of these aspects. Other activities in France and Germany were slow at the start of the game, being hampered by government red tape and inflexibility. In many ways, Europe poses a much more exciting future for DTH simply because of the wider array of ventures and economic conditions. The company that continues to be successful is *Société Européenne des Satellites* (SES), the commercial operator based in the Grand Duchy of Luxembourg. SES has established its orbit positions as the most popular real estate with the associated large base of existing TV receive antennas. This considers both markets—cable TV systems and DTH home receivers.

Table 6.3 indicates the quantity of satellite operators and their country of residence in Europe. It is beyond the scope of this book to go into detail for each and every one of these systems. Readers who need this detail can review the industry trade publications and technical reports published by various research organizations such as CIT Research [2]. Not shown in the table are the systems of INTELSAT, PanAmSat, and Orion. These provide some video distribution capabilities throughout western Europe but are hampered to some degree by the low-elevation angles due to the western location of the satellites.

We now review some of the key DTH systems and activities in Europe, followed by a few others around the world. These break down into government-

Table 6.3
Satellite Operations in Europe

Satellite Operator	Country	Number of Satellites	Coverage	Primary Services
SES/Astra	Luxembourg	5	European	Video, including DTH
Deutsche Telecom/ DFS Kopernikus	Germany	4	Germany	Video, DTH on TV Sat, other FSS
Eutelsat (including Hot Bird 1)	France	7	Europe	Video, including DTH, other FSS
Hispasat	Spain	2	Spain and north	Video, including DTH, other FSS
Italsat	Italy	1	Italy (Ka band only)	Video, other FSS
Sirius (including Tele-X)	Sweden	2	Sweden	Video, including DTH
Gorizont	Russia	11 (approx.)	Europe and global (C and Ku)	Video, other FSS
Telecom (including TDF)	France	3	France	Video, including DTH, other FSS
Thor	Norway	2	Nordic countries	DTH
Turksat	Turkey	1	Turkey and western Europe	Video, including DTH, other FSS

sponsored and commercial systems. Government-sponsored systems tend to be more innovated in a technology sense, pushing the state of the art. They usually have an industrial mission, that is, to promote domestic industry and gain a leg up in the international market. Commercial systems want to make money (usually), so their emphasis is on the market for the services. The information is very basic, so readers who need details for system design and service evaluations are advised to contact the appropriate satellite operator for current detailed information.

6.2.4.1 SES/ASTRA

It is always good to start off with a success story. In 1988, SES was in direct conflict with EUTELSAT and BSB and it was not clear at all if Astra would be successful. SES appeared to have a more marketable approach, using a medium-power satellite built in the United States with a sixteen-channel repeater. But EUTELSAT had a big head start since the first-generation satellites were delivering most of the satellite TV present in Europe.

SES/Astra is owned by several public and private investors, including the Luxembourg government and several banks, which provided the needed financial and political support. The idea came from an American, Clay T. (Tom) Whitehead, who also is credited with the successful Galaxy system operated by Hughes. Whitehead's original Coronet enterprise failed to get started and provided the framework for the eventual SES venture (Whitehead retains a small ownership interest). In 1995, SES launched its fifth satellite into service at 19.2 deg EL and currently provides over 100 channels of programming to all of the markets in western and eastern Europe.

Investors in SES include the following European financial and media companies:

- Aachener & Münchener;
- Beteiligungsgesellschaft AG;
- Banexi;
- Banque Generale du Luxembourg;
- Benson Holdings S.A.;
- Big Bank Luxembourg S.A.;
- Deutsche Bank Luxembourg S.A.;
- Deutsche Telekom AG;
- Dresdner Bank Luxembourg S.A.;
- Luxempart S.A.;
- Natinvest SAH;
- SG Capital Dévelopment;

- S.I.T.A. S.A.;
- Telfin S.A.;
- Thames Television PLC.

The Astra spacecraft are designed according the U.S. principles, which means that the emphasis is on performance and cost. This gave SES a cost advantage over its European rivals. U.S. manufacturers were also better able to deliver spacecraft on SES's critical schedule. Typical characteristics of Astra satellites are provided in Table 6.4.

By the time of launch on December 11, 1988, most of Astra 1a transponders had been leased. The last four transponders were taken for German-language channels. The anchor customer for this hot bird was Sky TV, controlled by News Corp. For the bulk of the channels, the marketing agent was BTI, which acquired rights prior to launch. Each transponder is leased according to a private deal, much the way channels are acquired in the United States. Table 6.5 presents the current programming lineup.

6.2.4.2 British Sky Broadcasting (BSkyB)

British Sky Broadcasting (BSkyB) is actually a contraction of BSB and Sky TV, brought about by the merger of the two organizations in 1992. BSkyB is a private

Table 6.4
Characteristics of Astra Spacecraft, Operated by Société Européenne des Satellites
(All Located at 19.2 deg EL)

Spacecraft	1a	1b	1c	1d	1e	1f
Launch date	12/88	3/91	5/93	10/94	10/95	4/96
Launch vehicle	Ariane	Ariane	Ariane	Ariane	Ariane	Proton
Manufacturer	GE Astro	GE Astro	Hughes	Hughes	Hughes	Hughes
Projected lifetime	10 years	12 years	15 years	12 years	14 years	15 years
Frequency range	14.25 to	14.0 to	13.0 to	12.75 to	17.3 to	17.3 to
Uplink	14.45 GHz	14.45 GHz	13.25 GHz	13.0 GHz	17.67 GHz	17.67 GHz
Downlink	11.2 to	11.2 to	10.95 to	10.7 to	11.7 to	11.7 to
	11.45 GHz	11.45 GHz	11.2 GHz	12.95 GHz	12.07 GHz	12.07 GHz
Transponder bandwidth	26 MHz	26 MHz	26 MHz	26 MHz	26 MHz	26 MHz
Polarization	Linear	Linear	Linear	Linear	Linear	Linear
Number of transponders	16	16	16	16	16	16
Power per transponder	45W	60W	63W	63W	85W	85W
Typical EIRP	52 dBW	52 dBW	52 dBW	52 dBW	52 dBW	52 dBW
Coverage area	Western Europe	Western Europe, Iberia	Western Europe, Iberia	Western Europe, Iberia	Western Europe, Iberia	Western Europe, Iberia

Table 6.5
Programming on Astra 1a, 1b, 1c, 1d, and 1e

Number	Service	Lessee	Encryption	Format
1	Screensport	WHSTV		PAL
2	RTL Plus	RTL Plus		PAL
3	TVS (Scansat)	Scansat	Eurocrypt	D2-MAC
4	Eurosport	Sky		PAL
5	Lifestyle	WHSTV		PAL
5	JSTV	JSTV		PAL
5	Childrens Ch.	Childrens Ch.		PAL
6	Sat1	Sat1		PAL
7	TV1000	Scansat	Eurocrypt	D2-MAC
8	Sky One	Sky		PAL
9	Teleclub	BetaFilm/		PAL
		Kirch		
10	3Sat	3Sat		PAL
11	Filmnet	Filmnet	Satbox	PAL
12	Sky News	Sky		PAL
13	RTL Veronique	Sky	Irdeto	PAL
14	Pro 7	Pro 7		PAL
15	MTV Europe	MTV		PAL
16	Sky Movies	Sky	Palcrypt	PAL
16	Shopping Ch.	Shopping Ch.		PAL

corporation with shares of ownership spread among several commercially oriented companies in the United Kingdom, Australia, and France but lead by News Corp. Originally, BSB was awarded the only U.K. license to use the BSS assignments from WARC-77. This venture was completely funded though investments by owners and by bank debt. It contracted with Hughes for the manufacture and placement in orbit of two satellites of the HS-376 spinner design (a conservative approach, including use of the world's most reliable launch vehicle, the Delta).

The venture had to use D-MAC according to its license, which followed existing EC policy for true DBS satellites in Europe. Delays in availability of D-MAC receivers delayed the rollout of service, which severely hampered the business. In the meantime, Sky TV was delivering both set-top boxes and programming. BSB emphasized that their ground receiving antennas are much smaller than Sky's and have promoted the use of flat plate arrays called "Squarials." The technology was not new, but they lost some ground with the first suppliers. With the merger, BSkyB dropped the use of the BSS frequencies and satellites and continues to grow using SES/Astra. One thing that can be said is that BSB did a lot to rekindle interest in DBS in the United States.

6.2.4.3 Telediffusion de France (TDF) and TV-Sat

France has made a strong technological entry into DBS through its *Telediffusion de France* (TDF) program. Several French aerospace and electronics firms contributed to ground and space elements. The French and German DBS projects were handled jointly to economize on the development of the spacecraft. These were manufactured by the Eurosatellite consortium, which was composed of the French and German companies, Aerospatiale (24%), Alcatel Espace (24%), Daimler Aerospace (24%), AEG (12%), ANT (12%), ETCA/ACEC (4%).

From the French government side, the project was supervised and funded by TDF and CNES. In France, TDF acts as a common carrier for broadcasters by operating the broadcasting stations and the microwave network. It was natural for them to take on the operation of a DBS activity. Similarly, CNES is the national resource for space technology and operations. The project achieved many technology goals, including demonstrating transponders at 200+W each. The satellite provides coverage of France and several French-speaking areas of Europe. In addition, major French-speaking cities in North Africa receive programming with antennas up to 2m in diameter. Both satellites were successfully put into orbit by Ariane and are operational.

Table 6.6 details the technical characteristics of the TDF spacecraft. Table 6.7 indicates the initial programming included on TDF.

In a recent article in *Cable and Satellite Europe*, it was mentioned that both TDF BSS satellites could be taken over by Eutelsat [3]. Eutelsat is the government-supported European regional satellite system that is largely responsible for extending TV coverage throughout western and eastern Europe (see discussion latter in this chapter). TDF 1 and 2, in particular, have attractions for Eutelsat because, like TV Sat, they use the BSS operating parameters. This is the band to be exploited for European DTH services in coming years. The two satellites are to be moved to the Eutelsat position of 13 deg EL to take place ahead of the operational Hot Birds.

TV Sat was originally controlled by the Deutsche Telekom and was built to the same general specifications as TDF by the same Eurosatellite consortium. The first satellite failed to become operational and has not been replaced. The only operating TV Sat satellite was sold to the NSD consortium and is providing TV services to Scandinavia.

6.2.4.4 EUTELSAT

As the leading European satellite operator, Eutelsat provides near universal coverage in support of the full array of FSS applications. The 44 member/owners of Eutelsat are the national telecommunication operators of the western and eastern European nations. When Eutelsat was formed in 1976, the direction

Table 6.6
Technical Characteristics of the TDF Spacecraft

Feature	Specifications
Launch date	
1	October 1988
2	July 1990
Launch vehicle	
1	Ariane 2 (single launch)
2	Ariane 4 (dual launch)
Start of service	1989
Projected lifetime	9 years
Orbital position	19 deg W
Manufacturer	Eurosatellite
Payload Characteristics of TDF	
Frequency range	
Uplink	17.3 to 17.7 GHz
Downlink	11.7 to 12.1 GHz
Transponder bandwidth	23 MHz
Polarization	Circular (right hand)
Number of transponders	5
Number of TWTAs	6
Power per transponder	230W
Typical EIRP	63 dBW
Coverage area	France (Germany on TDF 2)

Table 6.7
Partial Listing of the Initial Programming on TDF

Number	Service	Programming status	Format
1	La Sept	Operational	D2-MAC
2	Canal Plus	Planned	D2-MAC
3	Canal Plus Deutschland	Planned	D2-MAC
4	Canal Enfants	Planned	D2-MAC
5	(Failed)		

was clearly toward western Europe; but after the breakup of the eastern block, countries from east to west joined. The leading signatories, as they are called, and their respective ownership are British Telecom (20%), Deutsche Telekom (14.24%), France Telecom (11.23%), Telefonica of Spain (9.8%), Telespazio of Italy (8.39%), and PTT Nederland (7.43%).

Several of their satellites provide program distribution to cable systems, in competition with SES/Astra. However, unlike SES, all of Eutelsat's spacecraft are built by European consortia. The operating Eutelsat FSS satellites at the time of this writing are indicated in Table 6.8.

In the late 1980s, a study called Europesat was conducted to determine how to address the DTH requirements of several countries with a regional DTH system. One approach considered was to combine the assignments of several countries and implement a constellation of high-power BSS satellites at a single orbit slot. Originally, the administrations of many nations thought that it was good to have their own satellite positions. Some nations even thought the more satellites the better. Actually, the fewer the better, because you can aggregate programming and create a hot bird.

Eutelsat did not proceed directly with Europesat because of the perceived difficulty of combining all of the operating channels at a single orbit position. It took approximately 4 years for the industry to catch up, and in 1994 they introduced the Hot Bird series of satellites. These are intended to provide focus for cable TV and DTH services in the European region, primarily the west where incomes can support premium services. Hot Bird 1, built by Matra Marconi

Table 6.8
Characteristics of Eutelsat Satellites used for Video Distribution and Other FSS Services

Spacecraft	Eutelsat I, F5	Eutelsat II, F3	Eutelsat II, F1	Eutelsat II, F2	Eutelsat II, F4
Location	21.5 deg EL	16 deg EL	13 deg EL	10 deg EL	7 deg EL
Manufacturer	BAe, Matra, AEG	Aerosp., MBB, Alcatel	Aerosp., MBB, Alcatel	Aerosp., MBB, Alcatel	Aerosp., MBB, Alcatel
Launch date	7/21/88	12/7/91	8/30/90	1/15/91	7/9/92
Expected lifetime	7	10	10	10	10
No. of Transponders	14	16	16	16	16
Bandwidths	72 MHz	31, 36, 72 MHz	31, 36, 72 MHz	31, 36, 72 MHz	31, 36, 72 MHz
TWTA power	20W	50W	50W	50W	50W
Services	Broadcast TV	Cable, DTH	Cable, DTH	Broadcast TV	Broadcast TV
Leading channels	Canal Plus, Deutsche Welle TV	N. Africa, E. Europe	BBC, Eurosport, MTV, NBC Super, RTL 2	EBU	EBU

Space, began service in March 1995 and is collocated with Eutelsat II at 13 deg EL. With 16 medium-power transponders, it increases Eutelsat's capacity at this important position to 32 analog TV channels. Hot Bird 2 is expected to be launched in mid-1996 and placed at the same orbit position.

In consideration of the work done on the Europesat planning effort, Eutelsat will begin operation in the BSS bands with the launch of Hot Bird 3 some time in 1997. This is a 20-transponder high-power DTH satellite with the ability to cover Europe, central Asia, and the Gulf states. It is Eutelsat's intention to make this a digital DTH system, capable of transmitting to dishes of 45 cm in size.

6.2.4.5 Thor

The Thor series of DTH satellites operate in the BSS frequencies and provide service to several Nordic countries, including Norway, Sweden, Denmark, and Finland. The first satellite, Thor 1, was purchased from BSB in orbit as it was already functioning as a part of the defunct U.K. service (replaced by Sky abroad the Astra satellites).

Telenor Satellite Services AS of Oslo, Norway, has ordered an HS 376 satellite from Hughes Space and Communications International, Inc., to provide DTH programming to Scandinavia and northern Europe. The spacecraft is scheduled for launch in 1997. Called Thor IIA, the satellite is the 49th HS 376 model to be ordered from Hughes. An artist's conception of Thor IIA in orbit is shown in Figure 6.6.

Telenor is also party to the *Nordic Satellite Distribution* (NSD) joint venture company, which employs other satellites such as Intelsat 702. The Thor series have adequate power to transmit to 50-cm dishes. The Nordic market is moderate in size but hungry for programming, witnessed by the significant number of satellites and transponders serving the subregion.

6.2.5 Expansion of DTH in Asia

The first introduction of TV transmission to a small receiving antenna was conducted in 1977 in Indonesia over the Palapa A1 satellite. A cooperative effort of Perumtel (now PT Telkom) and Hughes Space and Communications, it demonstrated the viability of a form of direct video transmission. While the antenna was 4.5m in diameter, which is large by current standards, it nevertheless proved that low-cost installations could provide a quality product. The next major step was Aussat A with its medium-powered Ku-band FSS satellites, which could achieve the same result with an antenna of around 1m in diameter. However, it was not until the launch of AsiaSat 1 that a true commercial DTH service appeared in the region.

Figure 6.6 An artist's conception of the Thor IIA satellite in geostationary, operated by Telenor Satellite Services (photograph courtesy of Hughes Space and Communications).

The Asia-Pacific region is undergoing rapid expansion in commercial satellite terms, fueled by the demand for video transmission capacity. A summary of the satellite operators and the size of their respective constellations is provided in Table 6.9. As provided for Europe, the information given here for Asia is very basic, so readers who require details for system design and service evaluations are advised to contact the appropriate satellite operator for current detailed information.

6.2.5.1 Indonesia

The Indonesia satellite TV situation is the most extensive in the region, having provided the basis to deliver the national video signal across this extensive country of over 10,000 islands. As indicated in Table 6.9, there are three active satellite operators and five satellites in operation. The bulk of the capacity is at C band, which is consistent with two important factors. First, these satellites were originally intended to provide an effective telecommunications infrastructure for the country. The availability of regional C-band service for the expanding video market was more or less a fortunate accident. Second, this is a tropical region with some of the greatest thunderstorm activity in the world (see Figure 2.4). The result is that high rainfall rates will not impair the signal as much as they would have in a Ku-band system design.

The Palapa A series were launched successfully into service in 1976 and 1977, followed by the Palapa B series in the mid-1980s. Two Palapa B satellites are operated by the national telephone company, PT Telkom. Palapa B2P has become a focus of programming in the region and qualifies as a hot bird. Table 6.10 presents a listing of channels offered at the time of this writing. Readers should be aware that there is some volatility in the line-up of video services, as operators seek the most effective distribution satellites at the most attractive prices.

A second satellite operator called *Pasifik Satelit Nusantara* (PSN) appeared on the scene in 1991, gaining access to the orbit by purchasing a Palapa B satellite that was nearing end of life. The satellite operates in an inclined orbit to extend life to serve programmers in Indonesia and Taiwan. An uplink is installed on Batam, a small Indonesian Island about 50-km south of Singapore. PSN has more recently begun an ambitious MSS project called ACeS, which is discussed in Chapter 9.

The third satellite operator in Indonesia is called PT. Satelit Telekomunikasi Indonesia (Satelindo). Partially owned by Deutsche Telekom, Satelindo has three government telecommunications licenses that it is exploiting: an international PSTN gateway in Jakarta, a GSM cellular license for Indonesia, and the owner and operator of the Palapa C series of satellites. The first Palapa C satellite was launched in 1996 and provides C- and Ku-band transponders.

Table 6.9
Satellite Operations in the Asia-Pacific Region

Satellite/Operator	Location	Number of Satellites	Coverage	Primary Services
Palapa B/PT Telkom	Indonesia	2 (all C band)	Indonesia, ASEAN region	Video, including DTH, other FSS
Optus and Aussat/Optus Communications	Australia	4 (all Ku band)	Australia and New Zealand	Video, including DTH, other FSS
ChinaSat/MPT of PRC	Peoples Republic of China (Beijing)	4 (3 all C band, 1 C/Ku)	Peoples Republic of China	Video, other FSS
Insat/ISRO	India	3 (C/S bands)	India and India Ocean	Video, including DTH, other FSS
JCSat/Japan Satellite Systems	Japan	3 (all Ku band)	Japan, East Asia	Video, including DTH, other FSS
AsiaSat/Asia Satellite Telecom Co Ltd.	Hong Kong	2 (1 all C band, 1 C/Ku bands)	Asia and Middle East	Video, including DTH, other FSS
Superbird/Space Communications Corp.	Japan	3 (2 Ku/Ka bands, 1 all Ku band)	Japan	Video, including DTH, other FSS
Palapa C/Satelindo	Indonesia	2 (C/Ku bands)	Indonesia, ASEAN region, Asia	Video, including DTH, other FSS
APStar/APT Satellite	Hong Kong	2 (all C band)	China, Asia	Video, including DTH, other FSS
PSN/Pasifik Satelit Nusantara	Indonesia	1 (all C band)	Southeast Asia	Video
BSat/NHK-TAO	Japan	2 (all Ku band)	Japan	DTH
Taicom/ Shinawatra	Thailand	2 (C/Ku bands)	Thailand, East Asia	Video, including DTH, other FSS

Table 6.9 (continued)

Satellite/Operator	Location	Number of Satellites	Coverage	Primary Services
NStar/NTT	Japan	1 (Ku/Ka bands)	Japan	Video, other FSS
Measat/Binarang	Malaysia	2 (C/Ku bands)	Malaysia, India, and East Asia	Video, including DTH, other FSS
Koreasat/Korea Telecom	Korea	2 (all Ku band)	Korea	Video, including DTH, other FSS
Mabuhai/PLDT	Philippines	1 (C/Ku bands)	Philippines, Asia Pacific	Video, other FSS

Table 6.10
Video Services Available on the Palapa B2P Satellite, Located at
113 deg EL Position

Video Service/Channel	Origin
The Discovery Channel	United States
RCTI	Indonesian government TV channel
TV1 Canal Plus	France
ATVI	Australia
TV3	Singapore
Asia Business News	
ABS/CBN	Philippines
CNN International	United States
HBO Asia	Singapore (United States)
AN TEVE	Indonesia
GMA	Philippines
KBP People's Network	Singapore (United States)
TPI	
ESPN International	
SCTV	Indonesia
The Gold Network	Hong Kong
TVB	
Singapore International TV	Singapore

6.2.5.2 Japan

Japan was an early developer of DTH service with the BS series of satellites. The governmental broadcaster, NHK, has been delivering a DTH service to the Japanese audience since 1980, and it has become very popular. There are estimated to be over 7 million home dishes in Japan, and it clearly represents the most successful service of its kind in the world. Viewers pay an annual charge to NHK more or less in the form of a tax, akin to the policy of the BBC in the United Kingdom.

NHK provides the following three DTH channels using the BS spacecraft:

- *DBS-1:* a combined 24-hour satellite feed of news, sports, and general interest programming coming from Japan, the United States, and Europe;
- *DBS-2:* a cultural channel available about 23 hours per day, consisting mostly of Japanese material;
- *Hi-Vision (HDTV):* a demonstration HDTV channel utilizing the NHK standard, also referred to as MUSE; this channel is the only daily HDTV broadcast in the world.

NHK has been very successful with DBS-1, primarily because of the attractiveness of the programming. They have obtained the rights to rebroadcast CNN, ESPN, ABC, and other popular U.S. channels to the Japanese market. The feed is provided by Hughes Communications through multiple links from Europe and the United States and retransmitted over a Pacific INTELSAT satellite. It is then reuplinked from Japan to the BSS satellites. People in neighboring countries like Korea, Taiwan, and China have installed dishes just to receive this material, which is considered to be extremely current and objective.

A subscription DTH service is available over the BS satellites: *Japan Satellite Broadcasting* (JSB). This is a premium pay channel offered by a commercial company. The NHK and JSB services are made available at 110 deg EL, which is Japan's BSS orbit position allocated at WARC-BS. JSB was established as a company in 1984 and began operation in April 1991 using the service name "WOWOW." They employ the same analog signal format used by NHK with the addition of scrambling.

Satellite broadcasting moved further away from government domination with the creation of *Broadcasting Satellite System Corp.* (BSAT). This is a consortium composed of NHK, JSB, and other terrestrial broadcasting and financial partners. BSAT purchased an HS-376 BSS spacecraft from Hughes, which will be placed into the 110 deg EL orbit location in 1997. It will then replace the aging BS series operated by the *Telecommunications Advancement Organization* of Japan (TAO), the government-supported satellite operator. The four BSS channels are allocated between NHK, JBS, and new services to be instituted by the other partners. As many as two BSAT satellites will be operated jointly, allowing an expansion of DTH services in coming years.

Of the remaining satellite operations in Japan, two currently provide DTH service. These are JSat and SCC, utilizing the FSS allocations at Ku band. The services offer between 5 and 12 channels using analog format similar to the NHK services. However, in coming years, the capabilities will be expanded to DDTH and the channel capacity will be increased by as much as a factor of ten. A planning company has been formed by Hughes and Mitsubishi to implement a DIRECTV service in Japan, based more or less on the U.S. model.

6.2.5.3 AsiaSat

The best model of a video distribution system created by business interests is AsiaSat, currently operating two satellites in the Asia–North Africa region. The company was formed in Hong Kong by three partners, each with equal share: Cable and Wireless PLC, Hutchison Wampoa Ltd., and the *Chinese International Trust Investment Corporation* (CITIC). This is a very unique combination of western, Hong Kong, and Chinese interests. AsiaSat 1 was obtained on the cheap by taking over the Westar 6 satellite that was recovered by the Space Shuttle after a prior partial launch failure. The spacecraft antenna was modified for Asia coverage by the manufacturer, Hughes, and relaunched by the Long March rocket from China. The satellite has allowed Star TV to begin service and thereby become the leading Asia-based satellite TV service.

AsiaSat 2, built by Lockheed-Martin, was launched in November 1995, on the Long March 2E, and went into service in 1996. It is a hybrid C/Ku-band satellite with a total of 33 transponders. The C-band capacity is to be used by StarTV. However, the Ku capacity has excellent coverage of China, with EIRP sufficient to provide DTH service into antennas as small as 50 cm. Several Chinese government entities are planning to introduce a variety of video services over AsiaSat 2.

The only TV services available on AsiaSat 1 and 2 are those of Star TV, a Hong Kong-based programmer started by Li Ka Shing and his son, Richard Lee, and subsequently sold to Rupert Murdock's News Corp. The Star TV lineup consists of the services:

- Prime Sports;
- STAR Music;
- BBC Asia;
- ZEE TV;
- Star Plus;
- Chinese Channel;
- STAR Movies.

The coverage of AsiaSat 1 is split into two beams: the northern beam that focuses on China, Korea, and Japan; and the southern beam that extends through Southern Asia and the Middle East. Consequently, the Star services are repeated on the two beams. The exact number of DTH antennas currently pointed at AsiaSat 1 is impossible to know. This is because none of the services are encrypted and controlled. Users simply purchase the equipment as cheaply as possible (usually from domestic sources) and put them up. It is guessed that there are over one million cheap antennas in China alone, which is an amazing fact when you consider that it is illegal to have a dish. Service is widely used in India—owing to the popularity of ZEE TV. In this case, individuals subscribe to cable in their community. This is an unregulated business, where entrepreneurs invest the few dollars needed to string cables from residence to residence. Subscribers pay very little for access and nothing for the programming itself.

AsiaSat has purchased a third satellite, AsiaSat 3, from Hughes. In this case, Star TV is no longer an exclusive occupant from a video standpoint. This will allow AsiaSat to aggressively market the capacity of a high-power satellite throughout the region.

6.2.6 Expansion of DTH in Latin America

The final region to be examined in detail is Latin America. For years, Latin America has followed the United States in the development of terrestrial and satellite TV. Because U.S. cable TV satellite transmissions do not extend well below about the middle of Mexico, many groups and individuals in South America have still taken the trouble of installing 13m dishes just to be able to receive HBO and CNN. More recently, these programmers have introduced Latin American versions of these channels and offer them through PanAmSat, INTELSAT, and other regional operators.

The current array of satellite operators serving Latin America is provided in Table 6.11. There are currently two satellite operators in Latin America, plus the U.S. companies. The fact that the United States is so closely tied to Latin America means that, effectively, U.S. operators are Latin American operators. The best example is PanAmSat, which is a U.S. company that carries 50% ownership by Televisa of Mexico. They have done an excellent job of entering both the Latin American and Atlantic Ocean region markets.

From a DTH standpoint, the vital new addition to Latin America is Galaxy 3R, which was launched in November 1995. The satellite is part of the Hughes Communications Galaxy fleet; however, the DTH service is provided by a joint venture company called Galaxy Latin America (GLA). The partners include Hughes, Organisacion Diego Cisneros (ODC) of Venezuela, TV Abril of Brazil, and Multivision of Mexico. GLA is headquartered in Ft. Lauderdale, Florida, and provides a flavor of DIRECTV called, not surprisingly, DIRECTV Latin

Table 6.11
Satellite Operators That Focus on the Latin American Market

Satellite/Operator	Location	Number of Satellites	Coverage	Primary Services
Morelos-Solidaridad/ Telecomm Mexico	Mexico	3 (C/Ku bands)	1-Mexico; 2-Mexico, United States and South America	Video, including DTH, other FSS
Brasilsat/ Embratel	Brazil	2 (C band)	Brazil	Video, other FSS
PanAmSat	United States	3 (C/Ku bands)	Latin America, Europe, United States (eastern), United States (western), Pacific Rim	Video, including DTH, other FSS
Galaxy 3R/Hughes	United States	1 (C/Ku bands)	United States, Latin America	Video (United States), DTH (Latin America)

America. The services are uplinked from a broadcast center in Long Beach, CA. Additional broadcast centers will be operated from Mexico, Brazil, Argentina, and Venezuela. The frequencies are within the FSS portion of the band, but the system employs circular polarization. Sufficient EIRP is provided to allow the use of antennas typically 60 cm in diameter.

6.2.7 The INTELSAT System

Any discussion of worldwide video distribution satellites would not be complete without mention of the INTELSAT system. While their main objective is in the domain of providing trunking services, INTELSAT nevertheless is a resource for C- and Ku-band satellite capacity that many DTH operators are using or at least considered as an option. INTELSAT is a global consortium or cooperative that paved the way in satellite communication for an advancing world. To meet the growing international communication requirements of this decade and beyond, they continue to invest in new satellites and expand their fleets in the three ocean regions, namely, the *Atlantic Ocean region* (AOR), the *Pacific Ocean region* (POR), and the *Indian Ocean region* (IOR).

A summary of the types of spacecraft employed by INTELSAT is provided in Table 6.12. At the time of this writing, INTELSAT was the largest satellite operator in the world with 24 operational satellites, including five Intelsat VIs, built by Hughes, six Intelsat VIIs, built by Ford Aerospace (now SpaceSystems/ Loral), and six Intelsat VIIIs and one Intelsat K, built by GE Astro (now part of Lockheed-Martin) [4]. All satellites employ the FSS frequency allocations, which is indicative of the multipurpose nature of INTELSAT services.

6.3 RELATIVE COST OF SATELLITE DTH VERSUS CABLE

Cable TV is a very viable technology for providing a wide array of programming services to urban and suburban homes. In countries like the United States, Canada, Belgium, and Germany, the percentage of homes that actually take cable services is over 60%, in relation to the total that have access to cable service (i.e., the "homes passed" by cable). In only a few places in Asia does cable penetration approach this number, such as in major cities in Japan and

Table 6.12
General Characteristics of INTELSAT Operating Satellites

Spacecraft	Intelsat VI	Intelsat VII	Intelsat VII-A	Intelsat K	Intelsat VIII	Intelsat VIII-A
Manufacturer	Hughes	Ford Aerospace	Loral	GE Astro	GE Astro	GE Astro
Expected lifetime	13	10 to 15	10 to 15	10	14 to 18	14 to 18
No. of Transponders						
C band	38	26	26	0	36	28
Ku band	10	10	10	16	6	3
EIRP (dBW)						
C band	31.0	33.0	33.0	0	36.0	37.5
Ku band	44.7	45.4	47.0	47.0	47.0	47.0
Designations	601 to 605	701 to 705, 709	706 to 708	IS-K	801 to 804	805, 806

Korea and Singapore. For the highly populous countries in Latin America, Asia, Eastern Europe, and Africa, cable TV is simply much less economical in relation to digital DTH (DDTH).

In following analysis, it is assumed that both the cable TV and DDTH services provide in excess of 50 channels. The cost of providing cable service to a typical single-family home is approximately $2,000, which includes the cable, set-top box, its installation, as well as the headend equipment needed to receive satellite and terrestrial TV signals. Also required are administrative offices, maintenance facilities, and a studio to develop local advertising and program content. For a DDTH satellite system, the cost per subscriber includes a share of the cost of the satellites and uplinking facilities added to the cost of maintaining the subscriber base. For a system of three satellites and two uplinks, the investment is assumed to be approximately $600 million. The lower curve in Figure 6.7 shows the investment cost per subscriber, which is obtained by dividing the total investment by the number of subscribers. Added to this is the installed cost of the home-receiving system, which is around $700. In the DDTH model, the satellites, uplinking facility, and administrative systems are shared by between 1 and 10 million subscribers. The upper curve shows that the total investment per subscriber decreases to under $800 when a subscriber base of 10 million homes is reached. For both the cable TV and DDTH TV broadcast systems, the cost of program acquisition is assumed to be the same and is born by the subscriber through monthly access fees.

6.4 DTH SATELLITE DESIGN OPTIONS

In this section, we discuss various design approaches for the spacecraft that would support a DTH-based service business. The emphasis is on how to

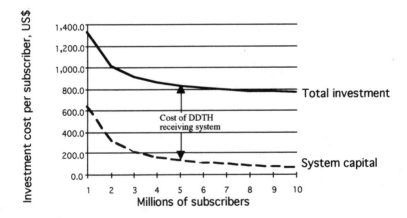

Figure 6.7 Investment cost per subscriber for a digital DTH system.

evaluate the technology alternatives and apply them effectively. Most of the necessary background material was covered previously. DTH is a delivery vehicle for programming, where the receiver is located with and probably owned by the end user. In the ideal case, specifics like the type of receiver, size of antenna, signal format, and satellite design are secondary. However, it is vital that the system developer understand these alternatives so that a poor technical choice at the beginning does not become the point of business failure. However, those who see this only as a high-technology battle will probably be the losers.

The emphasis should be on making it easy and relatively inexpensive for the subscriber, since their interest is in the programming and the cost of getting access to it. The quality of satellite-delivered digital video is as good as or superior to cable or over-the-air broadcasting; hence, the picture quality will probably not be a differentiating factor. Rather, subscribers could relate more to reliability of service along with the convenience factors cited previously.

In the main, the area where quality is seen as a negative is at C band where users have to contend with terrestrial interference. Ku-band systems, while generally free from terrestrial interference, are subject to rain fade, which produces occasional outages. Once this problem is solved through adequate link design and margin, subscribers will next be drawn by a desirable array of programming. Over the years in the U.S. market, this has come down to the range of channels, movies, and events that are delivered to cable systems and now provided over the operating digital DTH networks like DIRECTV and PrimeStar. Anything less in such a competitive environment will be rated below the leaders; the question is, what will prove the winner in coming years.

In Europe, the experience is exclusively at Ku band and the previous points about simplicity, cost, and programming have been validated over and over again. Sky TV succeeded over its higher technology competitor, BSB, precisely because of these aspects. Other activities in France and Germany were slower at the start of the game, being hampered by government red tape and inflexibility. In many ways, Europe poses a much more exciting future for DTH simply because of the wider array of ventures and economic conditions. Just where the "pot of gold" lies in DTH remains to be seen.

The Asian environment has many opportunities because of the primitive nature of the DTH industry in the region. Cable TV is a viable business in developed countries and city states; China, India, and Indonesia have large populations that are hungry for more and better entertainment. None of these countries are currently pursuing a BSS type of satellite or network. Importantly, money flows easily into major projects and business ventures. This has fueled the creation of several new satellite operators and the development of the largest satellite market in the world. Digital DTH systems, however, have not taken hold yet. The greater rainfall in the tropical parts of the region (where a high percentage of the population lives) is a significant factor in building a technically satisfactory system at Ku band. C-band FSS systems exist and serve a strong niche market, not unlike the early backyard dish segment in the United States. However, putting this together in a business the size of DIRECTV or BSkyB may take several years and false starts.

The story for Latin America assumes a different tone from the previous three. In the leading countries of Mexico, Brazil, and Argentina, pay TV is accepted and very popular among the middle class. Others enjoy a reasonably wide variety of standard broadcasting services. The stage is set for the introduction of DTH, with the launch of Galaxy 3R, a Ku-band FSS satellite, for DIRECTV Latin America and the pending start of a service by a consortium of large programming interests, namely, Globo of Brazil, Televisa of Mexico, and News Corp. The fact that these organizations are such powers in the area of TV content will have a big impact on the potential attractiveness of the services. It is interesting to note that both systems will use FSS satellites that are owned and operated by United States–based corporations. The BSS allocations continue to lie fallow. FSS satellite operators in the region, particularly Embratel in Brazil and Telecomm Mexico, have been inwardly directed up to the time of this writing. It will be interesting to see how they, too, will play in the DTH game in the wider region.

6.4.1 Medium-Power DTH Satellite Systems

Medium-power Ku-band satellites have a variety of uses, including video distribution, DTH, data broadcasting, audio distribution, point-to-point telecommunications, and VSAT interactive data networks. They are being used to create DTH businesses in the United States, Europe, Australia, Japan, and Latin America. A satellite operator who successfully creates a video hot bird from a medium-power Ku-band satellite or even a C-band satellite has a valuable asset upon which to expand a business. This is what happened with the Galaxy system in the United States, although this first involved cable TV rather than DTH. If the infrastructure is there because of the need for cable TV programming, then the DTH side can develop gradually. Building an exclusively DTH business has proven much more difficult for the pure satellite operator because of the lack of a flywheel from other revenues and easy access to programming. Another inherent problem for the medium-power DTH operator and TV programmer is that the cost of the receiving station is elevated due to lower satellite EIRP.

Technology improvements have made medium-power DTH more practical. Medium-power Ku band did not even exist prior to 1980. In the 1980s, the Americans and Japanese began producing low-cost receiving systems that made medium-power DTH possible. Perhaps it was the push of the U.S. C-band market that proved one could obtain a complete system for around $1,000. Also, the popularity of European services like Sky and RTL, available over

Astra, gave a strong push to Ku-band receiver and antenna production. Going to Ku band simply meant that enough potential manufacturers had to perceive a market. This is precisely what happened in Europe, where Ku-band FSS dominates.

To understand FSS architecture, it is important to explore the capabilities of the receiver. The functions of a typical FSS-style Ku-band receiver are listed in Table 6.13. This is based on the products from Pace, a leading United Kingdom–based supplier of these systems.

Medium power also means that there can be more transponders per satellite for the same total launch weight. Cost is primarily driven by weight, so medium-power DTH has an economic advantage in space. Ground costs would tend to be higher, as a result of the need for a larger dish (anywhere from 50% to 200% larger). The cost of increasing dish size is significant; but when considering the TVRO cost as a whole, the dish size has only a secondary economic effect. In terms of appearances, the larger dish could be a problem in some markets. The experience in Europe and Asia so far is that this is not the case, as consumers appear to accept the fact that to receive the programming one needs to have this type of dish.

Medium-power DTH can be difficult to implement from the standpoint of international coordination. DTH satellites that use the WARC-BS assignments do not need to be coordinated as long as they follow the plan. On the other hand, medium-power FSS satellites must be coordinated from the beginning. This process can take years to complete and poses risk to the business. The design may have to be changed to satisfy neighboring systems that could currently be using a lower power level. INTELSAT consultation must also be

Table 6.13
Functions of an Analog Ku-Band FSS Home Receiver

Frequency band	Ku
Number of channels tunable	500
Display type	Multifunctional vacuum fluorescent
Stereo system	Wegener Panda 1
Conditional access	VideoCrypt with dual viewing card
Tuner features	Dual LNB with 15-MHz and 27-MHz switchable bandwidth
VCR timer	8 event, 28 day
On-screen menus	Provided, multilingual
Satellite control	Automatic self-seeking and adjustment for a new satellite location; electromagnetic and electromechanical polarotor control

Source: Pace.

conducted, although this will not be a problem for domestic services and most regional systems. A detailed discussion of the coordination process is provided in Chapter 10.

Any technical compromises resulting from coordination could require receivers to have larger antennas, better polarization and/or sidelobe isolation, or narrower bandwidth (and a different frequency plan). The incompatibilities previously mentioned will likely impact the business, unless that satellite achieves hot bird status. Ancillary services such as data broadcasting may have to be curtailed to satisfy requirements of a coordination. This is because some carriers have power densities that could cause unacceptable interference. Therefore, system developers should maintain close contact with the government regulatory people who are pursuing the coordination process. Tradeoffs may have to be made along the way.

In summary, the principle advantages of the FSS approach are as follows:

- Existing satellite capacity can be utilized, provided that the EIRP is sufficient to allow reception by an appropriate receiver. Also, there must be enough bandwidth to support a channel capacity which will be attractive to users. In North America, the market demands that at least 100 channels be available at the same orbit position, whereas in Europe and Asia the number might be as low as 20 for a specific country market.
- The existing BSS assignments might already be taken by another operator. This is precisely the situation with respect to Sky versus BSB in the United Kingdom. Sky introduced their service on the Astra FSS satellite, ahead of BSB's launch of their compliant BSS satellite.
- The orbit slots and coverage patterns assigned by BSS WARC plans may not be optimum for a particular market or application. One of the big problems with the BSS coverages is that they are for a single country, whereas many services wish to cover a region where the same language is used. This philosophy has been challenged by Eutelsat, Hughes, and others who wish to use the BSS frequencies for regional or even global systems. This requires changing the plan, a process described in Chapter 10.
- Two-way services might be desired. BSS assignments are only for one-way broadcasting, whereas a new entrant might wish to introduce an interactive service that uses a two-way VSAT.

6.4.2 High-Power DTH Satellite Systems

DTH systems that adhere to the BSS plan have many characteristics in common. For one, the systems will generally not cause unacceptable interference to each other. This is provided by the plan and work of the ITU to maintain compliance. Reception is allowed in any country of transmissions from another country's satellite (installing of dishes may be prohibited, however). Antenna size is minimal, thanks to the availability of the highest practical power level. Introduction of DVB-based receivers should further encourage the launch of BSS satellites.

6.4.2.1 The A Priori Planning Approach of the ITU

The a priori plan of WARC-BS established both a planning process and a plan for two of the three ITU regions of the world. That this sort of thing could be accomplished in several weeks of deliberation is incredible. Today, several satellites operate according to the 1977 plan. Of course, there are even more satellites that do not. The fundamental difference is that BSS satellites use fixed orbit positions and frequency assignments, with their antenna foot prints covering typically one country each or a portion of that country. The transponders are used for video transmission, although sound and data broadcasting were also contemplated even back then. The planners established certain technical standards for the satellites based on their estimate of current and future technology. As stated previously, they overestimated the ability of satellites to carry high-powered transponders and underestimated the performance of future home receivers. The application of digital processing and advanced conditional access was not even considered.

The *protection ratio (C/I)* is the key interference test parameter. The single entry value corresponds to the interference due to one other frequency assignment. Multiple entry indicates the combined effect of all interference sources on the same channel. A single entry objective was set at 32 dB. The planners sought to hold the multiple entry *C/I* at 29 dB, which is a very high value. In the United States, we try to achieve 20 dB, but even here we often accept a poorer value. In accordance with the plan, the following are the assumed characteristics of each BSS satellite:

- High EIRP, typically 63 dBW;
- At least one satellite with five channels per country;
- Fixed-frequency plan with 23-MHz transponders;
- Circular polarization (left and right hand);
- Fixed orbital spacing between satellites:
  - 3 deg minimum separation,
  - 9 deg for cofrequency assignments,
  - Satellites that serve different countries are colocated where possible;
- Maximum spacecraft reflector size of ~2m;

- Elliptical beams without shaping;
- Mandated radiation levels into other countries.

Assumed home receiver characteristics include:

- Dish size no smaller than 0.9m;
- Receiver noise figure no lower than 9 dB;
- Relatively poor cross-polarization ability;
- Relatively poor angular discrimination.

6.4.2.2 Working Within and Around the Plan

In theory, the plan should have been able to satisfy all the needs for DTH services for several decades. However, some participants in WARC-BS saw a need to allow some flexibility in its implementation. The rules surrounding the plan have a few loopholes under which an ITU member nation can make technical changes. Any change, however, cannot raise interference by more than 0.25 dB (6%) into another planned satellite that complies with the plan.

There are a number of reasons for considering making a change to the plan. Some of these are:

- Change the downlink footprint:
  - More constant domestic coverage,
  - Increase service area;
- Reduce the number of satellites:
  - Decrease the investment of a domestic system,
  - Consolidate systems of several nations,
  - Build a better hot bird.

The key technology for altering the footprint is beam shaping onboard the satellite. It is impossible to create more energy than the power amplifier delivers to the antenna. The trick is to distribute that energy in a particularly effective way. The original plan assumed the use of very simple feed networks, that is, basically a single horn and an unshaped elliptical reflector. The same shaping techniques that increase the performance within the intended coverage area can be used to reduce the radiation into other countries. This can be needed if the calculated C/I into another member's system is unacceptable.

At the time of this writing, the Radio Regulatory Board was literally jammed with requests to modify BSS planned orbit assignments. This situation, as well as the general confusion over who is really proceeding, has caused the entire process to come under review. A forthcoming WRC will likely come up with a modified plan and process as a result.

6.4.3 Orbital Interference Limitations

DTH networks are designed to work with smaller receiver dishes that cannot reject adjacent satellite interference as well as the larger dishes found in video distribution. Operation in the BSS bands is made to be satisfactory because the ITU has assigned the satellites to orbit positions with adequate spacing. This assures a satisfactory service under expected conditions. In contrast, DTH systems that employ FSS satellites are not directly protected by the Radio Regulations, except through the coordination process. Typical frequency coordinations for FSS satellites allow satellites to be placed as closely as 2 deg apart, and there are even cases where a 1-deg spacing has been assumed. This causes downlink interference to be perhaps the largest single contributor of degradation to the service quality of DTH implemented in these more crowded bands.

The following paragraphs provide an introduction to the subject of interference design for DTH systems. This is more of an outline than a prescription for satisfactory operation. Any organization that is considering implementing a DTH system, either BSS or FSS, must gain a thorough understanding of the interference environment. They should provide adequate margins and conduct the frequency coordination activity through to successful completion of the process.

The first step in understanding this problem is to look at the basic mechanism that produces downlink interference. Figure 6.8 shows the ways that downlink interference can enter a DTH receiver. The small-diameter antenna on the ground has a main lobe and sidelobes on either side. These are exaggerated in the figure to make the example clear. In an actual case, the 3-dB beamwidth of a 45-cm (18-in) antenna is approximately 4 deg at 12.3 GHz. This means that the 3-dB point on this antenna lies 2 deg off the peak, in either direction. A satellite located in this direction would only have about 3 dB of isolation (by definition). If this satellite has the same EIRP as the desired satellite in the direction of the DTH receiver, then the C/I is the same 3 dB. However, if it happens to be twice as powerful and therefore has 3 dB more EIRP in this direction, then the C/I is 0 dB. Either value is, of course, unacceptable. The proper value must be determined for the specific modulation and frequency spacing, which is discussed later in this section.

The second adjacent satellite in Figure 6.8 is spaced far enough away to be located in the first sidelobe of the DTH receiving antenna. If we assume that the spacing places the interfering satellite at the peak of the first sidelobe, then the isolation is in the range of 15 dB to 25 dB, depending on the dish design. This value is the main component of the C/I, as discussed in the previous paragraph. For the example of the 45-cm antenna, this sidelobe peak occurs at an offset of 6.4 deg. This corresponds to an orbit spacing of about 6 deg.

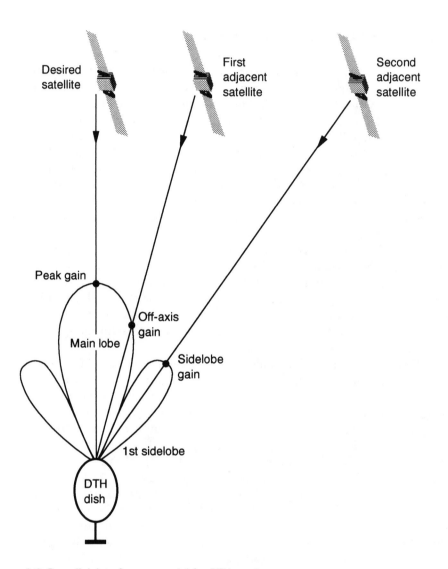

Figure 6.8 Downlink interference model for DTH service.

The orbit spacing, DTH antenna diameter, frequency, satellite power, and resulting *C/I* are all interrelated. Figure 6.9 presents an example of the results of a typical analysis for a DTH system operating in the BSS portion of Ku band. The receive antenna follows the expected main beam and sidelobe gain characteristic given in the ITU Radio Regulations, Appendix 30. The required value of *C/I* will depend on the makeup of the link budget for the particular

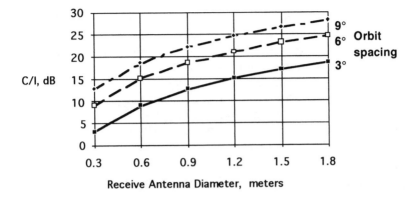

Figure 6.9 Typical calculations of C/I values as a function of the DTH receive antenna diameter at Ku band, for orbit spacings of 3 deg, 6 deg, and 9 deg.

application. In digital service, values in the range of 10 dB to 15 dB are typical. If we assume a value of 15 dB, then we obtain Table 6.14, which shows the relationship between orbit spacing and dish diameter.

These are the basic facts that govern the interference design of a DTH system on the receiving side. From an uplink standpoint, the broadcast center can utilize large enough antennas to reduce the uplink component to essentially zero. The overriding factor in the design is the downlink interference as received through the relatively small DTH antenna on the ground.

The consideration of orbital interference is now a critical part of designing a new DTH network. If we are applying the ITU BSS Plan as it is incorporated into the Radio Regulations, then the only concern is that the actual hardware complies with the plan's assumptions. The regulations provide for making modifications to the plan, provided that the change or addition does not increase the interference into any satellite network already in the plan by more than

Table 6.14
Relationship Between Orbit Spacing and Dish
Diameter, Given a *C/I* Value of 15 dB

Orbit Spacing (deg)	Dish Diameter (cm)
3	120
6	60
9	40

0.25 dB (i.e., 6%). This particular rule has been followed by many DTH operators and countries since there are good commercial reasons for making changes.

The rapid introduction of DTH systems around the world is increasing the difficulty of finding adequate FSS spectrum and orbit positions. For this one reason, BSS is likely to become more popular. Progress with digital DTH technology in general and DVB in particular is making it possible for literally any country to introduce satellite-delivered services within a year or less. Satellites with higher power than was contemplated when the plans were developed are in operation in the United States and will appear in other parts of the world.

6.5 DIGITAL COMPRESSION SYSTEMS AND SUPPLIERS

The discussion of DTH architecture has at its foundation the matter of the type of video processing and compression, along with the supply and cost of the set-top box. In this section, we briefly review the technical approaches to the uplink side of the network and review the capabilities of some of the key suppliers of the compression equipment. The market is still in its formative phase, so the names, owners, and ultimate suppliers of this critical element will change over the coming years.

There are several reasons why DTH operators are moving to compression [5]. One of the obvious is the need to increase the number of TV channels that a given transponder and satellite can support. We indicated earlier in the chapter that a key strategy in building a successful DTH business is the quantity and variety of programming offered. Digital compression has given the business the element that it needs to meet the market demand for the wide array of options. A second reason is that there are a limited number of satellites available in some regions, particularly Asia Pacific, and compression allows a new service to start even without a full transponder. On PanAmSat 1, the first of the series launched over the Atlantic, almost half of all video channels are compressed in nature. The cost per channel is also influenced in a favorable direction because the transponder can be shared by multiple users. When doing this, one needs to consider whether to use FDMA (on video per carrier), TDMA (multiple video signals sharing the same frequency on a burst basis), or TDM (some number of TV channels multiplexed onto one wideband carrier).

The basic arrangement of the uplink compression and downlink decompression systems is presented in Figure 6.10. The function of each of the blocks is described in previous paragraphs of this chapter and Chapter 5. We are concentrating here on the uplink side, which is contained within the broadcast center Earth station such as that shown in Figure 6.5 for DIRECTV. Included in the uplink baseband system, but not shown in Figure 6.10, is the equipment that digitizes and time division multiplexes the video, audio, and data information. Systems the employ FDMA use separate carriers for each video channel. In large networks, between 4 and 8 video channels and their associated audio

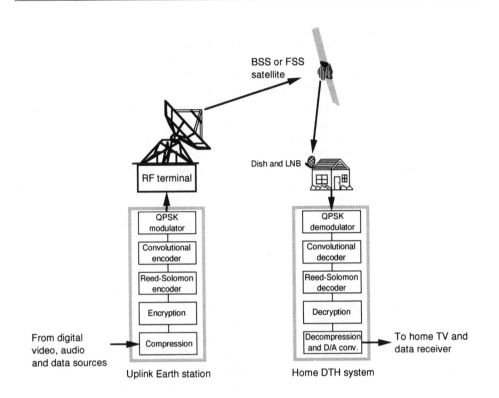

Figure 6.10 Digital DTH transmission system, indicating the baseband elements in the broadcast uplink and home receiver.

and data are combined using TDM onto a single carrier that would occupy the entire transponder bandwidth. The form of TDM could be either a fixed bit-by-bit multiplex or alternatively a statistical multiplex where data rates are adjusted based on content [6]. In either case, the carrier must be transmitted by a single Earth station as the multiplexing is done on the input side of the modulator.

The compression systems themselves fall into two categories: (1) those that comply with a standard, particularly MPEG-2 or DVB (this includes MPEG-2 as a component); or (2) those that use a proprietary algorithm and multiplexing scheme. Systems that started out in category (2) are quickly moving to MPEG-2 because of the rapidly decreasing cost of the receiving equipment. In the following subsections, we briefly review the offerings and capabilities of some of the leading suppliers.

6.5.1 Compression Laboratories, Inc.

As the leader in VTC systems and equipment, Compression Laboratories, Inc. (CLI) moved into the entertainment TV business by supplying the uplink compression equipment for DIRECTV and its DSS (which is based on MPEG-1). Their Magnitude product line of encoding and decoding equipment complies with key aspects of MPEG II, and in fact they lead the industry in putting it into operation for a major DTH network. Magnitude applications include cable headend delivery, DTH services, and VOD. It is related to their Spectrum Saver product line, which was introduced in Chapter 4 under distance education. Magnitude systems were installed by Thomson Consumer Products, the supplier of the baseband ground segment for DIRECTV and USSB, Keytech S.A., Bell Atlantic, National Technical University, Argentina Jockey Club, Wescott Communications, and *Electronic Data Systems* (EDS), for their Spectradyne subsidiary.

6.5.2 NTL

Based in the United Kingdom and part of News Corp., NTL has developed a DVB-compliant MPEG-2 compression system. They have produced uplink systems using their encoding and compression equipment combined with conditional access technology from the market leader, News Datacom (NDC—another News Corp. subsidiary), and modems from ComStream of San Diego, CA. NTL supplied the broadcast center equipment for the satellite operator of Thailand, Shinawatra Satellite Corp., and STAR TV in Hong Kong. Their strong support for DVB puts them in excellent position to supply compression systems that integrate well with the coming generations of DVB-compliant IRDs.

6.5.3 General Instrument

General Instrument (GI) of San Diego, CA, has the largest market share in satellite DTH systems by virtue of their pioneering work with VideoCipher and its digital counterpart, DigiCipher 1. According to GI, DigiCipher 1 has a strong leadership position in terms of the number of digital video compression systems sold. As of January 1995, there were 57 encoders in service in eight countries, providing 310 compressed digital video channels. Some of their leading customers include PrimeStar (United States), Multivision (Mexico), Televisa (Mexico), HBO (United States), HBO Olé (Venezuela), PanAmSat (United States), TV Avril (Brazil), and Telefonica (Spain). Like CLI, they are an active participant in the U.S. HDTV standard effort. Recently, they started to move from the proprietary approach of DigiCipher 1 to one that adheres to MPEG-2 and ultimately DVB. GI is licensing their technology to Scientific Atlanta (discussed below), AT&T, Zenith, Hewlett Packard, Microsoft, and Toshiba.

6.5.4 Scientific Atlanta

Scientific Atlanta (SA) has a solid base of experience in satellite video, having developed the B-MAC system that was adopted for Australia's distance education network and is so popular in private broadcasting applications by EDS and others. The first digital compression developed by SA was unique in its concept, using pel-by-pel compression rather than the DCT. More recently, they started producing MPEG 1 and MPEG 2 compliant products for cable TV and satellite transmission. Their DTH equipment is installed in the Middle East, Africa, and Asia. There are also plans to provide DVB IRD equipment to the European market.

References

[1] Elbert, B. R. "Digital Direct to Home TV Broadcasting Via Satellite," *Cable and Satellite China*, Beijing, China: International Institute for Research, 1995.

[2] *Satellite Communications in Europe 1995*, London, England: CIT Research, 1995.

[3] "Eutelsat Could Buy French DBS Twins," *Cable and Satellite Europe*, March 1995, p. 10.

[4] *A Comprehensive Guide to INTELSAT Satellite Beam Coverages*, Washington, DC: INTELSAT, 1995.

[5] Howes, K. J. P., "Digital Video Via Satellite," *Via Satellite*, March 1995, p. 34.

[6] Meyers, J. J., "Expansion of Digital Direct-to-Home Satellite Services and Implications for the Broadcasting and Communications Industries of South East Asia," *ASEAN Satellite and Cable Conference*, Kuala Lumpur, Malaysia, April 1996, International Institute for Research, Hong Kong, 1996.

Part III
Telephone and Data Communications

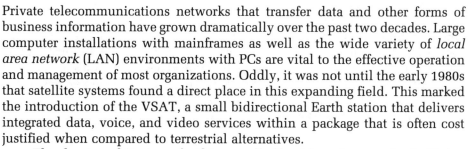

Data Communication and Very Small Aperture Terminal Networks

7

Private telecommunications networks that transfer data and other forms of business information have grown dramatically over the past two decades. Large computer installations with mainframes as well as the wide variety of *local area network* (LAN) environments with PCs are vital to the effective operation and management of most organizations. Oddly, it was not until the early 1980s that satellite systems found a direct place in this expanding field. This marked the introduction of the VSAT, a small bidirectional Earth station that delivers integrated data, voice, and video services within a package that is often cost justified when compared to terrestrial alternatives.

The classic architectures for this are called host-based processing (utilizing mainframe computers), peer-to-peer networks (usually employing minicomputers or large servers), and client/server networks (tying together personal computers, servers, and telecommunications networks). *Information technology* (IT) networks can serve basic administrative needs, like payroll processing and electronic mail (e-mail), or strategic needs, like a customer reservation system in the automobile rental business or a just-in-time inventory control system that ties a major customer to its network of suppliers. Today, terrestrial lines and switching in conjunction with VSATs provide a fast and effective mix to advance the competitive strategy of many medium to large businesses.

7.1 INTERACTIVE DATA NETWORKS

Data networks usually require a duplex connection so that information can be requested, selected, or exchanged. There is a wide variety of classes of data communications applications, leading to a very significant difference in the specific requirements for the type and amount of interactivity. Traditional host-based computer networks are perhaps the easiest to manage. More recently, peer-to-peer networks and client/server systems have replaced the host/mainframe approach in many organizations. On top of this, the specific nature of the data varies greatly. These factors make it impossible to generalize on the ideal architecture, data communications structure, or protocol. Instead, organizations must select the network architecture that best satisfies the needs of users and customers.

Modern data communications theory and practice is literally built upon the concept of protocol layering, where the most basic transmission requirement is at the bottom and more complex and sophisticated features are added one on top of another. The layering concept is embodied in the *open systems interconnection* (OSI) model and contained in relevant standards of the *International Standards Organization* (ISO) and ITU-T. As we move up the "stack," each layer above provides a standardized service, defined in the relevant protocol, to the layer immediately below. In this way, the details within the layer can be optimized for performance and isolated from the other layers. At the very top of the structure is the actual information processing application that requires the network in order to do its function. These principles are laid out in detail in our other work (see [1]); however, they are summarized in the following definition of the layers (which are normally arranged with layer 1 on the bottom, but listed here in ascending order for clarity).

- *Layer 1, physical:* provides the mechanism for transmitting raw bits over the communication medium (e.g., cable, wireless, and satellite). It specifies the functional, electrical, and procedural characteristics such as signal timing, voltage levels, connector type, and use of pins. The familiar RS-232 connector definition is a good example of the physical layer.
- *Layer 2, data link:* provides for the transfer of data between adjacent nodes or connection points either by a dedicated point-to-point line (e.g., a T1 private line or a satellite duplex link) or a medium capable of shared bandwidth (e.g., an Ethernet cable or satellite TDMA channel).
- *Layer 3, network:* responsible for routing information from end to end within the network, which would consist of multiple data link paths. This may involve decisions about the most effective route through the point-to-point links that comprise the network. A VSAT network may serve as one of these links and hence would have to interface properly with the network layer. Popular examples of the network layer are the X.25 protocol used internationally and the *Internet protocol* (IP) that is employed in the majority of router-based private networks.
- *Layer 4, transport:* provides another level of assurance that the information will properly traverse the network, from end user to end user. Two services are commonly available: connectionless, which transfers packets of data, one at a time, and connection-oriented, where a virtual circuit is first established before sending multiple packets that make up the entire conver-

sation. The familiar TCP layer of TCP/IP provides a connection-oriented service to computer applications.

- *Layer 5, session:* somewhat more complicated than layers 3 and 4 but provided to instill yet greater degrees of reliability and convenience of interface to applications. It manages the data exchange between computer systems in an orderly fashion to provide full-duplex or half-duplex conversations. One important service is that of reestablishing the connection in the event that the transport layer is interrupted for some reason.
- *Layer 6, presentation:* provides syntactic and semantic services to the application layer above. What this is saying is that the presentation layer is inserted to resolve the complexities between transport/network layers and the more simplistic needs of the actual application that employs the network in the first place. Some specialized services like encryption and data structure definition are considered to be part of the presentation layer.
- *Layer 7, application:* includes the actual data communication applications that are common in open systems, such as file transfer, virtual terminal, electronic mail, and remote database access. We refer to these as applications because they include not only the protocol elements that support specific types of information but also features and facilities that ultimately interact with the end user. Most nonexpert users will not use the application layer directly, instead relying on specialized software within the computer to improve the interface and functionality. For example, most subscribers to online information services use the e-mail package supplied by the provider. This package, in turn, will engage layer 7 electronic mail services to do the actual function of sending and receiving message traffic.

This was a brief introduction to the structure of modern data communications networks. The VSAT network is ideal for centralized computer networks, that is, those that employ a host computer. The majority of such installations are either from IBM or from a maker of compatible mainframes (Amdahl/Fujitsu and Hitachi). IBM offers the System Network Architecture (SNA) as the collection of data communication protocols. Following the layering concept, each computing or terminal device in the network has a unique address. SNA makes sure their information makes its way from source to destination so that users can employ a remote mainframe as well as smaller machines (minicomputers and servers) and send messages to other users. It by far represents the largest installed computer and software base; that is, most private line data networks employ SNA. The same can be said for VSATs because they are mainly being used as replacements for leased line networks.

A second important proprietary architecture is DECnet, from Digital Equipment Corporation, the largest manufacturer of networking minicomputers. DEC stole the lead in minis and created a highly functional peer-to-peer arrangement of computers. The network looks more like a mesh than a star. IBM has had to introduce peer-based computers, protocols, and applications and now offers some very effective alternatives to DEC. However, DEC still has a highly loyal customer base, particularly in technical computing. Because of the requirement for mesh connectivity, VSATs have not yet gained a strong hold in DEC computer networks.

Efforts to generate economy of scale in nonproprietary systems (i.e., the multivendor environment) have produced the OSI open systems architecture, sponsored both by ISO and the ITU-T. The most recognized subset of OSI is the X.25 network layer protocol. X.25 is found in many open systems and proprietary systems, including IBM and DEC. Both the terrestrial networks and VSAT developers have embraced X.25 because of its functionality and universal availability. It is being used more and more as a common format for exchanging data within and between data networks. Work continues on standardizing other aspects of networking above the basic network layer of X.25. Open systems applications such as file transfer, e-mail, and electronic data interchange (EDI) are available to solve complex networking problems in the international business environment. The fact that VSATs support X.25 infers that they will also be compatible, to a first order, with OSI applications.

A summary of the protocols in general use and their support over typical VSAT networks is provided in Table 7.1. Owing to the star topology of VSAT networks, the most common and successful protocol implementation is for SNA. Recently, the token-ring LAN (IEEE 802.5 standard) was added to accommodate the popularity of this technology for access by PCs, which largely have

Table 7.1
Network Protocols in Common Use for IT Networks and
Their Implementation Status Over VSATs

Protocol	Application	Availability on VSATs
SNA/SDLC	IBM host-based networks	Widely available
3270 BSC	IBM host-based networks (obsolete)	Available
Token-ring	IBM LAN installations	Available
Ethernet	LANs in general	Available
X.25	Wide area networks	Available
Asynchronous	Remote terminal access	Widely available
TCP/IP	Router-based WANs and Internet access	Available
DECnet	DEC peer-to-peer networks	Limited
Bit transparent	For unsupported protocols	Available

replaced IBM 3270 dumb terminals. Ethernet (IEEE 802.3 standard) is now a very popular access protocol, in support of the widest range of higher level protocols and networks that are now common in the LAN environment. Because of its worldwide availability, X.25 is well established on VSATs as a way to provide flexible interconnectivity and reliable data transfer. Remote access by a variety of terminals and devices is afforded by the asynchronous protocol, which can interface user equipment with unspecified protocols and a requirement for occasional network connectivity. The popularity of *wide area network* (WAN) routers and the Internet have caused TCP/IP to become an important addition to the range of protocols. As will be covered later in the chapter, TCP/IP and Ethernet are often combined to create a private internet that is capable of supporting a very wide range of applications and computer types. Support for DECnet is available on a limited basis. This protocol does not easily fit into the VSAT architecture but nevertheless may need to be included. Lastly, a bit transparent mode can transfer packets of other protocols since it allows information to be carried in a transparent manner. This then relies on the performance of the VSAT access and internal network layer to reliably transfer information.

Satellite links can support interactive data applications through two fundamentally different architectures: point-to-point connectivity (also called mesh topology) and point-to-multipoint connectivity (also called star topology). Point-to-point connectivity applies to either temporary connections or dedicated links to connect pairs of Earth stations. Temporary connections emulate dial-up service over the telephone network as well as connection-oriented data communications services. This type of service is common to the fixed telephony networks discussed in Chapter 8. TCP is effectively used on a point-to-point basis for common Internet services like FTP, SMTP, and access to the World Wide Web. Dedicated point-to-point links are useful for telephone trunks and broadband data lines that connect LAN segments together. Alternatively, point-to-multipoint connectivity includes a hub as the center of the star through which all communication passes and is controlled. In general, the details of data speed, protocol support, session management, and data protection can be tailored to the specific data communication application that the user wishes to employ.

7.1.1 Point-to-Point Connectivity (Mesh Topology)

The first satellite networks to be implemented were employed for point-to-point connectivity. Over time, this topology was less generally applied but remains an effective means of transferring information with minimum delay between pairs of points. As illustrated in Figure 7.1, node 1 in a point-to-point service conducts a full-duplex conversation with node 2 (shown with heavy

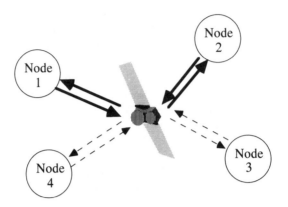

Figure 7.1 Satellite network arrangement for point-to-point connectivity (mesh topology). All stations must be capable of transmitting and receiving the same power level since connection between any pair is allowed.

arrows), and node 3 does likewise with node 4 (shown with broken arrows). In a true mesh network, all connectivities are possible and provided; that is, node 1 and node 3 communicate directly as do node 2 and node 4. The potential number of connections is equal to the permutation $N(N-1)/2$, where N is the number of nodes. The satellite provides separate bandwidth and power for each of the conversations, either by reserving spectrum (FDMA) or time (TDMA) for the duration of the particular interaction. The link must support transmissions between any pair of terminals; hence, the design must be balanced in terms of critical parameters like Earth station G/T and EIRP. The duration of a connection could be limited, as in the case of a dialup call or a virtual circuit, or a fixed duration, such as a permanently assigned channel or dedicated circuit. A mesh network is potentially the most versatile because it allows "any" to "any" communications. On the other hand, it must be managed since users can initiate connections on their own without central coordination.

Interactive mesh networks are implemented both at C and Ku bands, depending on the availability, pricing, and technical requirements of the particular application. The factors that lead to the selection were covered in Chapter 2. Link capacity is usually balanced, meaning that the same speed is used in both directions. This ranges from about 64 Kbps on the low end (for medium-speed data, voice, and low-quality VTC) up to the T1 or E1 speed (1.544 Mbps and 2.048 Mbps, respectively). In some cases, users have implemented 45-Mbps links as backups for terrestrial fiber optic transmission systems. Using the principles given in Chapter 2, it is possible to determine the proper antenna and HPA sizes on the ground as well as the required amount of satellite bandwidth and EIRP. Chapter 8 contains a more detailed discussion of the mesh topology as applied to fixed telephony services.

7.1.2 Point-to-Multipoint Connectivity (Star Topology With VSATs)

The point-to-multipoint connectivity is illustrated in Figure 7.2. The thick, shaded arrows represent the digital broadcast "outroute" from the hub to the remote nodes. It contains all outbound data in the network. The thin lines represent the "inroutes" from the individual remote nodes. Several VSATs share each inroute frequency and so must be separated in time (TDMA) or code (CDMA). All information transfer is between remote node and the hub; direct single-hop remote-to-remote data transfer is not possible with this topology.

The link design is imbalanced to simplify the VSAT design and thereby reduce its cost. An example of a typical link design for the star network is presented in Table 7.2. The antenna diameters of the VSAT and hub are assumed to be 1.2m and 6m, respectively. This results in a gain difference of 14 dB in favor of the hub. The imbalance in the link design is what allows small VSAT antennas to be used in the first place. The overall link C/N is the same for both the outroute and the inroute, assuring more or less balanced performance in the presence of rain attenuation. As a result of the star architecture, all communication passes through the hub, whether it is destined there or not. It is the

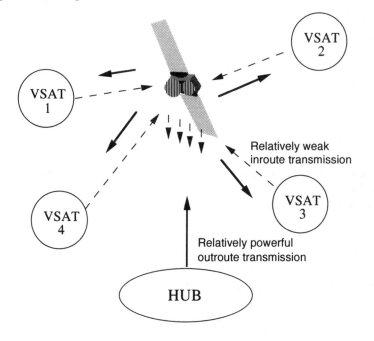

Figure 7.2 The star topology used in VSAT networks, where all communication and coordination is through the large hub Earth station.

Table 7.2
Example of a Ku-Band Link Budget for a VSAT and Hub Star Network

Link Component	Outroute	Inroute	Units
Uplink antenna	6.0	1.2	meters
Antenna gain	56.9	42.9	dBi
Transmit power	12.0	5.4	dBW
Uplink EIRP	68.9	48.3	dBW
Path losses	207.0	207.0	dB
Spacecraft *G/T*	2.0	2.0	dB/K
C/Tup	−136.1	−156.7	dBW/K
Bandwidth	500.0	100.0	kHz
	57.0	50.0	dB(Hz)
C/Nup	35.5	21.9	dB
Downlink EIRP	25.0	4.4	dBW
Path losses	205.7	205.7	dB
Downlink antenna	1.2	6.0	meters
Antenna gain	41.5	55.5	dBi
System noise temp	120.0	120.0	K
Earth station *G/T*	20.7	34.7	dB/K
C/Tdn	−160.0	−166.6	dBW/K
Bandwidth	57.0	50.0	dB(Hz)
C/Ndn	11.7	12.0	dB
C/Nth	11.6	11.6	dB

*The link has been designed to provide the same overall performance in the outroute and inroute directions

equivalent of a star topology used in many host-based data processing environments. Networks of this type are highly coordinated and can operate very efficiently. The disadvantage (which may actually not apply in many important cases) is that direct remote-to-remote communication is on a double-hop basis through the hub.

The point-to-multipoint architecture is very common in modern satellite data networks and is the basis of the success of the current *very small aperture terminal* (VSAT). A VSAT is a complete Earth station that can be installed on the user's premise and can provide business communication services. For the purposes of this book, we regard a VSAT as the remote Earth station in the star network architecture. An example of a typical VSAT with a 1.2m antenna is illustrated in Figure 7.3. The antenna feed has an RF head attached to it that contains the transmit-receive portion of the terminal. Cables connect between the RF head and the indoor unit, which is a complete baseband system contained in a cabinet about the size of a PC.

Figure 7.3 A typical 1.2m VSAT antenna installation and type of indoor unit that connects to the user computer or PC.

7.2 VSAT STAR NETWORKS

Businesses employ VSATs primarily as replacements for terrestrial data networks using analog private lines in a variety of industries, including retailing, automobile sales and distribution, banking, travel and lodging, and finance. Perhaps the first major installation was for Wal-Mart, a leading U.S. discount retailer with stores throughout the 48 contiguous states. Today, there are over 100,000 two-way VSATs installed in the United States and over 200,000 worldwide. While many organizations are considering VSATs instead of unreliable or expensive terrestrial alternatives, the technology should only be used if it makes sense. For example, a company needing to connect only five fixed locations to a data center would find that conventional VSATs would not be cost effective. Likewise, a large industrial organization that needs high-capacity links between major sites is not a candidate for existing VSATs. This would clearly be a better application for fiber optic links, if that were feasible.

7.2.1 Applications of Star Networks

Many centralized companies build their IT systems around the host computer that is typically located at the headquarters. This is an ideal starting point for VSAT network adoption. We offer the following examples from early experience with VSAT star networks.

Wal-Mart used VSATs to extend its reach to thousands of remote towns in rural America. Without VSATs, Wal-Mart would have great difficulty integrating their systems of distribution, credit verification, training, and financial management.

Chrysler Corporation, one of the Big Three U.S. auto makers, and regarded as an innovator in design and engineering, has provided every one of their 6,000 U.S. dealers with a VSAT to be used for order entry, financing, parts inventory management, distance education and training, and other applications that get added as time goes on. The VSAT network allows Chrysler to treat every one of its dealers equally, which is important because they are usually individually owned and operated by local business people.

Toyota of America's Lexus Automobiles, the most successful new luxury car line in the United States, uses their Lexus VSAT network as part of a strategy to give the customer the impression that there is one dealer and service organization operating in North America. A Lexus purchased in Torrance, CA, can be taken for service in Falls Church, VA, where all of the maintenance records are immediately available over the satellite link. All customer information is right in front of the service representative, just as if the car had been purchased in Falls Church instead of Torrance.

GE Information Services, a leading provider of virtual private data networks and online services to corporations, is very strong in the United States and holds a major position as a global *value-added network* (VAN) provider. As part of their interconnectivity strategy, GEIS can provide a VSAT to its customer in situations where public network access is unsatisfactory. The hub is located in Maryland and can reach any domestic user on a single hop. The bandwidth available over the satellite link is up to 64 Kbps, which is very competitive with narrowband ISDN service.

Holiday Inn, one of the largest up-scale hotel chains in the United States and worldwide, for its U.S. market, uses VSATs to connect between headquarters and the larger properties for access to the reservation systems and other applications for business management.

Banco Nacional de Mexico (Banamex), the largest financial organization in Mexico, relies on its VSAT network to reliably communicate with its branch offices. Tellers are supported with terminals that can connect back to the head-quarter's computers and databases. The system allows Banamex to provide consistent services to its clients whether they are in Mexico City or a small outlying town.

VSAT networks arose in the mid- to late 1980s as a result of electronic and software innovations that permitted all of the necessary features to be contained into an affordable package about the size of a personal computer. By affordable, we mean that the user pays a total network cost that is competitive with many terrestrial data network alternatives. This includes the cost of owner-ship of the VSATs and possibly the hub, along with the rental of adequate space segment capacity. The price of a VSAT itself started around $20K, dropping over the years to a level of around $6K at the time of this writing. The specific price of a VSAT depends on its configuration, particularly the quantity and type of port cards (which connect to user devices). In coming years, greater integration and large-scale production suggests that the VSAT of the twenty-first century will cost about the same as a typical PC. This will permit much larger scale application of two-way VSAT technology, as discussed at the end of this chapter.

7.2.2 VSAT Network Design

The VSAT antenna is typically in the range of 1.2m to 2.8m, the size depending on the satellite coverage performance (G/T and EIRP), the capacity requirements, and the frequency band. Common practice is to use Ku band because of the following factors.

- In many cases, Ku band is not shared with terrestrial microwave, and hence VSATs can be placed where needed to provide service. No frequency coordination is needed, and licenses can be granted on a network basis.
- Satellites at these frequencies usually have higher RF power and EIRP to allow the smallest antennas on the ground and to overcome the increased rain attenuation at this higher frequency.
- Adjacent satellite interference is less for the same Earth station antenna size because main lobe gain increases while sidelobe gain does not (for the same offset angle). This is taken advantage of by recognizing that a smaller Ku-band antenna has about the same interference potential as the larger antenna used at C band.

A smaller Ku-band antenna can be placed in an inconspicuous place such as on the roof of the building or behind a low screen wall. The transmit power level, being below 10W into the antenna, would not be a hazard to humans for these types of installations. C-band VSATs are available in areas without ade-quate Ku-band satellite capacity; however, the user must usually carry out the necessary frequency coordination prior to use in order to protect existing terrestrial microwave stations. These aspects are covered in Chapter 10.

Major satellite companies and telecommunications equipment manufacturers have diversified into VSAT equipment and services. Because potential customers are really users of telecommunications rather than technology managers, VSAT marketers must make a complete service offering that includes equipment, installation, maintenance, and, on occasion, hub and network operations. This requires that entrants in this market be prepared to become more vertically integrated than previously required for satellite operations and equipment sales.

The architecture of the typical VSAT star network is provided in Figure 7.4. This emphasizes the configuration of the ground equipment and ignores the techniques used for modulation and multiple access. It depicts how the user connects computers, PCs and other terminals, PBX and telephone systems, and video equipment used in private broadcasting. The hub of the star is shown on the right in the form of a complete Earth station facility with a relatively large antenna (typically 6m at Ku band and 9m at C band). The most common implementations of the star network use TDM on the outroute and TDMA as well as a derivative called ALOHA (to be discussed later) on the inroute. The

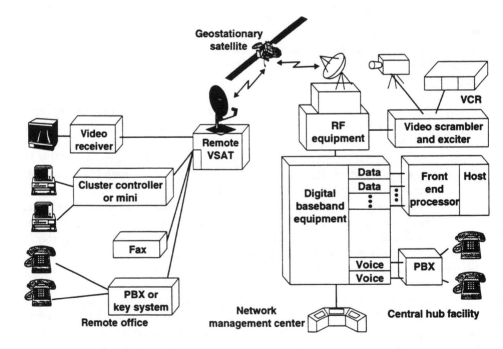

Figure 7.4 Architecture of a typical VSAT star network (courtesy of Hughes Network Systems).

hub and the entire network of VSATs can be managed from the single network management center attached to the hub.

7.2.2.1 Hub Equipment

The hub is the origination point for the outroute carrier that is received by all VSATs in the same network. It has a fixed capacity and transmits information as a constant stream of data using time division multiplexing. In addition to user data, the outroute also includes control information that allows the hub to manage each VSAT and thus control all network resources. Outroute data rates are in the range of 64 Kbps to 2 Mbps, where the rate depends on the total network capacity and the architecture of the particular vendor's equipment. A typical value is 512 Kbps, which is a good compromise between equipment cost, throughput efficiency, and space segment expense. If space segment charges dominate the evaluation of whether or not to use VSATs, then it might be preferable to reduce the outroute data rate to 128 Kbps or even 64 Kbps.

The RF equipment is fairly conventional, usually consisting of a redundant set of LNAs attached to the antenna feed and TWT power amplifiers in the range of 80W to 400W, depending on satellite G/T, capacity requirements, and a need for extra uplink rain margin. Up and down converters provide a 70-MHz IF interface to the indoor baseband equipment. High powers and voltages exist at this equipment, so it is important that it is protected from casual contact and the elements. Also, it should be possible to work on the equipment even during heavy rain (Murphy's Law would indicate that the time when you need to service the equipment also happens to be during a rainstorm). An added feature in Ku-band installations is UPC, which increases power during heavy rain to maintain the link.

The heart of the hub is the baseband equipment, which performs all of the protocol conversion, multiplexing, modulation, and multiple-access functions needed to manage traffic over the entire network. The customer's host computer connects directly to a data port in the same way as it would to a terrestrial data line operating at 64 Kbps or some multiple thereof. Hubs can also support telephone service within the confines of the star network. In this case, human speech is compressed from 64 Kbps down to either 16 Kbps or 5.6 Kbps, depending on the requirements of the customer. Either speed will work, but the higher speed provides a more natural-sounding service comparable to what is available on the public telephone network. Even still, purchasers of VSAT networks prefer the 5.6 Kbps rate as it reduces space segment costs. The hub provides demand assignment features, as will be discussed later, so that calls are set up and taken down as needed.

The third capability offered by the hub is video broadcasting using either live or recorded material. Generally, the video transmit capability is separated from the rest of the station because of the greater RF power required. Also, many companies rent time on video uplinks since they generally use this capability only on occasion.

The hub provides the central coordination and management point for the entire network. Control of the VSAT network is extremely important to the overall user or network operator. The basic approach is to use the hub as the coordinating point and to provide specialized network management functions over the same links between the hub and each remote VSAT. The inbound link provides the path for status information from the VSAT back to the management system at the hub. A network management workstation is connected to the hub equipment to allow the network operator to configure the network for service and to monitor its performance over time. This allows the operator to be proactive, meaning that the network is always under surveillance and every remote is monitored for its activity and potential for malfunction. If problem arises, then it is possible to isolate and resolve it in a relatively short time. The remote VSATs themselves do not require local operator support as long as the link is functioning.

7.2.2.2 Remote VSAT

A full-capability remote VSAT is shown at the left of Figure 7.4. The primary function is that of a communications multiplexer for the local PCs, terminals, and phones. We see that the customer's cluster controller (an IBM term) or end computer system such as a server is connected to a data port on the VSAT indoor unit. User terminals or PCs are then connected to this device. We consider the connection to a LAN in the next section. In this particular case, the host computer polls the remote devices through the link to the cluster controller as if they were connected to it over a terrestrial network.

The baseband equipment and the VSAT perform the necessary protocol conversion, an important part of which is to fool the end computer systems into thinking that the polling is happening without the satellite delay. The term used for this is protocol *spoofing*, which operates in the following way. The host computer uses a polling technique to allow each remote device to obtain service. The remote device does not, on its own, initiate contact with the host. If we were to allow this dialog to be conducted over the satellite, then each poll would take at least four satellite hops to complete, or a total of one second. The hub baseband equipment simulates the polling response of the remotes, while the VSAT indoor unit simulates the polling command. In between, the protocol over the satellite link takes care of reliable data transfer, delivering the actual information from source to destination. The delay encountered is principally that of a single hop (about one-quarter second) or at most a double hop (about a half of a second). Spoofing along with reliable data transfer are the techniques that make the satellite network perform as well or better than the terrestrial telephone network.

Other services can be implemented at the VSAT to correspond to the capabilities offered back at the hub. If telephone service is offered, then the VSAT can connect to a PBX or key system to allow placing calls over the satellite to the PBX at the hub. VSAT-to-VSAT calls are generally avoided as they experience a double hop. The video receiver would be included if the site is to be part of the private broadcasting network. This is a TV receive-only system, so any response would use the narrow bandwidth of either the data or telephone facilities of the network.

The typical VSAT is designed to be unattended, requiring only limited attention by customer personnel, who only need to worry about maintaining power and occasionally inspecting the minimal display on the indoor unit. Problems can arise within either the indoor unit or the RF head, which can render either the satellite link or the local connection insufficient to maintain service. These problems are relatively infrequent because of the simplicity of the electronic design and the structure of the satellite network. This situation is usually detected by the hub Earth station and corrected by a competent field technician that would be sent to the site.

7.2.3 Integrator of PCs, Local Area Networks, and Internets

Data communication networks have grown far beyond the realm of the host computer, propelled by the popularity of three facilities that are now common-place: the PC, the LAN, and the Internet set of protocols. The latter is actually more than a set of protocols, since it now comprises many public and private networks worldwide. VSATs have worked to keep pace with these develop-ments, motivated by the extensive use of IT facilities in literally every organiza-tion in every developed and developing country. We consider in the next subsections how the three facilities can be supported by VSATs.

7.2.3.1 Personal Computer Integration With the VSAT

The PC is the ideal direct user interface with the VSAT in applications where online information delivery is required. Typical telephone networks have a real throughput of about 20 Kbps, not considering data compression found in 28.8-Kbps modems. This was once adequate for applications such as online service connection, dialup terminal access to e-mail, and fax. With the advent of the World Wide Web and the increasing demand for the transfer of large files for graphics and database replications, the analog telephone network ceases to be adequate. The marketplace is ready for VSAT networks that have typical throughputs in the range of 64 Kbps to 2 Mbps and more. This bandwidth can be delivered to a single user, being able to download a 1-MB file in seconds rather than minutes. It is more common, however, to multiplex together several channels and blocks of information for delivery of data to thousands of subscribers. This implements what has been called videotex but will become more interactive over time.

A classic example of using the point-to-multipoint feature and the PC is shown in Figure 7.5. The data files or streams are uplinked from a hub Earth station that is shared by all users. A public videotex service would have the hub owned and operated by a service provider or common carrier. Information is delivered to the hub over backhaul circuit(s) from database(s) or other information sources (e.g., a stock market ticker). Subscribers purchase and install a low-cost VSAT, which, to be effective, need only receive the high-speed forward link broadcast from the hub. Some existing and potential applications of this medium are listed in Table 7.3.

The interface for most data applications between the PC and the VSAT is through the standard RS-232 connection. A separate modem or terminal adapter is not required because the information is already bidirectional and at baseband. Client software in the PC can select information from the stream based on a variety of criteria selected by the user. This can be displayed in real time. If it is a file transfer, then the block of data is loaded into RAM and possibly saved to the hard drive. This type of application has actually been around for several years but lacked both the throughput afforded by a megabit-per-second data

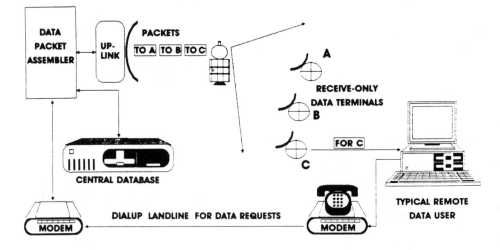

Figure 7.5 Using a data broadcast to integrate the PC into a VSAT data communications network.

Table 7.3
Typical Applications for Integration of the PC With the VSAT, Utilizing Data Broadcasting

Application	Example	Availability
Data broadcasting	Stock market ticker; commodity markets	Existing in U.S. and other major economies
	Wire services and electronic newspapers	Dow Jones, Reuters, and others available
	Airline flight schedules, arrivals, and departures	Used privately by airlines; available to the public over wire networks
	Traffic information	In some large cities over radio towers
Interactivity through PSTN	E-mail	Private networks
	World Wide Web access	An ideal application
	File transfer	Private networks
	Download of computer games	Developing application

rate and the demand for the service. The fact that microprocessor speeds and other capabilities have expanded over the 1990s also contributes to the growing demand for wider bandwidths than the typical telephone network is capable of supplying.

The quality of transmission can be very high because of the simplicity of the radio link from the satellite. The error rate can be maintained at less than 10^{-10} using forward error correction, which means that there would be one bit error per gigabyte of data received, on average. This is enough data to fill a typical hard disk drive found on a new PC at the time of this writing. The typical telephone modem link does not perform this well but does provide automatic retransmission request. The basic satellite delivery system operates much in the same manner as the digital DTH systems discussed in Chapters 5 and 6. This suggests that DTH operators are in a position to offer advance data delivery services to PCs at any time they care to enter the market.

Readers are probably wondering how some interactivity can be added. Figure 7.5 offers the simplest and least costly solution—the PSTN. The way we use the PSTN is as a request channel, which means that the amount of information is relatively small. Also, most of the time is taken up with the user either thinking or slowly typing. As a result, there are several good reasons for using the PSTN where it is available, such as:

- Low cost of access, including service charges and the cost of a standard modem (most PC owners have a modem for fax and online access);
- Low cost of return data transfer (assuming short requests);

- Relatively low delay or latency;
- Wide availability in developed countries and major cities in developing countries.

Data broadcasting has been around for a decade or more, pioneered by Equatorial Communications, Inc. Perhaps the most familiar service available in the United States at the time of this writing is DirecPC, an offering of hardware, software, and satellite service delivery from Hughes Network Systems (HNS). As the leader in VSAT technology and market share, HNS was in the best position to bring a high-performance data broadcasting service into being. It implements a 12-Mbps outroute transmission from a Ku-band satellite operated by sister company, Hughes Communications, Inc. (HCI). A description of the equipment provided for the service is shown in Table 7.4.

Two distinct services are made available on a pay-as-you-go basis, namely, Digital Package Delivery and Turbo Internet. Both are provided with full DES encryption security and conditional access features. From the antenna, the received carrier is delivered to an internally mounted PC board that demodulates, decodes, and demultiplexes the data. This makes it immediately available within the PC to the processor and memory. The natural integration of the

Table 7.4
Specifications of the DirecPC Equipment to be Installed at the User Site

Element	Feature	Characteristic
Receive antenna	Diameter	60 cm (24 inches)
	Frequency range	11.7 to 12.2 GHz
	Polarization	Linear
	Cable connection	Single RG6/U coax
PC internal card	Type	ISA bus card, 16-bit ISA standard
	Bit rate	11.79 Mbps
	Input frequency range from antenna cable	950 to 1450 MHz
	Security	DES-based conditional access
Software	Operating system	MS-Windows 3.1 or Windows 95
	Software packages	Antenna pointing utility Package Explorer Turbo Internet Device Driver Digital Package Delivery
PC hardware requirement	Type	IBM-compatible
	Processor	Intel 486/66 MHz
	Configuration	Multimedia

DirecPC data stream and the PC hardware and software produces an efficient system for the delivery of broadband data applications in a readily usable form.

With the Digital Package Delivery application, you may download files at about 100 times the speed available over the public telephone network. Any size file may be transferred on a broadcast basis to an unlimited number of DirecPC receive points. The inroute side of the link uses the terrestrial network and is accessed through internal software and a conventional analog modem. This takes advantage of the natural integration of satellite, PC processing, and modem/terrestrial access. The satellite port access speed is 12 Mbps, making this suitable for software download and maintenance, electronic documents, photos, videos, movies, and multimedia training. It can be adapted to an electronic kiosk where customers can request a download of software or other data (music and image) where it is immediately copied to CD and delivered to the customer.

Perhaps more exciting is Turbo Internet, a 400-Kbps satellite connection that provides access to the highly popular World Wide Web that is 14 times faster than a standard 28.8-Kbps modem and telephone line. This service can download Web pages and data faster than either an ISDN line or fractional T1 service operating at 384 Kbps. The process for using Turbo Internet is as follows:

- Dial your Internet access provider (HNS provides this as an option, but there are literally hundreds of possible providers in the United States alone).
- All keyboard and mouse actions are transferred over the terrestrial line.
- All download requests are fulfilled over the DirecPC satellite link instead of the telephone line.

From this discussion it is easy to see that PC integration is possible and attractive on a standalone basis. We have already broken the consumer price barrier with the DTH antenna and will see how IRD mass production techniques will be applied to the indoor unit. The next steps involve the identification and implementation of the data communication applications that appeal to the mass market. This could involve the highly recognized World Wide Web or perhaps the long-awaited electronic newspaper.

7.2.3.2 Integrating LANs With VSAT Networks

The LAN represents the foundation of nearly every IT strategy because of its versatility and cost/effectiveness. LANs are found in small companies comprising only a few employees and all the way up to the largest organizations engaged in global business. The challenge that IT managers continue to face is how to interconnect LANs into an enterprise data communication system akin to what has been possible with the largest mainframes from industry leader IBM. The door is open to VSAT networks, provided they can meet very stringent usability, performance, and economic objectives.

The internetworking of LANs into the WAN is discussed in enterprise environments in our work [1] as well as other references such as [2]. These systems are heavy users of the Internet series of protocols, that is, TCP/IP. In contrast to proprietary architectures and protocols, like IBM's SNA and Digital Equipment Corporation's DECnet, TCP/IP is completely open and vendor independent. The protocols themselves are in the public domain, and much of the networking software is available free of charge over the Internet or by license for a very low cost. The cost and complexity come in through the actual applications that are sold by leading vendors like IBM/Lotus Development, Novell, and Microsoft.

As a review, the fundamental pieces of networking equipment that are used by all organizations to interconnect LANs are the bridge and the router. These devices are produced by several suppliers in the United States, particularly Cisco Systems, Bay Networks, 3Com, IBM, Hewlett Packard, and Digital Equipment Corp. Users of IT capabilities implement client/server architectures using these devices to build a private WAN or to access much larger networks such as the Internet, GEIS, Infonet, or Sprintnet. Through the use of stored-program processors, the bridge and router intelligently transfer information between LANs on more or less a seamless basis. Management of a WAN is provided through a series of standards for network management, including the *simple network management protocol* (SNMP). Organizations purchase network management workstations from companies like Hewlett Packard's OpenView and IBM's NetView that use SNMP to gather relevant information and control many of the devices and systems in the WAN environment. Support for SNMP is growing rapidly but remains a future possibility with respect to the mainstream of satellite communication networks.

We first review the application of the bridge and router before going into detail on how VSATs will support a WAN strategy. The bridge, shown in a typical application in Figure 7.6, provides the most basic connection between pairs of LANs at the link layer (layer 2) of the OSI model. The bridge does just what its name implies; it provides a point-to-point connection for packets of data that must be delivered from one LAN segment to the other (in both directions at the same time). The intelligence in the bridge reads the link layer address (called the medium access control (MAC) address) to determine if the destination is local or remote. If it is remote, then the packet is transmitted over the link to the bridge on the other end where it is delivered onto the remote LAN segment. From there, it is carried to the appropriate station.

Packets that a particular bridge recognizes to be local are not retransmitted. One of the most popular techniques for accomplishing this is called *learning*,

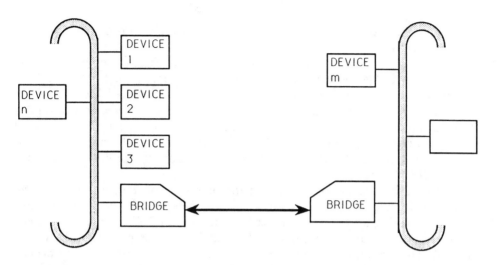

Figure 7.6 A bridge is a device used to connect two or more LAN segments together, using the link layer protocol of the data communications architecture.

wherein the bridge learns what stations are active and builds an internal table of the corresponding local addresses. This is simply done by recording the sending address of packet traffic over time (it may only take a few seconds to identify, say, 90% of locally active stations). Packets that are addressed to local stations are not applied to the transfer link, while any "unfamiliar" addresses are so transferred. This function, also called filtering, reduces outbound traffic and loading of the remote LAN and provides a security feature as well (i.e., blocking certain local addresses or remote addresses from being connected over the bridged link).

There are a number of operational consequences of the typical implementation of a bridge, using OSI layer 2. Because the bridge is a link layer device and only looks at the corresponding Link layer addresses, it is specific to the particular LAN design (i.e., it can only connect LANs that are alike). Thus, an Ethernet bridge must connect between Ethernet LANs; likewise for token-ring and likewise for fiber distributed data interface (FDDI) LANs and backbone systems. This focus on layer 2 has the important benefit that protocols like TCP/IP and SNA are essentially invisible to the bridge. It merely transfers the MAC layer packets with the higher layer protocol treated like user data.

Modern bridges can connect several LANs together to form a ring or near-mesh network. However, they still operate at the MAC layer and therefore cannot "see" the higher layer protocols. This precludes the pure bridge from being used as an access point to a TCP/IP internet, a DECnet, or an SNA network. The router, on the other hand, is the appropriate node device that provides the access protocol support and the ability to manage a major portion of the network environment as well.

A typical application of the router is shown in Figure 7.7. While looking a lot like the bridge application in Figure 7.6, the important feature of the router is that it can create a versatile WAN environment. This will include private links as well as public data network services. The router operates at layer 3 of the OSI model, which means that it transfers packets based on the network layer protocol. Examples of such protocols include, of course, IP (from TCP/IP), DECnet, and IPX (from SPX/IPX, which are the protocols employed in Novell NetWare).

The link layer is not a factor in the operation of the router; hence, the router can transfer packets between different LAN types. For example, the router on the left in Figure 7.7 can connect to an Ethernet LAN and route IP packets to the WAN. The connection to the WAN might be through FDDI or by way of a T1/E1 point-to-point cable connection. The WAN will route the IP packets to the appropriate router, suggested on the right of the figure. This could be a 64-Kbps line using a V.35 connection. The router, in turn, might transfer the IP packets through a token ring LAN and its associated MAC layer protocol.

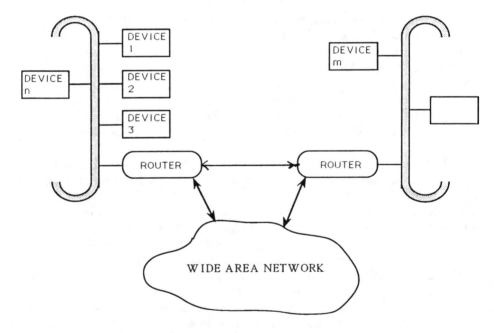

Figure 7.7 Routers are used to interconnect LANs to WANs and as the switching nodes within the WANs themselves. They employ the network layer protocols to assure reliable data transfer from end to end.

The most common routers found in business applications are called multi-protocol routers because they have software that permits the device to simultaneously route two or more protocols. A typical multiprotocol router could service different applications over the LAN/WAN environment even though they individually use different protocols. For example, the two LANs in Figure 7.7 might have Novell NetWare servers and clients that can access each other through the SPX/IPX protocols, while some UNIX workstations on each side might employ TCP/IP to exchange graphics files. The routers treat the protocol data individually, placing the packets on the shared access line to the WAN. Figure 7.7 is quite simplified, and it is more common to build up a mesh type of WAN using routers at each LAN end of the network. The router selects a particular output line based on the type of protocol and the destination address of the packet.

We are now prepared to examine how a VSAT star network can create an appropriate WAN environment for interconnection of standard LANs. Figure 7.8 shows how a VSAT star network would be used to interconnect both token ring and Ethernet LANs into a WAN environment. As will be evident, the functionality provided is similar to that which could be obtained with bridge connections. All of the remote LANs and their associated indoor VSAT units fulfill the role of a WAN with a star topology; that is, the satellite network emulates a star with dedicated connections between the remote bridges and the bridge at the host LAN.

As a specific example of LAN-to-LAN implementation, we consider LANAdvantage, which is a product from HNS that operates at the MAC sublayer of the link layer. This approach allows the VSAT network to transfer the network layer and higher packet data on a completely transparent basis. Thus, LANAdvantage can support the widest array of network architectures and protocols that are found in client/server environments. The performance for TCP/IP and SNA applications is emphasized in that nearly the full throughput of the inroute and outroute are made available. This is further enhanced by filtering software at the remote and hub that can recognize IP and SNA addresses.

LANAdvantage is a specific implementation of LAN connectivity with a VSAT star network. To give a feel for the capabilities of this type of product, we give more details of the services and options that it offers.

MAC sublayer support for both token ring (IEEE 802.5) and Ethernet (IEEE 802.3) at both the hub and remote sites allows SNA/SDLC data to pass from a token-ring LAN over the satellite to an Ethernet-connected host at the hub. The same can be said of TCP/IP services that would otherwise have required routers to achieve the same result from a terrestrial WAN.

Message switching enables efficient support of remote-to-remote communications for remote traffic on the same hub. In this way, the star topology is

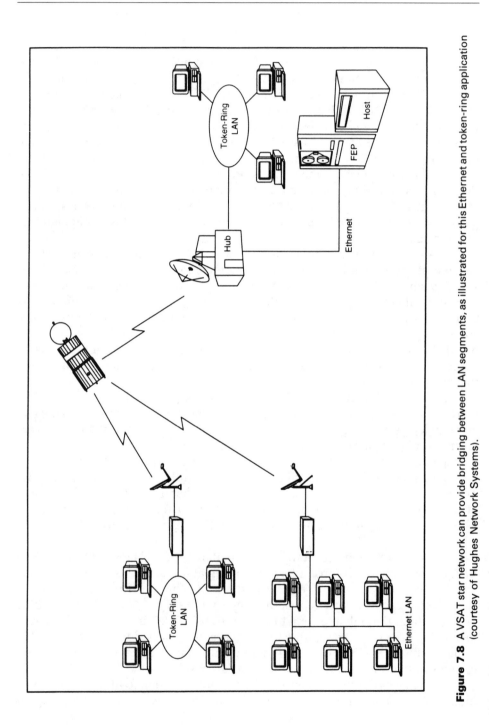

Figure 7.8 A VSAT star network can provide bridging between LAN segments, as illustrated for this Ethernet and token-ring application (courtesy of Hughes Network Systems).

made to perform the functions of a full-mesh network cloud. The extra delay from a second hop and hub queuing could be acceptable since the data are transported at the inroute speed (128 Kbps) with minimal overhead.

Protocol filtering determines that only those protocols marked as permitted in the protocol list to pass from hub to remote LAN or from remote LAN to remote LAN. It provides a simple way to implement a "firewall" that prevents local users from making outside connections and conversely protects a given location from intrusion over the VSAT network. Protocol filtering may be enabled or disabled on each direction, independently.

Packet filtering only forwards those frames over the satellite link whose destinations are not known to exist on the attached LAN. This is the basic function of a bridge, as discussed earlier in this section. Frames with specific source addresses can also be specified to be forwarded or discarded. Filtering can be based on learned addresses or addresses that are configured (i.e., downloaded by the operator).

Address learning means that addresses on the LAN segment are monitored and learned, reducing network reconfiguration requirements when stations are moved. Address learning is inherent in the learning type of bridge, so there is no need to purchase these devices if LANAdvantage is employed throughout the network environment. Learning can be enabled or disabled.

Address aging means that addresses are removed from the address table when no packets have been generated by the station within a specific time-out period, again simplifying network reconfiguration. Aging can be enabled or disabled.

The question may arise about how this star network might support applications that need communication between pairs of remote LANs, without passing through the host. This must be accomplished in a manner that does not overload the outroute from the hub. The approach is for the indoor unit at the hub to redirect such packets to the single-destination LAN (not all remote LANs simultaneously). In the typical WAN environment, this is the job of the router, not the bridge. Hence, the star VSAT network emulates the operation of the router even though this particular equipment need not be provided.

In cases where remote-to-remote traffic is heavy and where the double-hop delay is unacceptable, then it is better to implement a mesh topology. This type of arrangement, as discussed at the beginning of this chapter, implements direct point-to-point links that do not pass through a hub. Most mesh networks over a satellite now employ circuit switching, so the connection must be first requested before it is available for data transfer. The time needed to make the connection is typically of the order of several seconds due to the existence of multiple satellite delays and the necessary processing time to arrange the connection. Once the circuit is established, data can be exchanged with the delay of a single hop.

7.2.4 VSATs in Business TV

VSAT networks offer a special capability to introduce business TV into the environment at a relatively low cost. In Chapter 4, we reviewed the role of business TV as a means to tie organizations together for a variety of uses. Principle among these are distance education and teleconferencing. The latter requires a two-way interactive link, which is feasible with certain classes of VSATs, that is, the mess network architecture using TDMA or FDMA. Otherwise, all VSAT networks, star or mesh, can deliver video for private broadcasting and distance education with relative ease.

Business video has been touted and advertised as the medium of business advantage. In reality, it is used for some very specific purposes. It is estimated that there are about 100 private business video networks in operation around the world, primarily in the United States, Europe, and Japan. These reach a collective total of about 100,000 receive points. The rate of growth has been about 15% per year.

Most business TV installations have been additions to VSAT data networks, making use of existing antenna installations. However, there have been a few networks installed for business TV activity alone. Typically, the demand for it has come from executive management where the basis to proceed may have been to improve control. This does not mean that all business video applications are not economically justified. It will depend on the manner in which the network is implemented and operated. Once the network is installed, however, users become very cost conscious. This has given rise to the business TV service provider, the largest being EDS.

7.2.4.1 Video Teleconferencing

This application allows two or more groups to conduct a meeting. The medium provides duplex video transmission, so the groups get the impression of a face-to-face meeting in real time. The duplex link is a digitally compressed video channel using P*64 or switched 56-Kbps service. The first meeting rooms used in VTC service in the United States were specially designed and cost upwards of $1M. Today, the rooms are replaced with roll-around units that allow virtually any conference room to be employed for the VTC. With the work going on with the MPEG 4 standard, it is entirely possible that VSAT star networks might become capable of allowing two-way VTCs to be conducted without reverting to a mesh topology. An example of a VTC linkage is shown in Figure 4.8.

An adaptation of VTC is being applied to the field of medicine as a means to assist physicians in the field who need expert advice from specialists. This field of telemedicine employs a variety of telecommunication technologies to assist in the proper diagnosis and treatment of victims of accidents and rare sicknesses who otherwise would have had to be transported to a major city where the specialists reside. Paramedics in the United States may already practice telemedicine to the extent that they transmit EKGs from the scene of the accident to a hospital. This has been extended to allow a patient in Saudi Arabia to employ the expertise of a specialist of Columbia Presbyterian Hospital in New York.

At the top of the hierarchy is VTC since it allows the specialist to actually see the patient and speak to the on-site physician. The reverse direction is important so that the physician can better interact with the expert and also view medical image data after evaluation or processing. Hughes Electronics and the Telecomm Mexico cooperated in a telemedicine trial using VTC between a clinic in Chiapas and a hospital in Mexico City. This allowed the local physician to work with the specialist of his own nationality who was located at the best medical facility in their country. While telemedicine always seems to impress those who experiment with it, the lack of adequate telecommunication facilities remains a roadblock to extensive use. This is an area where VSATs can fill the gap once the applications are refined to the point where a cost-effective architecture can be introduced to the medical community.

7.2.4.2 Private Broadcasting

This is a private satellite video distribution network, constructed much like that used in cable TV and broadcasting for entertainment. Many of these networks were created to allow the CEO to talk to organization members and to hold nationwide press conferences. More important for the business, some companies have found ways to use their VSATs to deliver specialized broadcasts that contribute to corporate effectiveness. Examples include retailers (JC Penney), fast food servers (Domino's), computer manufacturers (HP and Tandem), and automobile manufacturers (GM, Chrysler, and Toyota).

An added feature is audio return using the satellite inroute link. This works similar to audience call-ins for radio and TV talk shows, but the line may stay up for the duration of a teleconference. This author participated in a private nationwide panel discussion with four other industry executives. We assembled in the studio of the Prudential Insurance company in New York and were uplinked to Westar IV. As many as 12 sites around the country received the signal for viewing at coordinated meetings of the *Society of Satellite Professionals International* (SSPI). After each speaker made his or her presentation, the floor was opened up for questions. People at the remote locations were in contact through the telephone network (the satellite network did not include return links).

These applications are only a sample of what is possible, based on what is actually going on in industry. One or more of these can be combined and implemented to create a unique capability that could mean success in a business context. This involves experimentation by the organizations, perhaps starting out with one or two applications that justify the value of installing the VSAT network. Then, more applications can be introduced as the original applications are modified or replaced. The idea is to continue to innovate.

7.3 CAPACITY PLANNING AND SIZING

The greatest challenge that the developers of VSATs overcame was to build a versatile and reliable data communication systems that could compete with existing terrestrial technology. This was achieved and still holds true, even as telephone networks are converted to digital technology. However, implementers of VSAT networks have their own challenge to face, that of properly sizing the network and maintaining its performance as requirements change. The purpose of this section is to provide some guidelines and examples of how to approach the problem of capacity planning and network sizing. This is not an exact science, as will be apparent from the discussion.

7.3.1 Collecting Requirements for the VSAT Network

We are taking the systems approach to this problem, which consists of determining the requirements, sizing the network, and determining overall performance against the requirements. The typical requirements for a data communication network fall into the following categories.

1. The *number of users and locations* that are to be serviced defines the geographic properties (physical distribution) and the associations (linkages) of the requirements. The output of this part of the exercise is the topology of the network.

2. The *traffic or information volume* that is offered to the network by each user or by an expected volume of users considers the particular amount of information as well as the timing of its occurrence (also referred to as its temporal nature).

3. *Throughput* is the amount of useful data that is transferred per unit of time, measured in bits per second, packets per second, or the like. This provides an estimate of the aggregate bandwidth required from the VSATs, hub and satellite.

4. *Time delay* (also called *latency*) is the amount of time required to transfer a specific amount of data. This is composed of line transfer time, access protocol time, propagation time, and node processing time. Interestingly, satellite system users focus on propagation time, which is only one of the

factors. If this is a circuit-switched type of network, then another time to be considered is the *call setup time*. All of these times are the sum of a number of contributors, some of which are constant and some of which are variable or random in nature.

5. *Response time* is measured from when a user initiates a request to when the response is displayed on the user's terminal device. This applies mostly to data communication networks where users employ PCs or other types of display terminals.

Each of these requirements has a definite impact on the capacity, complexity, and cost of the network, VSAT or otherwise. Requirements 1 and 2 are the basic inputs to the network design, defining the structure, distribution, and timing of the information to be carried and/or distributed among users. Believe it or not, this is the hardest thing to determine because in most situations it is simply not known with any precision. We will discuss these requirements in more detail later in this section. The remaining three requirements are performance measures that can be measured from the operating network. Throughput measures the actual quantity of data flowing between users over the network. What we are observing here is actual user data that reaches the end of the link for the particular computer application. In the process of carrying data, the network introduces overhead information that is needed to control and manage the network operation. This is not part of throughput. Another nonthroughput contributor is data that must be retransmitted due to congestion or errors on particular links.

The logical way to design a data network is to begin with the requirements for data transfer among the various users and points of operation. The best situation for this is if you are starting with an existing star network with a working host computer and many remote users. Typically, this would use terrestrial leased lines that connect from the remote sites back to the host. The architecture popularized by IBM actually connects the remotes together in a daisy chain that is referred to as a multidrop line. This is cheaper to implement and employs polling to coordinate the dialog between user terminal device and the host. The data requirements in such a star network can be collected at the host where all of the data can be observed for each remote and each leased line connection. Statistics on the information packets, their delays, and the overall response times are collected using monitoring software that is loaded into the host.

In a peer-to-peer or client/server environment, the network is far less structured and there is no single point of concentration where data characteristics can be gathered. Remote hosts and servers allow some of this information to be collected, as can monitoring devices on LAN segments and routers used to interconnect LANs. This information may be available at a network management station that employs SNMP. If necessary, technicians can make manual measurements using a protocol analyzer such as the Sniffer by Network General or the Lanalyzer by Novell. Information such as quantity of packets and their lengths, throughput, retransmissions, number of virtual circuits, and transfer delays are definitely of value in sizing the VSAT network. It is likely, however, that this information will only address part of the requirement because there will likely be new applications and computer systems. This brings us to the rather rough estimating process outlined in the next section.

7.3.2 Estimating Delay and Response Time

Factors 3 and 4 are quality measures that can have a direct impact on user acceptance. Values of acceptable time delay and response time must be established ahead of time so that the network components can be sized properly. Figure 7.9 shows the major elements of the link between user terminal and host computer over a VSAT star network. This is at a high level and should not be regarded as a specific design (we provide such an example later in this chapter). We can see that there are several contributors and each adds to the delay as experienced by the user of the PC client workstation. At the start of an exchange between client and host, the user enters data and hits the return key. This causes a block or frame of data to be applied to the local access line between the PC and the VSAT indoor unit. The speed of this line determines the time delay for transfer of the entire packet (we can ignore propagation time since the distance is usually less than 1 km). Thence, the data are repackaged within the indoor unit and formatted for the satellite return link. A pair of arrows indicates the relationship between this VSAT and the satellite access protocol. VSATs typically employ either the ALOHA access technique or

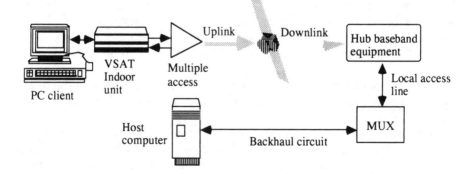

Figure 7.9 Major contributors to delay and response time in a VSAT star network.

TDMA, as discussed later. This almost always introduces some time delay dependent on the number of other VSATs sharing the same transmission channel over the satellite.

The uplink and the downlink introduce nearly constant propagation delay, amounting to approximately 260 ms total (for a single hop). The delay of the satellite repeater is typically measured in nanoseconds and therefore can be ignored in these estimates. An exception would be an advanced processing type of repeater that requires a certain amount of time to route the packets. This would be specified by the satellite manufacturer and would be expected to be a constant number much less than the propagation delay. The next significant contributor to delay is the processing within the hub baseband equipment, resulting in a restored block of data at the local access line. Most hub installations multiplex several streams of such data together, a process that can add a small but measurable delay. Finally, there is the transfer and propagation time associated with any backhaul circuit between the hub and the host computer system. Processing time within the host is not supposed to be counted as part of the response time of the network, but it is often added since it will, of course, be experienced by the user at the PC client.

An example of a typical time budget for this end-to-end connection is provided in Table 7.5. We have made the assumption that the user wishes to transfer a block of 1,000 bytes, which is 8000 bits. Assumed line transmission speeds and processing delays are indicated in the table.

An approach like that in Table 7.5 provides a way to estimate the total response time. In this example, we looked at one direction only; also, there are other contributors, such as an end-device queuing delay, that add to the total. There would be a separate estimate for the forward direction compiled in the same manner. This estimate shows that the transmission delay contribution occurs only once and is determined by the lowest speed in the link (64 Kbps at the original access point, producing a delay equal to $64,000/8,000 = 125$ ms). The uplink/downlink contributes a total of 260 ms of propagation delay, to which is added only 3 ms for the backhaul circuit (assumed to be 750 km in length). Processing delays for the equipment in the system are estimates but can be measured on real devices to improve accuracy. The multiple access delay of 25 ms is based on a TDMA frame length of 50 ms, dividing it in half to represent the average holding time before transmitting a burst. This is comparable to the delay experienced in a token ring LAN, where stations on the cable must wait their turn in the cycle. Finally, we have assumed a host processing time of 100 ms, which could be an underestimate in the case of a heavily loaded computer system.

The estimation process shows that there are many contributors to total delay. If we merely doubled this estimate to account for the round-trip situation, then the response time is 1.27 sec. This would satisfy a requirement of 2 sec,

Table 7.5
An Example of a Delay and Response Time Budget for a Terminal to Host Application
Over a VSAT Network

Element	Line Speed (Kbps)	Intrinsic Delay (ms)	Propagation Delay (ms)	Actual Contribution (ms)
PC client		0		0
Access line	64.0	125		125
VSAT indoor unit		50		50
Multiple access protocol		25		25
Uplink	128.0	62.5	130	130
Satellite repeater			0	0
Downlink	128.0	62.5	130	130
Hub baseband equipment		50		50
Local access line	64.0	125		0
Multiplexer		25		25
Backhaul circuit (750 km)	384.0	20.8	3	3
Host computer		100		100
TOTAL				638

which is a typical value used in industry. What we see is that the satellite propagation delay is about 40% of the total. It would be possible to reduce the total delay by increasing the basic line speed or by reducing the processing delay in the host.

7.3.3 VSAT Access Protocols

Capacity estimation in a VSAT network is a complex problem driven by the fact that many of the key inputs to the problem are not well known ahead of time. Also, even if you have a good estimate of the required number of users, their locations, and the data throughput needs, there is still the problem of considering how these users will interact with each other and with the system at any given time. We typically work with an average of some sort and allow for some peak during a busy hour or seasonal peak of activity (like just before Christmas in the retailing industry or summer in air travel and hotels). But this does not consider the statistical peaking that occurs from an instantaneous heavy load after some unexpected event or if some particular users activate an application that produces a surge in the network.

What we have are some basic methods of estimating the load and from that the capacity. This should be taken as a starting point for sizing and not used as if it was an accurate prediction. Fortunately, VSAT networks have a number of built-in features that tend to forgive for estimating errors. For example, the overall capacity of the network is determined by the amount of satellite bandwidth that is both allocated to the network and that can be accessed by the Earth stations. The fact that this pool of capacity is available to all means that any given user has potentially a large reserve that can be utilized under a critical load. This is constrained by the equipment that is available to the user, such as the bandwidth of the particular uplink chain through which the data must flow to reach the hub. The hub, on the other hand, has a much larger reserve of bandwidth, but this to is not unlimited as we shall see.

The basic technology of the VSAT star network exists on a foundation of many years of experience with digital satellite communication. We use a large hub Earth station to receive all direct transmissions from VSATs, which are sent in packet form. The hub transmits one or more continuous streams as a data broadcast to the entire network. Individual VSATs select packets addressed to them from the broadcast. A group of VSATs transmits their information to the satellite on the same frequency, but not overlapping in time. The use of burst transmission in the inroute and broadcast TDM for the outroute is illustrated in Figure 7.10; the outroute is shown as a continuous data stream, and the inroute is discontinuous, allowing individual VSATs to timeshare the particular narrowband channel.

VSAT and other satellite data networks make use of *automatic retransmission request* (ARQ) protocols that operate on top of the multiple access system. This technique can actually be found in all high-performance networks, including TCP/IP, X.25, and IBM's SNA. In the case of satellite networks, the parameters are adjusted so that the round-trip delay of about 0.5 sec does not cause the throughput to reduce to zero. We employ a block-oriented protocol that allows the receiving end to detect errors and limit the requests for retransmission to only those blocks that contain errors. Another name for this technique is *look back N* because the receiving end checks for back blocks and tells the end to look back in the frame and only resend the errored block.

Combined with FEC, the *look back N* technique produces nearly perfect data transfer without a significant increase in average delay. The version of this protocol that is most popular employs the *modulo 128* scheme. That is, $N = 128$ blocks for this frame of data—the receiver collects the full 128 blocks before requesting the retransmission of the errored blocks. The number of bits in the block is set as part of the design of the link, and could be as short as

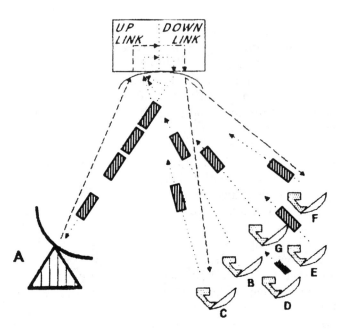

Figure 7.10 Multiple access technique used to collect VSAT inroute transmissions at the hub. This combines TDMA and ALOHA, depending on the requirements of the application.

1,000 bits or as long as several kilobytes. The low cost of computer memory makes this approach very affordable even in low cost VSAT networks.

7.3.3.1 Operation of the TDMA Access Protocol

The inroute is shared by multiple VSATs that transmit their data in bursts. TDMA is one of two basic multiple access methods used for this purpose. As indicated in Figure 7.11, the inroute transmissions from the VSATs are coordinated and highly synchronized so as to prevent overlap and a resulting loss of information. The master frame in this example is 45 ms, which is a multiple of 360 times the frame rate requirement for telephone service. Each station (numbered 1 through 10) is allotted a fixed interval of time in which to transmit data. The frame repeats every 45 ms, producing and average delay per inroute burst due to multiple access of 45/2 = 22.5 ms. Obviously, the shorter the frame, the less the average delay. This also reduces the amount of storage needed in the VSAT and hub to accumulate data and voice samples for inclusion in the next burst. The start of the frame may be identified by a sync

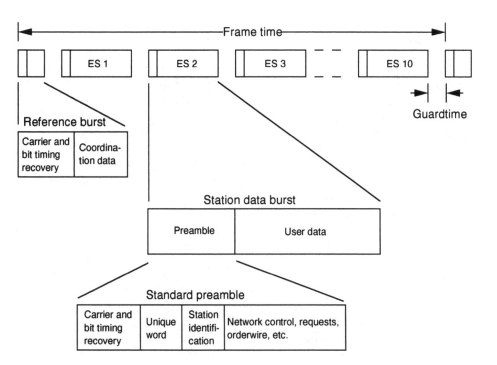

Figure 7.11 An example of a TDMA time frame format, indicating burst transmissions from ten different Earth stations all operating on the same frequency. Stations synchronize their bursts to a reference transmitted by the hub.

burst that is transmitted by a master station for reference. In the actual case of the VSAT inroute, no sync burst is needed because the hub controls transmission via the outroute. Additional delay and loss of throughput results from the addition of the preamble, as shown at the bottom of Figure 7.10. However, this overhead is needed to permit the hub to lock onto the incoming burst, restore the bit stream, identify the source VSAT, and subsequently recover the original data packets.

TDMA represents an optimum approach for timesharing the bandwidth of a radio channel. Full-transponder TDMA was originally developed in the late 1960s to use the bandwidth and power of the transponder in the most efficient manner. The only loss of capacity relative to a single continuous carrier is due to guardtimes and overhead. Later, multiple-carrier TDMA was introduced to save on Earth station transmit power and to allow a transponder to support TDMA and other incompatible services at the same time. It can be applied in VSAT star networks, as discussed above, but is also very effective in mesh networks where each station operates on an equal basis.

7.3.3.2 Operation of the ALOHA Access Protocol

The other approach for separating the inroute transmissions in time is the ALOHA protocol. The scheme is simpler in that the transmissions are uncoordinated; however, the complexity occurs because there are occasional overlaps that result in lost communication. This is overcome by retransmissions from the affected VSATs. For example, an ALOHA channel with three users is shown in Figure 7.12. The first three lines provide timelines for each user's uplink; the fourth link depicts the downlink showing how the ALOHA packets appear after passing through the satellite repeater. Each user remains in an idle state until there is data to be transmitted. In this example, user 1 is the first to need the channel and so transmits the block of data without waiting. User 2 happens to be next, independently of what happens at users 1 and 3. From the downlink timeline, we see that user 1 and user 2 do not overlap and hence get through in the clear.

The next packets from users 1 and 3 have reached the satellite at approximately the same time and so have produced a collision. In the event of such a time overlap, the signals jam each other and the information is lost (indicated by the presence of a dark in the downlink). Neither packet is received at the hub—a condition that is detected by these users because of nonacknowledge by the hub. The way that packets are ultimately transferred is through automatic retransmissions, as shown at the ends of the curved arrows in Figure 7.12. The delay between the original and retransmitted packets is selected randomly by each user to reduce the possibility of a second collision. The result of this protocol is that the delay is as small as it can possibly be for a packet that does not experience a collision. For one that does, the delay is lengthy since it includes at least two round-trip delays plus the delays of the random offset as well as from processing within the hub and VSAT.

The bandwidth of the inroute provides the most fundamental limit of capacity at the VSAT level. If we could perfectly allocate capacity to a group of VSATs, then a 128-Kbps inroute could serve any number of VSATs as long as the total throughput does not exceed 128 Kbps. TDMA offers the best prospect for high throughput and predictable delay because the bursts are ordered in time. Even still, there must be guardtime between bursts to allow stations to keep their burst separate. This fixed guardtime causes the throughput to decrease more or less proportionally to the number of bursts (stations) on the same frequency. In TDMA, we do not expect to exceed about 90% utilization, which gives a throughput of 115.2 Kbps. If we estimate that each VSAT needs to transmit about 19.2 Kbps on the average, then this inroute could support six VSATs. The fact that we have six independent users on the same frequency means that we can probably allow a significant amount of occasional peaking by individual stations since the total bandwidth can be shared dynamically

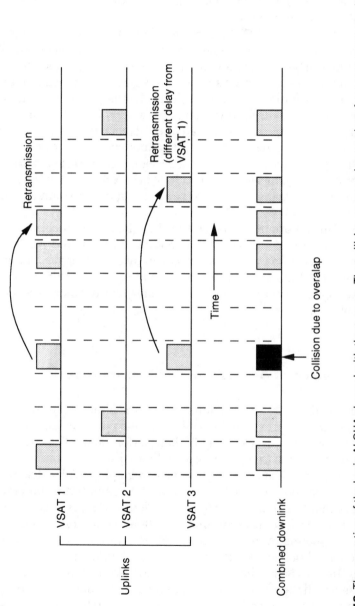

Figure 7.12 The operation of the basic ALOHA channel with three users. The collision occurs when packets from two users overlap at the satellite.

(e.g., when one user is peaking some others are either idle or at least operating below their average throughput).

The situation with ALOHA is quite different because the bursts are not coordinated. In any multiple access scheme where there is no such coordination, the throughput will be substantially reduced. The reason for this is that conflicts will occur on a statistical basis, as we see in Figure 7.12. Since the originating VSAT users are operating independently of each other, combining their activity into the same channel will follow what is called a Poisson arrival process. An ALOHA channel behaves in a statistically predictable way, as will be presented in the next section. The result is that we cannot even approach a throughput of 90% without experiencing a continuous sequence of collisions that reduce the throughput to zero. Because of this, the ALOHA protocol offers a relatively low throughput of perhaps 10% in exchange for simplicity and a short access delay.

There is an important refinement in the ALOHA protocol that increases throughput in a VSAT network. This is referred to as slotted ALOHA, where the inroute packets are partially synchronized by a reference from the hub. As shown in Figure 7.12, a collision is either total or not at all; a partial overlap of two packets cannot occur. As will be seen later, slotting increases throughput by a factor of two.

7.3.4 Comparison of Access Protocol Performance

TDMA and ALOHA are very established as VSAT access protocols for the inroute. As can be seen from the descriptions in the previous section, they differ in their operation and most certainly in their performance. The consequence of this is that users should select the protocol based on the requirements of the application. As we indicated in a previous paragraph, ALOHA has the benefit of offering the transfer of a data packet with the minimum possible delay. This requires that the channel be lightly loaded. TDMA is best applied to an application that requires near-continuous data transmission. It is the function of the VSAT indoor equipment to take a continuous data stream, such as for a telephone channel or file transfer, and break it up into a periodic sequence of bursts. The time-division channel can be provided on a preassigned basis to be maintained until the network is reprogrammed. Alternatively, the connection can be established on demand, through a process called bandwidth reservation. This would employ the ALOHA protocol to request the bandwidth, which is subsequently provided via the TDMA protocol. At the conclusion of the call, a release request is transmitted and the bandwidth is released.

7.3.4.1 Use of an Analytical Model

The analysis of throughput and time delay in ALOHA and TDMA is complicated by the fact that user data requirements are usually unknown or at least highly variable. There are some theoretical relationships that give us a good understanding of how these protocols tend to perform. For example, Figure 7.13 indicates the average delay experienced by each of three types of links, namely, a standard ALOHA channel, a slotted ALOHA channel, and a TDMA channel used in reservation service. A slotted ALOHA channel differs in that the packets are transmitted within defined slots, according to timing provided by the hub Earth station. This means that a collision only occurs if the packets are actually transmitted in the same slot, as adjacent slots are nonconflicting. The upshot of this is that, on average, the time delay of slotted ALOHA is one-half that of standard ALOHA.

The average packet delay is plotted against the throughput fraction, which is the ratio of data that arrives at the destination divided by the maximum data communication capacity of the channel. Say that the channel operates at

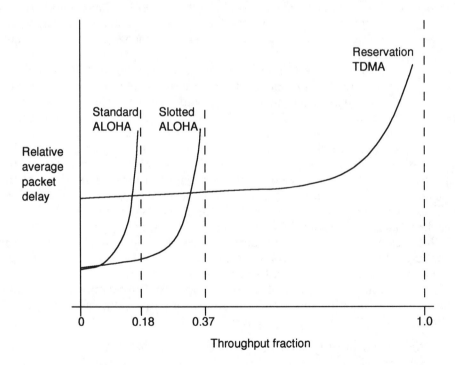

Figure 7.13 Comparison of the contention delays for three VSAT access protocols: standard ALOHA, slotted ALOHA, and TDMA with transaction reservation.

128 Kbps; then a throughput fraction of 0.10 means that the actual usable data transfer is 12.8 Kbps. Figure 7.13 displays how delay increases as one attempts to increase throughput toward the ideal of 1. The importance of the throughput fraction is that it allows the network designer to determine how many inroutes are needed to satisfy the data transfer requirements.

The ALOHA channel has the lowest delay at zero throughput, a condition under which there are no collisions (that is, our VSAT has the channel to itself and wishes only to transmit one packet). At this point, there is a fixed delay that results from the basic transmission speed of the channel. Based on simple Poisson statistics, the standard ALOHA channel experiences infinite delay at a throughput of 0.18 due to a continuous procession of collisions and the resulting loss of data transfer. The channel effectively breaks down. For the slotted ALOHA channel, this point of instability occurs at 0.37, which is about twice the throughput of the unslotted ALOHA. Operation with a satisfactory frequency of collisions and average delay would probably correspond to throughput values of 0.1 and 0.2 for the standard ALOHA and slotted ALOHA channels, respectively. While not clear in the figure, the standard ALOHA curve starts at a value of delay below that of the slotted ALOHA channel and crosses above it well before a throughput of 0.18 is reached. For this reason, VSAT networks employ the slotted ALOHA scheme in recognition of its greater throughput potential.

These results are well known; as a consequence, the ALOHA technique is recognized as effective for light traffic. Furthermore, the individual messages themselves should be uniform in length and relatively short. This further reduces the potential for collision. Slotted ALOHA seems to be preferred because it has roughly twice the throughput for the same average delay.

For the case where the throughput is less than 10%, the relative number of retransmissions is essentially equal to the throughput. An example of the application of this rule of thumb, derived from the Poisson statistics of the packets, is that if the throughput is 6%, then, on average, 6% of the packets will experience a collision. Above 10%, the percentage increases rapidly and reaches 100% at a throughput of 37%. Any variation in packet length will further degrade this result. The performance of the channel in this case is dominated by the largest packet length, which means that capacity will be lost as compared to using a standard shorter packet.

The frequent occurrence of retransmissions is a strong indication that the throughput is pushing too high. The following approaches can be applied, either individually or as a combination:

- Reduce the quantity of packets per second from the VSAT with the greatest demand;
- Reduce the length of the longest packets or force the use of a single packet size;

- Reduce the number of VSATs sharing the same inroute—that is, add another inroute to reduce the average load.

The performance of TDMA is also indicated in Figure 7.13 for the case of a bandwidth reservation type of service. In this case, there is a fixed delay corresponding to the time required to establish the connection. Delay increases slowly with throughput to a knee in the curve at a throughput of approximately 0.9, which is considerably higher than that for the ALOHA channels. Increasing throughput further produces added delay, in this case because the bandwidth reservation process employs ALOHA as the request mechanism. A mesh-type TDMA network would have delay increase linearly to a point somewhere just below 1.0, which accounts for the guardtimes allowed for the number of individual bursts along with the additional overhead indicated in Figure 7.10.

7.3.4.2 Use of a Computer Simulation Model

Analytical curves like those in Figure 7.13 provide some clarity into the operation of the protocols employed in VSAT networks. The difficulty comes in using the results for network design purposes. A more accurate way of proceeding is to build a discrete-time simulation model in a digital computer and then to simulate the operation of the network under the expected packet loading. There are a number of software tools that can be applied to this type of problem, such as COMNET III from CACI of La Jolla, CA, and BONeS by COMDISCO of Foster City, CA. Another approach is for a computer programmer to build a simulation model in a third-generation language like C or Ada.

An example of the results of such a simulation is shown in Figure 7.14 for a typical VSAT star network employing the slotted ALOHA protocol on the inroute. The graph displays the round-trip delay from VSAT to hub to VSAT and therefore represents the response time minus the host processing time. To add complexity, it is assumed that both short and long messages are to be transferred. The graphs are cumulative distributions of user message delay in milliseconds, shown for the short message, the long message, and the combined average.

The graphs indicate that message delay is not a constant but in fact varies statistically across the total quantity of packets carried over the link. We see that the minimum delay is about 750 ms and 1200 ms for short and long messages, respectively. On the other hand, the maximum delay experienced is 3500 ms (e.g., 3.5 sec). Between these limits, we can choose the 50% cumulative percentage as a point of reference. At this mean value, half of all the packets take less time and half take more. The nominal values of delay at the 50% point are 1,200 ms, 1,400 ms, and 2,000 ms for the short, combined, and long

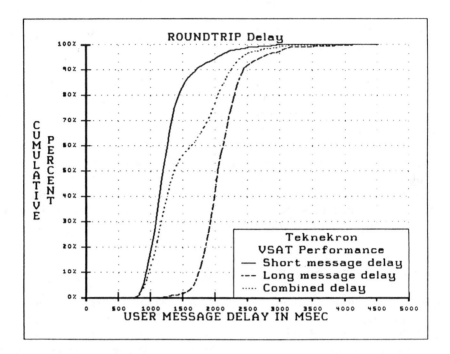

Figure 7.14 An example of the results of a computer simulation of the performance of a VSAT network using the ALOHA access method (courtesy of TCSI, Berkeley, CA).

message conditions. The useful result is that this network would meet an objective of a 2-sec response time on an average basis for the worst case type of message.

These results are interesting in that they report on the dynamic behavior of a network. We have to keep in mind, however, that they were obtained for a specific network configuration for a given traffic input. Change the configuration or traffic, and the results will change, possibly dramatically. The ideal kind of situation is one where we can closely model the network and the traffic and then test the sensitivity of the delay to changes in assumptions. This allows us to test different hypotheses about network loading and examine how a change will impact the network. One useful type of study would be to look at how delay increases as we add more VSATs to the network or if we were to increase traffic loading by 50% in the original network.

7.3.4.3 Integrating ALOHA and TDMA Protocols

As we have discussed, the ALOHA protocol has certain benefits in applications that require short packet delivery with minimum delay. On the other hand,

TDMA is better suited for the transmission of longer blocks of information, such as voice conversation and file transfer. The bandwidth reservation mode, which uses ALOHA as to request bandwidth and TDMA to provide stream transmission, combines both.

The two have been integrated in VSAT star networks through the use of a combined time frame, as illustrated in Figure 7.15. The frame period of 45 ms in this example is 360 times the minimum time required for speech transmission; therefore, each speech burst must contain 360 voice samples. The figure indicates how this frame is allocated among the protocols, with the first segment available for pure TDMA, referred to as stream (continuous real time). Bursts that appear in this segment will repeat indefinitely or until the network is reconfigured on a long-term basis. A second TDMA segment is allocated for demand assigned bursts, that is, for bandwidth reservations. Here, the bursts have been allocated through a dynamic process in response to requests from remote VSATs.

ALOHA transmissions are confined to the third block of time within the frame. Within this segment, VSATs can transmit packets using the slotted ALOHA protocol. This means that there will be instances of collisions since VSATs transmit in an uncoordinated way. The only means that we have to control the frequency of collisions is to limit the number of VSATs that share this particular block of time (which limits the maximum possible throughput).

One important outcome of using a common time frame for all services is that every user will experience the same average time delay of one-half of the time frame (22.5 ms in this example). This is only about 10% of the one-way satellite hop delay and therefore might be considered to be acceptable. An exception, of course, is for ALOHA users who are subject to an additional delay due to collisions and retransmissions. The benefit of the common time frame is that a single inroute can be shared very effectively and the bandwidth allocated as needed over the life of the network. It is possible to move applications from one protocol to another as a way to better meet user requirements. Also, the number of VSATs assigned to a particular segment can be adjusted up and

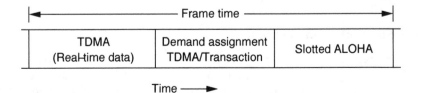

Figure 7.15 The time frame for an integrated VSAT inroute, providing ALOHA, TDMA, and demand assignment/transaction reservation services.

down. These are functions for the network control center that is associated with the hub Earth station.

The fact that the access protocols need to be tailored to the application has caused two additional modes to be introduced in the market. The first is the dynamic selection of access method based on the traffic demand. The activity level of the data channel is monitored at the remote VSAT and at the hub. If the situation requires a change, then the hub automatically causes a switch in access protocol. For example, if the particular user is employing slotted ALOHA and the requirements shift so that more collisions are experienced, then the network can automatically transfer to a bandwidth reservation mode of operation. The user application would only experience an improvement in performance, recognizable as a drop in response time.

The switching between access protocols would be based on a criterion to be selected by the operator. Two such criteria are (1) switching based on the rate of traffic flow and (2) switching based on the size of the message being applied to the remote. In traffic rate switching, the remote VSAT senses the amount of data in the queue buffer that holds information that is to be placed into packets for transmission to the hub. When the amount of that data exceeds a threshold, the switch is made from ALOHA to bandwidth reservation. Later, when the amount of data drops below a second threshold, the protocol is switched back to ALOHA. In message size switching, the VSAT uses the measure of the size of the message to make a determination on the protocol. As long as short messages are coming into the remote, the ALOHA protocol is employed. However, when long messages begin to appear, the protocol is switched to bandwidth reservation or continuous TDMA, as appropriate. The selection criterion is a programmable parameter.

The other technique is to make a bandwidth reservation on the fly without having to employ the ALOHA protocol. This reduces the setup time and therefore improves the operation of the particular application. Referred to as *piggybacking*, the request for bandwidth is attached to another user packet that is already going from the remote to the hub. The hub recognizes the request and then returns the reservation information to the remote. This is invisible to the user application.

7.3.4.4 Structure of the Outroute from the Hub

The outroute is typically transmitted as a continuous stream of data using a TDM format. It is not TDMA because only one station is providing this uplink and there is no need to employ burst transmission. This also simplifies the design of the remote modems in the VSATs since they can take more time to lock on to the transmission at the initiation of service (or when there is a break due to a link fade or equipment change). As we indicated earlier in this chapter, the outroute speed is typically some multiple of the inroute speed to take advantage of the larger antenna size and greater potential transmit power of the hub. For example, in one design the outroute operates at 512 Kbps while the inroute operates at 128 Kbps. There might then be multiple inroutes to balance each outroute, determined by the specific data transfer requirements of the network.

There could be a situation when one outroute is insufficient to provide the forward link for the expected quantity of VSATs and their expected data throughput requirements. In this case, additional outroutes would have to be provided. The general architecture of the system usually requires that there be one set of equipment at the hub for each outroute. This set of equipment is referred to as a "network," not to be confused with the overall network itself. A network of this type could support one outroute at, say, 512 Kbps and a quantity of inroutes, say, 4 at 128 Kbps each. There might be instances where the traffic is imbalanced, that is, where the total of the inroute data rates is greater than that of the outroute. Also, the number of VSATs that can share an inroute will depend on the aggregate data throughput requirements, including the extent to which the ALOHA and TDMA access protocols are used. This can change on a dynamic basis and therefore would have to be adjusted as the needs vary.

There are architectures where the hub can be downsized to reduce cost and therefore the price of entry for the user. (Another way to accomplish the same result is to share the hub among several networks, as discussed later in this chapter.) Reducing the outroute speed has a direct effect on the bandwidth and power that must be obtained from the satellite provider. Keeping the hub and VSAT antennas the same size will allow the commitment for satellite capacity to be reduced by the ratio of outroute speed. It also means that the hub antenna might be smaller in size, say, 2.8m instead of 6m. For some applications, a smaller antenna might be required by limitations of the hub site. Also, having a lower capacity on the outroute, such as 128 Kbps, would allow a smaller high-power amplifier to be used. This will depend on the size of the hub antenna and the need to maintain service in the local rain environment. Reducing the outroute speed also decreases the number of VSATs that can be supported in the network. Therefore, for a small network with perhaps only 100 remotes, the smaller hub and outroute speed could be a good match.

7.3.5 Sizing the Remote VSAT

The design of the typical VSAT is arranged so that it can be configured to satisfy a range of data transfer requirements and thereby achieve an acceptable response time. Whether a satisfactory result can be obtained depends on this match; not every application will fit the particular architecture and throughput capabilities of the typical VSAT network. Cases where there is a good fit usually adhere to the following:

- The network is clearly a star topology, where there is a central host computer that supports a large quantity of remote locations. This depends on having a centralized data processing structure, where remotes depend on the host for information access and network management.
- The throughput requirements are consistent with the basic data transfer capabilities of the VSATs and the hub. This must be verified through the process of sizing.
- Requirements are currently being met using terrestrial telephone lines or low- to medium-speed data lines (up to 19.2 Kbps but probably not as high as 64 Kbps). The ability of the VSAT to meet the requirements of a remote location are constrained by the maximum throughput of the inroute data rate, which typically is not greater than about 256 Kbps.

The elements that must be sized in a remote VSAT are indicated in Figure 7.16. The user terminal is indicated to the left and would normally be connected to the host by a terrestrial circuit of some kind. In substituting the VSAT, we need to consider the bandwidths and performance of each critical element. This is the first limiting aspect of the design and should be selected in a way that does not degrade performance. Manufacturers offer a variety of port speeds that mirror the interfaces found on typical terminals. Examples include 2.4, 4.8, 9.6, and 19.2 Kbps; 48, 56, and 64 Kbps. The higher the speed, generally the higher the cost. Electrical interface connections can adhere to the RS-232, RS-422, and V.35 specifications. Fortunately, the VSAT is typically within the same building as the terminal; hence, we can use a much higher data rate than we could normally afford on a long-distance basis. For example, if the throughput requirements are estimated to be an average of 20 Kbps, then we

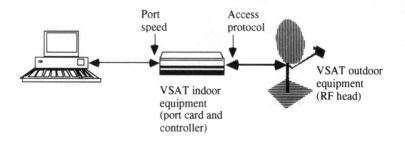

Figure 7.16 The elements of a remote VSAT that have the most direct relationship to network sizing.

can employ a port speed of 64 Kbps or a LAN connection at 10 Mbps to prevent this part of the system from ever impacting performance. The utilization on this line will be relatively low. The cost of doing this is probably relatively minor since all that we are purchasing in this case is cable and connectors. Note that no modems are used because this is a direct digital connection using interfaces like RS-232, Ethernet, or token ring.

The example in Figure 7.16 is for a single user that is directly connected to the port card of the VSAT. There is no contention for access, and therefore the user will not experience any delay for this portion of the system. In many installations, several user devices share the same access port using polling to control contention. This is consistent with the protocol used by the hub to determine which remote devices are in need of service. As we discussed previously, the host computer is spoofed by the hub baseband equipment into thinking that it is exchanging polling information with each remote device. Instead, polling is recreated at the remote and information is transferred over the satellite on a nonpolled basis using ALOHA or TDMA, as appropriate. Polling between the VSAT indoor equipment and multiple terminals on the access line is accomplished on a more or less seamless basis as long as the bandwidth at the port is sufficient for satisfactory service. Utilization in a polling scheme can exceed 90% as in the case of TDMA because the VSAT controller is able to force an orderly entry of information from the user devices. It is just a matter of ensuring that the total access speed is sufficient for the expected demand from the particular number of users who share this port.

The delay that the port card introduces then involves two components: the data transfer time (determined only by the line speed) and the queuing delay (determined by the number of users sharing the line and the contention, control method). A dedicated port involves no contention, while a polled port must be evaluated carefully and perhaps should not exceed 60% utilization to avoid significant additional delay. Later in this chapter, we consider a port connection to a LAN.

There are, in fact, lots of options for port cards and control approaches. Because of efficiencies now possible in microcircuitry and software, multiple access lines can be accommodated with a single port card. The particular configuration must be determined at the time the equipment is ordered. Suppliers can help the user determine the appropriate types and capacities. Changing this part of the network is certainly feasible but can be difficult after a large quantity of VSATs are already installed. For this reason, it is a good policy to install a pilot network with a smaller quantity of remotes before committing to a very large quantity of VSATs and the cost of the associated port cards.

The other side of the problem in VSAT sizing relates to the access protocol and satellite link data rates. We discussed earlier in this chapter that the inroute may use one of three protocols: ALOHA, TDMA, and bandwidth reservation.

The selection of the particular protocol depends on the nature of the application. Also, there are features in the network that allow this protocol to be changed dynamically to match user demand. The speed with which the data are transferred over the satellite is fixed by the equipment installed at the VSAT and the hub and hence cannot be changed very easily (or at least that was the situation at the time of this writing).

Typical capabilities that can be provided by the outroute and inroute are summarized in Table 7.6. Most of our discussion has centered around the first type of network, that is, the unbalanced star. The majority of networks in operation at the time of this writing are of this type. It allows the remote to transmit with relatively low power (because of the lower data rate) while maintaining a relatively high data rate on the outroute from the hub. This tends to reduce cost and improve the total throughput of the network. The balanced star networks reduce the investment cost of one or more of the elements but may increase space segment charges and degrade performance as compared to the unbalanced star. This may still be the right choice for a particular network implementation.

The mesh networks provided in the table are implemented with higher data rates than observed in the star networks. This is because of the reliance on TDMA, wherein all stations share the same channel. The data rate is selected by the most demanding application. For example, if the network is to support LANs with relatively low throughput requirements, then the low data rate case could suffice. An example of when high data rates are necessary, even as high as 10 Mbps, would occur if we need to support a broadband application like video teleconferencing.

Coming back to the first case of an unbalanced star network, the inroute design has a direct impact on the configuration of the remote VSAT. The RF equipment, shown with the antenna in Figure 7.16, must support the transmission between the VSAT and the hub. The capacity of the link is determined

Table 7.6
Typical Examples of Inroute and Outroute Data Rates
in Use Over VSAT Networks

Type of Network	Inroute Rate (Kbps)	Outroute Rate (Kbps)
Star, unbalanced	128	512
Star, balanced	128	128
Star, low data rate	64	64
Mesh, low rate	256	256
Mesh, medium rate	1,544	1,544
Mesh, high rate	10,000	10,000

by the access protocol and the associated throughput versus delay characteristic discussed earlier in this chapter. The link itself can be evaluated using standard link budget calculations, telling us the required EIRP and *G/T* of the station (see Table 7.2). From this, we can select an appropriate antenna size and corresponding HPA. The typical approach is to make the antenna as small as possible consistent with the available solid-state power amplifier that can come with it. These are usually sold as a system. Readers should note that the RF head tends to determine the overall reliability of the VSAT, primarily because of the rather hostile environment to which it is exposed. For this reason, a well-proven design with many similar units already in use is the best choice.

In the balanced network, the lower data rate case has the benefit of reducing the required transmit power and antenna size. This would reduce the equipment and installation cost of each remote. Networks of thousands of remotes are good candidates for this type of installation since the cost of equipment tends to dominate the cost of the network. The inefficiency in using a space segment and hub are probably overcome by the economy of scale on the remote VSAT side. In networks of much fewer stations, it probably makes sense to allow the size of the VSAT to grow so that other savings can be introduced. All of these factors and tradeoffs should be examined for the particular supplier and network prior to making an commitment.

7.3.6 Sizing Hub Facilities

The sizing of the hub equipment is particularly important to VSAT network design because it relates to the most costly single element. Quantities of port access cards, modulators and demodulators, and TDM/TDMA baseband processing equipment must be determined and placed on order with the supplier. However, this is a complex problem that warrants careful consideration and significant involvement by the supplier since they understand the unique architecture and capabilities of their products. The fortunate aspect of the design of the hub is that it is very modular in nature, being composed of a variety of racks of equipment that are configured for the particular network requirement. It is relatively simple to add, remove, or rearrange equipment within the hub, thus allowing the capacity to be managed and new capabilities introduced from time to time. The one part of the hub that might be more difficult to alter is the RF terminal that typically is located in an external shelter or cabinet. Adding equipment or increasing RF power may require a major modification of this part of the hub, and even a complete replacement could be a consequence.

For all of these reasons, it is usually best to allow the supplier to participate in the sizing of the network. This can be done using the power of competition to obtain the best estimate, both in the terms of the quantity of equipment and the ultimate price. This must also consider the possible consequence of a low estimate and the resulting undercapacity. On the other hand, additional equipment could be brought after the network is put into service to resolve cases where response times are excessive.

The critical elements of the hub related to sizing are shown in the simplified block diagram given in Figure 7.17. In its role as the central point of the star network, the hub coordinates all transmissions over the satellite and provides the connection to the host computer. These elements must be sized to support the maximum capacity of the hub, which also determines the capacity of the network. In this example, the host computer employs a *front-end processor* (FEP) to transfer data from applications within the host to the external world. From a network standpoint, the FEP establishes all of the sessions between host and remote user device and assures reliable transmission. Generally speaking, the FEP assumes that the network is implemented with fixed point-to-point connections using terrestrial telecommunications services.

The backhaul circuit, which is either a short direct connection in the case of a locally connected hub or a dedicated circuit using private or public transmission facilities, can interface directly with the FEP precisely because it is of the type that the FEP normally employs. The hub contains one or more port cards that interface the backhaul circuit and thereby reach the FEP. Communication between the port and the FEP follows the protocol used by the host in its normal operation. For example, in the case of an IBM SNA network, the protocol would typically be *synchronous data link control* (SDLC), a block-oriented protocol with powerful error control features. Since this protocol is different from that employed over the satellite, the port provides spoofing to make the FEP believe that it is connected to a standard network.

The capacity of the backhaul circuit is chosen so as not to restrict the flow of data on the outroute and inroutes. Thus, it must support every application that the host delivers to the network. Congestion here will directly impact all users and hence should be avoided. It is a good practice to overdesign the backhaul circuit in terms of its bandwidth, for example, by providing double the expected throughput. This also allows for peaking of traffic during expected and unexpected increases in demand and for growth. Another consideration is the provision of backup service in case of failure. A typical strategy is to provide an alternate path using possibly a lower speed data line or even a point-to-point satellite link. The link can be used for additional capacity and thus provides for unexpected peaking of demand.

The hub baseband equipment is difficult to size because it is really a system composed of several functional elements. Port speeds and quantities are selected based on the needs of the backhaul circuits and host FEPs. Typical interface options include RS-232 (for speeds under 56 Kbps), RS-449 (for 64 Kbps), V.35 (for 56 Kbps up to approximately 256 Kbps), T1 and E1, and LAN interfaces such as IEEE 802.3 (Ethernet) and 802.5 (token ring).

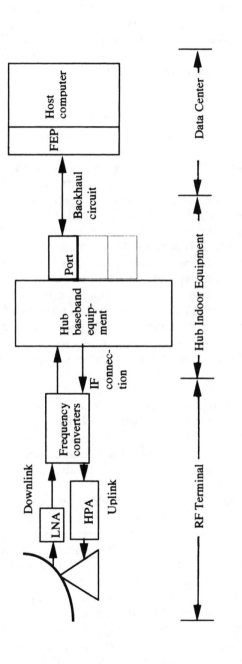

Figure 7.17 The elements in the hub that determine network size and performance.

Port cards and the supporting buffering equipment make the digital traffic available to the outroute equipment and accept traffic from the inroute equipment. The buffering is necessary to allow the packetized information to be placed into the time division frame that comprises the outroute transmission. This introduces an amount of delay that consists of:

- Processing delay;
- Buffering delay;
- Outroute access delay (nominally half the frame time).

The outroute capacity is dictated by the speed of transmission. If we assume this to be 512 Kbps, then each 45-ms frame contains $512,000 \times 0.045 = 23,040$ bits or 2880 bytes. The frame, illustrated in Figure 7.18, is divided among the VSATs that receive this outroute. The header indicates the start of the frame and provides a timing reference for all remote VSATs. Real-time packets are assembled into the first segment of the frame; these correspond to the TDMA stream inroutes that repeat periodically. Non-real-time packets in the second segment are dynamic and respond to ALOHA and bandwidth reservation allocations to the remotes. The breakdown need not be constant as services can be quite dynamic in nature. The data can be packed as tightly as the accessing traffic allow. Obviously, maximum throughput occurs when the traffic is preassigned and not subject to change. If we are responding to requests from remote VSATs who are using ALOHA on a totally random basis, then the makeup of the outroute frame will change constantly.

A typical packet structure is provided in Figure 7.18 for reference. In this instance, it is assumed that the user data can be contained within the variable portion of the packet. The architecture of the network is constrained so that there is a maximum length, as indicated in the figure. If the user data input from the host exceeds the maximum, then a segmentation process in the hub breaks the user data packet down into smaller pieces that fit within the maximum size allowed by the network.

The complexity of the architecture and data structure of the VSAT network along with the wide range of possible user configurations at the remote and hub make sizing a very difficult process. There are a variety of tools and techniques available to the network designer to appropriately match the demand from either end. The supplier may be able to offer some of these tools to facilitate the sizing process. Furthermore, the dynamic and flexible nature of the VSAT and hub allow user demands to change as they may (something that is expected). This means that after the hub and VSATs are installed, there are many options to be explored. This includes letting the network itself choose the best multiple access technique in response to variation in user demand. If the capability of the installed equipment is exceeded, then it is still possible to add ports and

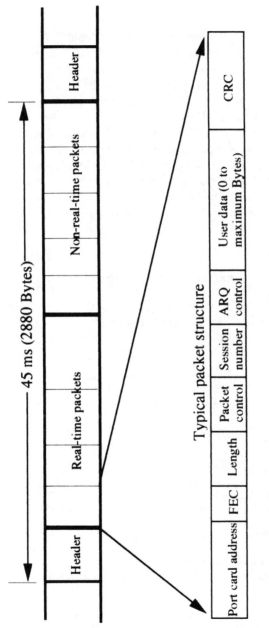

Figure 7.18 The outroute frame and packet structure for a typical VSAT network.

hub baseband equipment elements to bring the limiting element up to sufficient capability.

Sizing of the RF terminal, as indicated to the left in Figure 7.17, is based on the standard principles for Earth station design. An RF engineer would perform a basic link calculation as outlined in Table 7.2 to determine the uplink EIRP for the outroute carrier. In this case, the receiving station is the basic VSAT remote terminal. From the uplink EIRP, the RF engineer works backward through the hub antenna gain to determine the power output from the HPA. The design might have to allow for multiple outroutes, either for the initial service or for future growth. The result is that the HPA should be sized for multiple carriers with acceptable intermodulation distortion. The HPA would be selected to have greater power than the requirement, calculated as the product of the power per outroute carrier times the number of carriers times a backoff factor. A typical value of the backoff factor is in the range of 4 to 6 (e.g., 6 dB to 8 dB). More power margin should be provided if the hub is to use UPC. An example of an RF terminal sizing is given in Table 7.7.

7.3.7 Allocating Transponder Bandwidth

A given VSAT network will only require a fraction of a satellite transponder that allows one transponder to support several such networks. This is shown

Table 7.7
Example of the Sizing of the Ku-Band Hub HPA
Used to Uplink a Maximum of Six Outroute
Carriers With an EIRP per Carrier of 50 dBW

Factor	Value
EIRP per carrier	50.0 dBW
Antenna gain (6m)	56.7 dB
Waveguide loss	3.0 dB
HPA power per carrier	-3.7 dBW
	0.43W
Multicarrier backoff	7 dB
	(5, as a factor)
Number of carriers (maximum)	6
Total carrier power	12.9W
Uplink power control range	12 dB
	15.8
HPA size requirement	203.8W
HPA size selection	250W

in Figure 7.19, where several outroute carriers and their associated inroute carriers occupy the bandwidth of one transponder. This packing efficiency permits a given network to go into operation at a relatively modest cost because transponder capacity can usually be purchased based on the percentage of the transponder that is needed. As in the case of a multicarrier Earth station HPA, sharing the transponder power must allow for multicarrier backoff of approximately 5 dB to control intermodulation distortion. As a further refinement, the outroute carriers in Figure 7.19 can be unequally spaced to avoid some of the stronger intermodulation products. This particular technique is also practiced in constant-carrier FDMA networks, such as that displayed in Figure 4.7.

The amount of satellite transponder capacity that is required for a VSAT network is determined by the combined bandwidth and power for the outroute and inroute carriers. It is a very straightforward process to add up the required bandwidth from the bandwidth of each type of carrier. Because ground receivers cannot separate carriers that are too close to each other, it is appropriate to allow a guardband of 10% to 15%. A typical example is provided in Table 7.8.

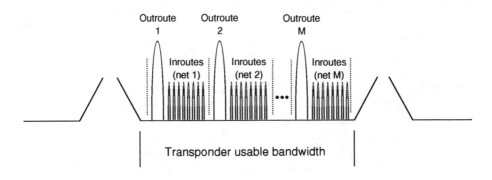

Figure 7.19 Allocation of bandwidth for multiple VSAT networks in a transponder.

Table 7.8
Example of an Estimate of the Total Capacity Required for a VSAT Network

Component	Unit Bandwidth (kHz)	Unit Power (%)	Quantity	Total Bandwidth (kHz)	Total Power (%)
Outroute (512 Kbps)	310	8	1	310	8
Inroute (128 Kbps)	154	2	6	924	12
Guardband allocation	123		1	123	
Total Network				1357	20

The outroute data rate is 512 Kbps and employs four-phase PSK, while the inroute data rate is 128 Kbps and employs two-phase PSK. The power allocation includes backoff and is based on the percentage of the total transponder power. We are assuming the transponder bandwidth is 54 MHz. The amount of power used is approximately 20% of the total available, while the amount of bandwidth is 2.6%. Under this condition, the satellite operator would charge for service based on 20%, which is the bigger fraction. The power of the transponder would therefore be used up before the bandwidth, which is a typical result in multicarrier operation.

7.4 HUB IMPLEMENTATIONS

The network itself will have both financial and strategic benefits for a lot of organizations. That a VSAT can provide a range of services through a relatively inconspicuous antenna is well established. Many companies consider the VSAT itself to be attractive and would like to take advantage of its cost effectiveness and flexibility. The star topology incorporates the hub to coordinate transmissions and provide access to a central host computer. However, the hub represents a major investment as well as a commitment to operate a complex telecommunication facility for the duration of the network lifetime.

Users of VSAT star networks have two avenues open to them, that is, the dedicated hub and the shared hub. The dedicated hub is owned by the network user and operated solely for the benefit of the user. On the other hand, the shared hub is owned and operated by a service provider who makes the investment and operates the hub for the benefit of multiple user networks. The following paragraphs compare these alternatives in general terms. Anyone considering the choice of dedicated versus shared should conduct a detailed examination that considers the operational, business, and financial aspects of the question.

7.4.1 Use of a Dedicated Hub

Early adopters of VSAT networks were large companies who were willing and able to risk being pioneers. Not only did they have to overcome the early barriers to integrating satellite communication into the conventional terrestrial networks of the time, but they also had to make the full financial commitment to hub ownership and operation. Companies like Wal-Mart, the world's largest retail chain, and E. D. Jones, a regional stock brokerage in the United States, built their businesses on VSAT technology. Taking responsibility for the hub was an expected commitment along the way to achieving a competitive advantage over their rivals. Interestingly, the feature of VSATs that attracted them the most was their ability to deliver a consistent set of telecommunications applications independent of location. Wal-Mart placed their stores in rural towns where there was no competition. Likewise, E. D. Jones served stock brokerages in the midwestern United States where the latest telecommunications technology was being installed at a slower pace than in major cities like New York and Seattle.

The basic arrangement of a dedicated hub serves a single data center, as that shown in Figure 7.17. In the ideal case, the host computer and hub share the same facility or are within a short cable distance of each other. This removes the backhaul circuit that could otherwise impact throughput and response time. In addition, the service is not susceptible to failure of a backhaul circuit provided by an outside telecommunication operator. The hub Earth station site may have to be located away from the host computer due to practical considerations such as difficulty with placement of the antenna or availability of adequate room for the equipment. Also, the organization may maintain two host locations for redundancy reasons. The backhaul circuit would be required in these instances.

The following are some general guidelines in favor of a dedicated hub.

The total number of VSATs exceeds 400. This is not a hard and fast rule but rather is based on economics. The actual breakpoint number is determined by a variety of parameters, in particular, the total investment in the hub and the cost of obtaining competing data communication services over the terrestrial network. There are instances where a dedicated hub was installed for a network of only 60 to 100 remote sites. Likewise, networks much larger than 400 remotes have been placed on shared hubs. If an organization wants to consider this question on an economic basis, then it should run a thorough evaluation to see which direction offers the lowest cost over the expected period of operation.

Applications are very strategic. A strategic application is one that gives a company a significant competitive advantage in its market. Telecommunications can be an ingredient in achieving this aim, as we discuss in our other work [3]. The VSAT network is truly a private network infrastructure that is independent of the public network. Furthermore, it can deliver features and resources in ways different from what is available to the general public and, in particular, competitors (unless they implement their own VSAT networks). This would allow a company to extend their business through telecommunications in a unique way and thereby gain an advantage. This strategy is employed by Wal-Mart to grow rapidly in rural communities.

The highest level of reliability and availability are required. A dedicated hub can be made as reliable as resources and money will allow. The facility is not shared with other users, and hence there is no conflict over how resources are allocated and what steps are taken to assure the availability of services. For example, all of the equipment is available to one organization, who can decide which particular service or end user will be maintained in service even during a partial interruption. Also, there will tend to be less potential conflict between the services that run through the common equipment within the hub.

Adequate technical staff is available. Design and installation of a VSAT network, including the hub, can be contracted out to the supplier. After the network is up and running, it is time for the user organization to take over the operation and maintenance and to evolve the network as applications are modified or added. This type of activity can be very technical and complex in nature. You are, after all, a service provider to your internal and external users. Some organizations prefer to stay out of this line of activity as they do not consider it to be a core competence. On the other hand, through recruiting and training, the organization can raise its standard to become a competent VSAT network operator.

The backhaul circuit is unacceptable. The dedicated hub configuration shown in Figure 7.17 uses a backhaul circuit to connect the hub to the host computer. While this is the general case that allows for separation of the two elements, there are instances where backhaul circuits of adequate quality and reliability are simply not available. This could be in a developing country or in a remote area of an advanced economy. The dedicated hub allows the host and hub to be physically adjacent to each other. In this case, the backhaul circuit is a piece of cable that would be provided along with the installation of the hub.

Video transmission is frequently required. We discussed earlier in this chapter how business TV can be added to the network simply by installing a video receiver with each VSAT and arranging for an appropriate uplink facility. The latter need not be part of the hub but could be obtained on an occasional basis from a video service provider (e.g., a teleport). For large VSAT networks that can justify their own video uplink, the hub should probably be of the dedicated type. This gives the user complete control over all aspects of its operation, removing the possibility of conflict over resources.

The dedicated hub is probably the most popular manner in which the VSAT network is implemented. It is the vehicle for maximum control over the network and its direction. Clearly, the cost, organizational, and strategic implications should be examined thoroughly before making this kind of commitment. If good shared hub facilities are available, then this option must be examined as well. This is the subject of the next section of this chapter.

7.4.2 Use of a Shared Hub

A shared hub is a telecommunication facility that is operated as a business for the benefit of customers who wish to reduce their costs and operational commitments. Generally speaking, the cost of using the shared hub would be less than that of implementing a dedicated hub and operating it for the benefit of one organization. The idea is that there is an economy of scale in loading a hub with several independent networks, each with their own respective community of remote VSATs. Since the hub would have one RF terminal and antenna, the remotes would all have to be within the coverage area of the same satellite.

Figure 7.20 indicates how a single hub can be shared by three customers, each with their own independent host computer. From the customer's point of view, their network is an independent entity support by their host, shown at the right of the figure. Not shown are the remote VSATs themselves, which would be located around the satellite coverage area. Each customer's host is connected to the hub via a dedicated backhaul circuit (one per customer). This circuit would be maintained by a terrestrial communications service provider, probably the public telephone operator, or could be a private point-to-point link using cable or microwave. There have been instances where the backhaul circuit is provided over a GEO satellite as well, although this introduces an additional propagation delay of one-quarter second.

The hub operator has the responsibility of satisfying the special data communication and application needs of each individual user. Their respective networks may employ different types of hosts, protocols, and computer applications. This increases complexity and might introduce some conflict within the operation or performance of the hub. For these reasons, the hub operator will have to maintain an experienced staff and be on call to resolve difficulties as they arise.

The following are the general guidelines for determining if the shared hub approach is of potential value to a customer.

There are fewer than 200 VSATs. As with the case of the dedicated hub, there is no hard and fast rule based on the maximum quantity of VSATs. However, if there are less than 200, then it is often difficult to justify the investment in a dedicated hub. The reason for this is that the hub is a major telecommunication facility, costing in the neighborhood of $1 million. To make this effective, you should divide this cost over a sufficient quantity of remotes to make it pay. The breakpoint has been reduced in some cases by compressing the hub into what is called a minihub. This is a hub with a lower outroute speed, say, 64 or 128 Kbps, allowing a reduced space segment charge and investment in RF terminal and baseband equipment. The other side of the coin is that a minihub cannot support much growth in traffic or network size. For this reason, it may be prudent to start on a shared hub and then transfer (migrate) to a dedicated hub when requirements dictate.

Technical staff are unavailable. This is the converse to the situation posed for the dedicated hub. Organizations that do not wish to extend their activities into the highly technical operation of a hub should best leave the effort to an experienced provider. Then, they would not be concerned with the recruitment, training, and management of a hub staff. The cost of doing this can be significant

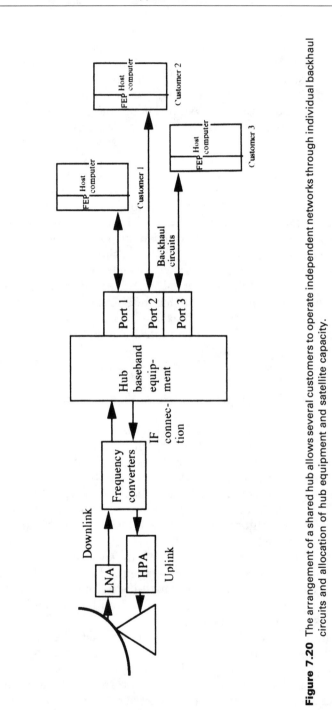

Figure 7.20 The arrangement of a shared hub allows several customers to operate independent networks through individual backhaul circuits and allocation of hub equipment and satellite capacity.

but is usually recovered in large networks. On the other hand, some owners of dedicated hubs prefer to hire outside contractors to run the hub (a process called outsourcing the hub maintenance).

Reliable backhaul services are available. Shared hub service only makes sense where reliable backhual services are available. This is because of the criticality of the backhaul circuit, which must have near 100% availability and high-circuit quality in terms of error rate. Fiber optic or coaxial cable links are preferred. If these facilities are available at reasonable cost, then the user should at least evaluate this option.

Technology is changing rapidly. The investment in the hub covers the cost of hardware and software. The software part can be modified and upgraded over time to match the needs of users and to correct for problems and bugs that are detected in this and other hubs. The hardware portion is less flexible and can become obsolete when new equipment arrangements and options are introduced by the vendor. If the customer believes that technology is changing rapidly, then it would be prudent to allow the shared hub operator to make the investment and take the associated risk. They are also in a position to move equipment around among users as their specific needs change.

Capital commitment not possible. The same principle applies to the investment in the hub. Organizations sometimes find themselves in a position where they cannot make investments outside of the core business. Money might more appropriately be spent on adding a new production line or retail outlet as opposed to investing in a telecommunications facility.

7.5 SUPPLIERS OF VSAT NETWORKS

The market for VSAT equipment is diverse in terms of the number and types of suppliers. There are no standards at this time, and it would be necessary to stick with a single supplier of the hub and remote VSAT stations. As of the time of this writing, there was no effort to open up the technology so as to permit multivendor VSAT networks. A key issue for the prospective buyer is the long-term viability of the chosen supplier. On the other hand, if the supplier can deliver the desired capability and the user is technically able to maintain the system on a continued basis, then long-term supplier viability is less of an issue. Perhaps more important in this case is the performance and reliability of the equipment coupled with the available support from the supplier during the time that the network is being brought into use. There are, after all, many things that can go wrong or at least must be adjusted during installation and for some time thereafter.

To give the reader an idea of the scope of resources, Table 7.9 provides a selected listing of VSAT suppliers that were addressing this market at the time of this writing. There are some very familiar names on the list, such as Alcatel, AT&T, GE, Hughes, and Scientific-Atlanta. Others are less well known but are still very strong players in VSAT equipment and service, such as ComStream, Qualcomm, and STMI.

The process of selecting a supplier is as complex as one wants to make it. A good strategy is to make sure that you understand your requirements very well before looking into suppliers. The next step is to gather data on the suppliers and to learn as much as possible about what can and cannot be done with currently available equipment, software, and services. This can take several months, particularly if the network is to be installed outside of a major developed country. A consideration also is the quality of service and support available

Table 7.9
Selected VSAT Network Suppliers That Are Active in the World Market

Supplier Name	Location	Status
Alcatel Telspace	Nanterre, France	Partnered with STMI
AT&T Tridom	Marietta, GA, USA	Equipment and space segment
Cancom	Mississauga, ONT, Canada	Service provider
ComStream, a Spar Company	San Diego, CA, USA	Equipment
GE Capital Spacenet	McLean, VA, USA	Equipment and space segment
Gilat Satellite Networks	McLean, VA, USA	Equipment
Hughes Network Systems	Germantown, MD, USA	Equipment, service and space segment
Matra Marconi Space	UK and France	Equipment and service
Microspace Communications Corp.	Raleigh, NC, USA	Equipment and service
NEC America	Herndon, VA, USA	Equipment and service
PanAmSat	Greenwich, CN, USA	Service and space segment
Qualcomm	San Diego, CA, USA	Equipment
Satellite Network Systems	St. Paul, MN, USA	Service
Satellite Technology Management, Inc.	Irvine, CA, USA	Equipment and service
Scientific-Atlanta	Melbourne, FL, USA	Equipment and service
Stanford Telecom	Sunnyvale, CA, USA	Equipment
Telesat Canada	Gloucester, ONT, Canada	Service and space segment
Titan Information Systems Corp.	San Diego, CA, USA	Equipment
Vitacom	Mountain View, CA, USA	Equipment and service
Wegener Communications Inc.	Duluth, GA, USA	Equipment

for the supplier. Since the VSATs and hub are an integrated system that must handle a range of services and computing facilities, a safe approach is to rely on the larger companies like Hughes, AT&T, and Scientific-Atlanta. They have addressed more customer situations than the other companies and hence have the software and expertise that can get your network into service as quickly as possible. On the other hand, a smaller "boutique" supplier like STMI or Wegener may have the precise product or software that your network requires and may make the most attractive office in financial terms.

7.6 ADVANCED VSAT NETWORKS AND TECHNOLOGY

The state-of-the-art in VSAT technology is driven by advances in microcircuitry and manufacturing technology, which should result in a VSAT costing substantially less than what was on the market at the time of this writing. But it is a simple case of which was first, the chicken or the egg? The investments in producing truly low-cost VSATs depend on the market being there for the product. This is precisely the same situation that needed to develop for digital DTH receivers and networks. The technology is there to have the same result, but the applications currently lag.

The hardware and software of the VSAT can be improved and made less expensive over time. However, much of the overall advance in VSAT network architecture will come as the satellite repeater moves from a simple bent-pipe to an onboard processor like what is carried on modern MSS satellites. The processor can adjust gains and data rates between the hub and the remote. For example, instead of having a high data rate on the outroute and a low data rate on the inroute, the roles could be reversed with a remodulating type of repeater. With greater bandwidth on the inroute, the network could support greater information transfer for data collection and image transfer. There is also the potential for full mesh connectivity, using the satellite as the switching point in the sky.

This will happen and the consumer VSAT will become a reality before long (possibly by the time you read this book). The other aspect is the performance of the satellite itself. We see many organizations advancing plans for advanced satellite-based digital communication networks capable of delivering megabit-per-second speeds directly to users. Hub Earth stations may or may not be required, depending on the application. It would be possible to implement such networks at Ka band, relying on multibeam coverage and high power to overcome rain fading and to provide an adequate signal to antennas as small as 60 cm in diameter. The technology base is there to do these things today; all we need is the business base to make it a success.

Fixed Telephony Satellite Networks

8

Satellite links have long been used to carry telephone calls to extend the reach of public networks. This role has changed over the years due to the adoption of digital transmission and fiber optic systems throughout developed countries. During the 1990s, undersea fiber has been introduced on a global basis and so the original *raison d'être* for satellite telephony has diminished.

Another trend that has been underway is the rapid economic expansion of many underdeveloped countries throughout Asia and Latin America as well as increased inhabitancy of rural areas of developed countries. Other regions like Africa and the Commonwealth of Independent States (CIS) should also offer an opportunity for development in coming years. The cost and time required to extend the terrestrial network to these areas amount to a large barrier to development. There is now a growing awareness of and market for satellite-based telephony networks. These have proven themselves in years past, and the technology used in low-cost VSATs and cellular systems is being applied to greatly improve the flexibility and cost-effectiveness of these *fixed telephony satellite* (FTS) networks.

The focus of this chapter is on the application of satellite networks to public telephone service, emphasizing its attractiveness in rural and underdeveloped areas. By fixed telephony we mean that the Earth stations are located on the ground and serve stationary users. The idea is to use a satellite network as if it was implemented in the classical manner, using local loops, switches, and trunks. The intriguing difference is that the switching function is provided by the satellite, either using SCPC or digital routing. Most of the technologies and methodologies have been covered previously, so we concentrate here on the special situations and requirements for this particular application. In addition, the following chapter on mobile satellite service covers a closely related area and should be considered to be part of the solution set for extending public telephone services to literally every corner of the planet.

8.1 ROLE OF SATELLITES IN TELEPHONE SERVICES

A FTS network is designed to fit into the conventional *public-switched tele-phone network* (PSTN) architecture. As shown in Figure 8.1, the PSTN is arranged as a hierarchy of telephone exchanges, which we simply refer to as switches. The local exchange is the first point of entry into the PSTN, serving each subscriber on a separate local loop. This loop (indicated with the number 5 in the figure) would normally be implemented with twisted pair wire and extend a distance of a few kilometers. A radio-based technology called fixed wireless is an alternative in rural areas and overbuilt cities to extend local

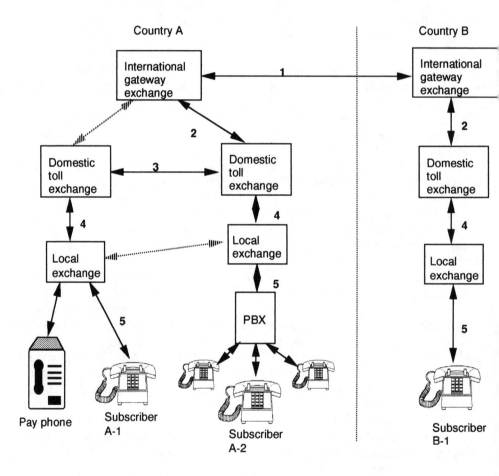

Figure 8.1 A typical PSTN hierarchy, showing local subscribers, a PBX, local offices, toll offices, and international gateways. Satellite links and FTS can be applied for any and all links, both on a permanently assigned and demand-assigned basis.

telephone service quickly and cheaply. FTS addresses this situation by extending the local loop hundreds or thousands of kilometers.

Many businesses employ privately owned telephone switches called PBXs to concentrate telephone usage within a building or campus environment. This improves the effectiveness of telephones and reduces the number of required local loop lines, as indicated in Figure 8.1. A FTS capability could be used to interconnect the PBX with the local exchange or to connect several PBXs together into a private network. The benefit of doing this is that FTS links can be implemented for efficient calling even if the PSTN is not in existence within the area of interest. Also, such lines can bring a remote location into an existing private network and thereby improve integration.

FTS links can be implemented within the PSTN itself, particularly in regions where the construction of terrestrial lines is impractical or too expensive. An example would be in highly mountainous areas or over barren deserts. Such trunks extend between a local office and a domestic toll office (4) and between toll offices (3). These would typically be dedicated links of the T1 or E1 type (or higher) due to traffic carrying requirements at this level of the hierarchy. Access to the international gateway (2) is also a candidate for a dedicated satellite link. Figure 8.1 shows PSTN connections between two countries (1) that can use the satellite capacity of the INTELSAT system on a global basis. Regional satellite systems like EUTELSAT have also been applied in this manner.

8.1.1 Domestic, Regional, and International Services

That satellite links continue to be used in telephone networks should be no surprise. The complimentary nature of the telephone hierarchy and FTS provides an excellent match of need and capability. Potential connections and their applicability to FTS are indicated in Table 8.1. The number in the left column corresponds to the link shown in Figure 8.1 and is comparable to the class of exchange where the link terminates. Domestic FTS was introduced in the United States as a means for new competitors to enter the long-distance telephone market. This died rather quickly after fiber optic systems were installed between major cities. This is because the cost per telephone channel of a high-capacity fiber cable is 50% to 90% less expensive than a comparable satellite link, provided that the fiber capacity is fully utilized. However, FTS is still important in private networks and for addressing the telephone needs in Alaska and U.S. territories in the Pacific Ocean. On the other hand, Indonesia, China, and Russia are highly dependent on FTS to tie together the far-flung reaches of their respective countries. While cable and microwave are effective transmission means, FTS proved from the beginning that it was more cost effective and practical.

Table 8.1
Application of Satellite Links for Circuits Use in Telephony Networks

Class of Circuit	Purpose	Typical FTS Application
5	Connect subscriber (or PBX) to local exchange	Demand assigned
4	Connect local exchange to toll office	Demand assigned or preassigned
3	Connect toll office to toll office	Demand assigned or preassigned
2	Connect toll office to international gateway	Preassigned
1	Connect international gateway to international gateway	Preassigned

We see that two modes are available to achieve connectivity in domestic FTS systems: demand assigned and preassigned. In Chapter 7, we introduced the concept of demand-assigned bandwidth in VSAT networks as a very useful technique for improving the loading of the satellite network. This was done with TDM/TDMA and bandwidth reservation. The approach in FTS is to use SCPC transmissions, where each voice channel is assigned to a particular frequency for the duration of the call. Frequencies are maintained in a pool of transponder bandwidth and managed by a central network control system. This is the same function as a telephone exchange, except that it is performed at a national level (e.g., as if every user were connected directly to the country's national switch). FTS can therefore bypass all of the exchanges in Figure 8.1, except the international gateway.

Regional FTS networks are somewhat limited in their implementation. The best example is EUTELSAT, the largest regional satellite operator, which happens to be owned by the *public telecommunication authorities* (PTAs) that it serves. Being a system by and for the PTAs of Europe, EUTELSAT long has been the centerpiece for regional telephony and TV service. With the extension of EUTELSAT into Eastern Europe, domestic FTS has become a popular area of growth. A brief review of EUTELSAT satellites is provided in Table 6.8.

INTELSAT continues to be a central source of satellite capacity for international FTS services. Virtually all service is digital in nature, making the quality nearly comparable to that available over fiber optic links. The primary mode of transmission is dedicated using fixed carriers with multiplexed telephone channels. Some of the communication uses high-speed TDMA at 120 Mbps. This efficiently uses transponder bandwidth yet allows several Earth stations and their associated international gateways to access each other in a common network. Refer to Table 6.12 for a summary of the capabilities of the INTELSAT satellites that were in operation at the time of this writing.

8.1.2 Trunking Versus Thin Route

Satellite links used in FTS fall into the category of either trunking or thin route. As suggested in Figure 8.1, a trunking application provides links between telephone exchanges at the higher levels of the hierarchy. They tend to support multiple channels of telephone transmission, using T1, E1, or greater bandwidth. Thin-route telephony responds to the need for only a few voice channels of transmission. With demand assignment, these channels can be used to connect to multiple destinations so that specific links do not have to be arranged ahead of time.

The telephone traffic unit of measure is the Erlang, which is equal to an average load of one channel during the busy hour [1]. The traffic requirement between a pair of thin-route service locations can be a small fraction of one Erlang. In mathematical terms, the number of Erlangs of load is $A = \lambda/\mu$, where $1/\lambda$ is the average idle time between calls and $1/\mu$ is the average call holding time. Another way of looking at A is that it is the average utilization in terms of the number of channels. Because the calls are coming from outside sources (presumably subscribers), we are concerned about there being enough channels to serve the peak load as well as the average. To determine the peak load, we must first specify grade of service, which is measured by the probability of blockage, P_b. A standard value of P_b in telephone networks is 1% or 0.01. In words, this states that there will be one blocked call out of 100 attempts (e.g., 99 successes). Cellular networks are designed for three times this value, in recognition of the higher infrastructure cost and value of the spectrum.

The following formula relates the grade of service to the average utilization and number of available channels (the peak cannot exceed the quantity available). Any demand that exceeds the peak cannot be served, resulting in blocked calls. The probability of this occurring is

$$P_b = \frac{A^N/N!}{\sum_{n=0}^{N} A^n/N!} \tag{8.1}$$

where N is the number of available channels in the same group. In standard telephone networks, a group would consist of a bundle of channels that connect the same pair of telephone exchanges together. The more channels in the group, the greater the relative efficiency of handling calls. When we look at a demand-assigned SCPC network, the group consists of all of the available frequency channels on the satellite. This achieves the maximum bundling efficiency.

This concept also applies to cellular telephone networks, but the irony is that satellite-demand-assigned FTS existed long before modern cellular telephony was even conceived. In cellular networks, the bundling occurs at the base station for a specific cell. Users in cell A, for example, who experience a lack of channels cannot use available channels in cell B because the associated tower is out of range. This is not true for the satellite, since all channels are available to all users.

The topology of FTS networks is almost always full mesh in nature, as compared to the star topology of the VSAT. By avoiding a double hop, FTS is able to deliver service at a quality that is comparable to the terrestrial PSTN. When employing GEO, the ITU has recommended that the double hop be avoided. This particular rule can be broken in MEO and LEO satellite networks (covered in Chapter 9) because their lower altitude allows a double hop to meet typical delay standards. To avoid a double hop in GEO systems, a domestic FTS that is destined for another country should be connected over an international terrestrial link (not INTELSAT).

8.1.3 Interfacing to the Terrestrial Telephone Network

The satellite network can be designed for efficient operation and good commercial quality. However, the most critical problem encountered in the application of FTS is the proper interfacing of the service to the terrestrial PSTN. Telephones worldwide are nearly the same, as far as the twisted wire connection (the local loop, 1, in Figure 8.1). The problems come when using FTS for the higher level connections (2, 3, 4, and 5) where information is coded and signaling features are included. There are more than five interface standards in use around the world, and many countries have multiple interfaces in operation within their domestic borders. This makes the interface into a potential bottleneck and quagmire.

An understanding of the importance and complexity can be viewed in Figure 8.2, which indicates the range of interface options encountered in FTS networks. Support of thin-route service and private networks is indicated at the bottom in terms of the interface to end-user equipment. This is comparable to what is possible with VSAT star networks (covered in Chapter 7). The fact that FTS also supports direct connections to telephones, computers, modems, fax machines, and the PBX reflects the importance of satellite communications in business and government. In many cases, the satellite network provides the primary means of communication and therefore must be at least as flexible as the telephone network.

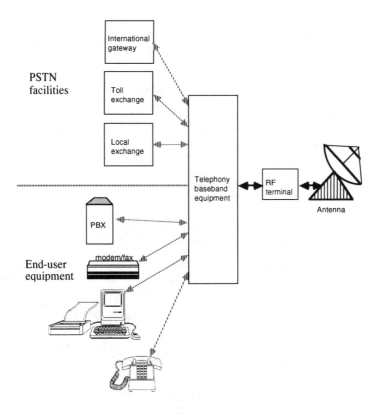

Figure 8.2 The range of interfaces to be employed in FTS applications, considering the connection to user devices as well as various options for the interface to the PSTN.

The use of FTS in the PSTN is indicated at the top of Figure 8.2. Each type of telephone exchange corresponds to a level of the PSTN hierarchy that is shown in Figure 8.1, covering local, toll, and international gateway interfaces. A brief summary of some of the appropriate technical interface standards that apply in each case is provided in Table 8.2. This is a summary and not a design guide because the details vary significantly from situation to situation.

The telephone interface brings with it unique electrical, bandwidth, and signaling characteristics that must be reflected in the FTS Earth station and network operation. The interface to the local exchange is like that to the PBX and could be considered to be the same in many cases. However, there are many options here, based on the local telephone network standard and the types of switches in existence in the particular country. One may even find manual switches (e.g., switchboards and operators) in some underdeveloped areas. More recently, the local exchange interface has moved to the digital era,

Table 8.2
Typical Interface Standards for Connection From an FTS Network
to Public and Private Facilities

Facility	Type of Access	Standard	Requirement
Private; end-user	Telephone	two-wire, touch-tone (DTMF)	Response to off-hook/on-hook and tone
	PC data	RS-232	Typically 28.8 Kbps
	ISDN	Basic rate interface (BRI)	Total 144 Kbps; Terminal adapter within PC
	Modem/fax	Same as telephone	Bypass of speech processing or compression
	PBX	Four-wire E&M; alternatively T1 or E1	Private trunk
PSTN facilities	Local exchange	Two-wire touch-tone (DTMF); alternatively four-wire E&M, T1 or E1	Range of options considers a wide variety of local conditions
	Toll exchange	Four-wire E&M, T1, or E1	PSTN trunks must comply with national standard
	International gateway	Domestic access, T1 or E1	Must comply with national standard
		International connection, T1, E1, or SDH	Must comply with international network

supporting international standards like the E1, CCITT signaling system no. 7 (SS7), and the ISDN basic rate interface.

The toll exchange interface is almost always digital in nature and follows a national standard. The exchanges themselves are supplied by world class manufacturers, and many have been installed recently as part of major network expansions. Therefore, this is perhaps the easiest interface for the FTS implementer. Because this is at a rather high level of the hierarchy, we can expect the capacity requirement to be in multiples of, say, 10 channels.

Interconnection to an international gateway can occur either on the domestic side or on the international side. On the domestic side, the FTS interface is basically the same as that for the toll exchange, following the domestic standard for transmission and signaling. Interfacing on the international side means that the FTS link is part of the global network such as that of INTELSAT. This implies that the international transmission plan and signaling system is in use. Operators of these types of Earth stations and networks comply with the appropriate ITU standards and usually come under the control of an international body. By following these rules and standards, the operator of the FTS facility will have many of its questions answered and should be able to avoid the risk of incompatibility. An exception exists when an independent operator wishes to compete with existing FTS operators and hence must find a means of entry into the market. The consequence of this is that the only standard that exists is that required to make the connection at each end.

To clarify some of the abbreviations of Table 8.2, the two-wire type of interface is that normally found on subscriber telephones. The common standard today is the RJ-11 modular jack, which actually has provision for six wires. Voice frequency information passes over one pair in both directions at the same time. The Earth station detects when the phone goes off-hook and on-hook and provides ringing as well. Within the FTS baseband equipment, a device called a hybrid separates the send and receive energy so that it can be connected to the corresponding parts of the Earth station (e.g., send must be routed to the uplink and receive to the downlink). The hybrid is a transformation device that can introduce an undesired return path for audio signals. This results in an echo that can be heard at the distant end over the satellite link. Because even a low level of echo on a satellite link is very objectionable, it must be removed through active techniques. The ITU-T has specified the echo cancellor in its Recommendation G.165. The advent of the digital echo cancellor has made it possible for satellite telephony to become competitive with terrestrial phone service. Every FTS installation will include echo cancellors either as part of the Earth station or within the associated telephone exchange where it can service long-distance terrestrial circuits as well. In integrated digital networks, the echo cancellation function is often performed on a bulk basis on T1/E1 circuits.

The usable bandwidth within the analog voice channel extends between 300 Hz and 3,400 Hz, which is an international standard. It is capable of transferring speech, low- and medium-speed computer data with the use of modems, and subscriber signaling through the touch-tone pad. The touch-tone signaling approach is officially referred to as the *discrete tone/multifrequency* (DTMF) code. In this scheme, the telephone number digits can be sent over the active voice channel using pairs of audible frequencies.

In the case of a four-wire connection, the send and receive are separated from the origination point, eliminating the need for the hybrid. In a totally private network, it might be possible to use four-wire telephones and other devices to provide the best possible transmission quality and eliminate the potential for echo. The other aspect of a four-wire transmission to the exchange is that on-hook and off-hook indications are conveyed over a separate pair of wires call E (for ear) and M (for mouth). The M lead interface to the Earth station accepts line and address signaling by detecting DC signaling state changes. The complementary function is performed by the E lead, which generates output line and address signaling. This also improves network performance and simplifies interface design. It still is possible to use DTMF to transfer dialing digits and other network information.

The signaling system allows the network to transfer dialing digits, supervisory information, and customer care information. The R1 system is generally employed in North and South America, while the R2 system is the standard for Europe and much of the rest of the world. Both depend on DTMF to transfer dialing digits and the E and M leads to service the circuit. However, they perform DTMF signaling differently in that the standard version of R2 employs the compelled mode. According to this protocol, the sending point waits to end its transmission until the receiving end has responded. This introduces long time delays for each signal sent. Semicompelled R2 signaling was first implemented in Indonesia's Palapa A FTS network as a cure to this compelling problem. This uses pulses of supervisory signals that are not held for response from the distant end. Whenever DTMF signaling is used in the R2 signaling system, it should adhere to the semicompelled scheme.

With the rapid digitization of the PSTN worldwide, there is a trend toward the use of SS7 instead R1 or R2. This is a natural outgrowth of the popularity of the intelligent network, a forerunner to the more comprehensive standard called *integrated services digital networks* (ISDN). Intelligent networks offer services like the use of calling cards or credit cards, virtual private networks for voice and data, and flat hierarchies that need not follow the traditional structure shown in Figure 8.1. The nature of SS7 is that it is transmitted outside of the voice channel, a technique referred to as out-of-band signaling. Furthermore, the signaling information for several voice circuits is combined onto one SS7 channel, which runs at a multiple of 64 Kbps. It would be possible to add SS7 as a signaling interface to a demand-assigned network, which is something that had not been available at the time of this writing. On the other hand, the out-of-band or common channel signaling is standard in a demand-assigned SCPC network, as will be discussed later in this chapter.

Readers are probably interested in newer service options like ISDN. In Table 8.2 we suggest that an ISDN interface can be provided to subscribers via the FTS network. This is fundamentally straightforward when digital transmission is used; however, the problem arises with the specifics of implementation. The concept of providing two digital dial-up circuits at 64 Kbps each is not new and is in fact available within many developed countries. The issue has to do with the specifics of the data protocols and signaling schemes, which have not been settled in many countries (including the United States at the time of this writing).

The table covers a very wide range of options, and it is doubtful that one type of FTS interface equipment will satisfy all them without modification or reconfiguration. The approach taken by suppliers of FTS Earth stations is to make the interface portion as modular as possible. In addition, there is a high degree of software control needed to accommodate the protocols involved with the interface to the PSTN. For the international gateway situation, there are fewer standards to contend with, so the interface can be satisfied in a straightforward manner.

8.2 DEMAND ASSIGNMENT SCPC NETWORK ARCHITECTURE

The capability of FTS is greatly enhanced through the demand assignment feature that has been adopted in thin-route networks around the world. Rather than identify specific frequency channels or time slots ahead of actual demand, the required bandwidth is made available only for the duration of the call. In Chapter 7, we discussed the use of demand assignment in conjunction with the bandwidth reservation mode of TDMA. This is applied to FTS networks as well, but the more popular approach is to use FDMA in the form of SCPC. Channel pairs are assigned in response to call demand. This is precisely the same technique that is applied in terrestrial analog cellular systems like AMPS and TACS.

8.2.1 Demand-Assigned Network Topology

Most demand-assigned SCPC networks consist of various types of Earth stations that are capable of mesh connectivity. An example of such an arrangement is shown in Figure 8.3. In response to the variety of interface situations identified in Figure 8.2, there are two classes of Earth stations, namely, the remote terminal and the gateway. The remote terminal serves private or public end users who require one or more telephone access lines into the FTS network. The gateway is a much larger Earth station that serves all remotes by providing access to the PSTN. The network is controlled from a central point, indicated by the *network control system* (NCS) Earth station. This could be located at a gateway or operated as an independent facility.

The use of the term gateway here should not be confused with an international gateway, which is a telephone exchange that connects calls to other countries. Rather, it is a *gateway Earth station* (GES) used to complete calls from remote terminals to terrestrial subscribers and for allowing terrestrial subscribers to call subscribers that can only be reached via the FTS network. There is also the potential for placing calls between gateway Earth stations so that the FTS network can serve at higher levels of the telephone network hierarchy (Figure 8.1).

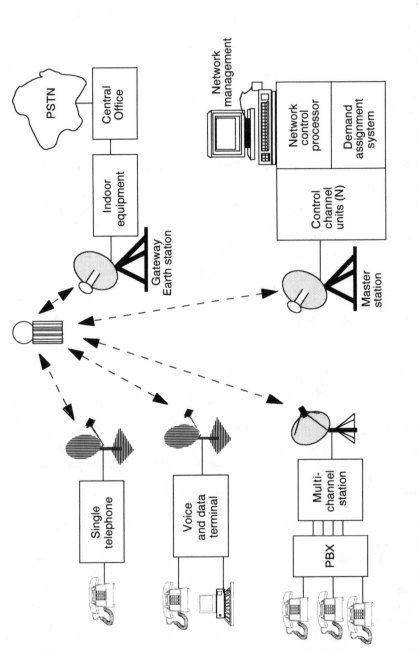

Figure 8.3 Typical arrangement of a demand assignment FTS network with central network control and management (courtesy of Hughes Network Systems).

The demand-assigned FTS network allows any station to establish an SCPC circuit to any other station. Thus, we have a fully interconnected mesh network activated on demand. Reflecting the compact design of the VSAT, a typical remote terminal is composed of an outdoor antenna and RF head and an indoor unit that contains the electronics and interface equipment. The following telephone interface options are suggested in Figure 8.3:

- Two-wire telephone;
- Data communication device (PC or terminal);
- Fax machine (not shown);
- PBX, whether two-wire or four-wire;
- Host computer.

The GES typically uses a larger antenna and RF amplifier to be able to service more simultaneous calls. Sufficient SCPC channel units would be provided to support the maximum calling demand of the particular GES. The channel units would interface to the local or toll exchange on a four-wire or T1/E1 basis. The latter requires a multiplexer to take the individual analog outputs of the channels units and combine them into a high-speed digital stream. Alternatively, the digital outputs of the modems within the channel units can be transformed into the T1/E1 directly using a device called a trans-multiplexer (also referred to as a transmux).

The third element of the topology is the NCS, which implements the demand assignment feature. It has become a common practice to put the intelligence of network control into a central facility that is operated by a service provider. In this way, one entity can manage all aspects of the network and collect the necessary call accounting information for billing purposes. There is an Earth station associated with the NCS since all control and information gathering is performed over the same satellite transponder that contains the traffic channels. This is performed with control channels that are accessed by the remote terminals and gateway Earth stations in order to reach the NCS. A supply of special control channel units at the NCS are available at all times to support the calling demands of the various Earth stations of the network. Also, software and computers at the NCS process the data transferred over these links and make frequency assignments to the remote stations and gateway Earth stations. Attached to the NCS is a network management station that allows operations personnel to configure the network for service and deal with various problems that arise from time to time.

8.2.2 Use of Satellite Capacity

We have indicated that the majority of demand-assigned FTS networks employing mesh topology use the SCPC mode of satellite transmission. The

SCPC approach has all carriers operating at the same power level at the satellite, producing a uniform spectrum across the transponder bandwidth. The NCS assigns channels in pairs to allow two Earth stations to make a duplex connection for the duration of a call. In addition to these traffic channels, there are the control channels that are used to pass call setup and administrative messages between the NCS and the Earth stations (remote and gateway). The architecture of the control channel network is similar to the VSAT star, with outbound control channels between the NCS and the Earth stations and inbound control channels between the remote Earth stations and the NCS. As with the bandwidth reservation mode of the VSAT, the outbound channels use TDM and the inbound channels use ALOHA.

The manner in which spectrum is provided and used in a demand-assigned SCPC system is shown in Figure 8.4. In a typical network, there might be a total of 2,500 channels that are available to service calls and for network control (the figure shows a lesser number for illustrative purposes). Total transponder power must be divided among the carriers and an allowance made for output backoff to control intermodulation distortion. The precise number of channels is determined by the required EIRP per carrier (from the link budget) and the associated bandwidth per channel. A smaller allocation of bandwidth is made for the control channels between the NCS and the Earth stations.

There are a number of factors that the network designer can control in the use of the bandwidth and power of the transponder. The required power per carrier is an important characteristic because it is the basis of determining the total amount of transponder capacity and the resulting cost of the space segment. In Table 8.3, the total number of traffic and control channels is estimated for a typical network at Ku band. The transponder in this example has sufficient power to support the equivalent of 1,000 full-time SCPC channels. The carriers are assumed to be voice activated, which means that the individual

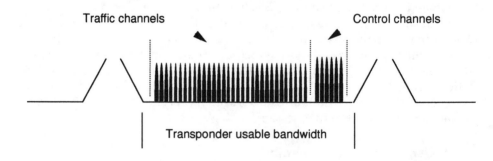

Figure 8.4 Arrangement of demand-assigned SCPC channels within the usable bandwidth of a transponder.

Table 8.3
Capacity Calculation for a Typical
Demand-Assigned FTS Network

Factor	Value
EIRP per carrier (from link budget)	10 dBW
Transponder saturated EIRP	45 dBW
Required output backoff	5 dB
Number of channels	30 dB
	1,000
Voice activity	40%
Channel capacity	2,500
Control channels (10%)	250
Traffic channels	2,250

transmitters are turned on and off in response to speech. An average duty cycle of 40% is accepted as an industry standard. Therefore, the transponder can support 1/0.4 times the number of active channels, which amounts to 2,500. There must be sufficient bandwidth for this number of channels at full loading. Of these, an allocation of 10% is made for control channels, leaving 2,250 for revenue-bearing telephone traffic. This assumes that each control channel can support approximately 112 call requests during the same time interval.

This example is somewhat simplified in that we assume that all carriers are operated in the same manner and that the channels are uniformly loaded across the transponder bandwidth. In a real network, carrier powers and bandwidths vary between types of Earth stations and channel capacity requirements. Also, there are variations in power due to link fading and equipment power output misadjustment. All of these factors tend to reduce the usable capacity. A thorough analysis for the type of network should be performed in cooperation with the network supplier. They most likely know the detailed assumptions and operating parameters for their system and can provide specific examples for the planned network.

8.2.3 Fixed Telephony Earth Station Design

The typical FTS remote Earth station, shown in Figure 8.5, incorporates the VSAT RF terminal with SCPC transmission features to provide mesh connectivity. The dotted box indicates the equipment that resides within the Earth station. In this example, the Earth station interfaces with an analog telephone on a two-wire basis. This analog interface emulates the local exchange by providing such functions as:

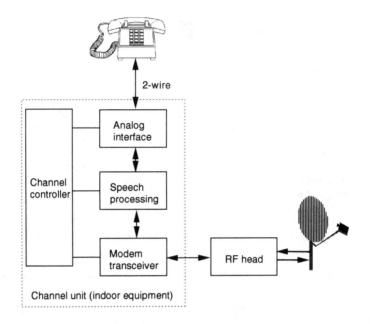

Figure 8.5 Typical equipment arrangement for an SCPC Earth station in a demand-assigned FTS network. The channel unit contains the baseband and control elements and the RF head is similar to the type used in a conventional USAT. The station provides a two-wire interface to a standard telephone instrument.

- Provides DC current for the phone;
- Detects on-hook/off-hook conditions;
- Responds to DTMF signaling;
- Provides both DC and ringing functions;
- Provides two-wire to four-wire conversion with a hybrid.

The speech processing function within the channel unit is critical to the operation and quality of performance of the FTS service. Within it, analog speech is converted into digital data through the A/D process. This bit stream is compressed through one of a number of techniques that produces a lower data rate for satellite transmission. In addition, error correction is introduced to improve overall performance. The output of the speech processor would be in the range of 5.6 Kbps to 16 Kbps, depending on the requirements of the service. Both ends of the connection must use the same processing. We consider various speech compression techniques in the context of mobile communications in Chapter 9.

The channel unit, as illustrated in Figure 8.6, is the core of the FTS Earth station. There is one channel unit for each active conversation. These devices are fairly complicated and, as a consequence, represent a significant part of the expense of the Earth station. A station with a large capacity will, of necessity, have a large investment in channel units. From the telephone interface, analog voice band information is digitized, compressed, and modulated for transmission. The reverse direction is provided so that the channel unit is a balanced device. Within the channel unit, the speech processor provides the following important functions:

- Echo cancellation to control the echo produced in the two-wire-to-four-wire conversion process, which should satisfy the requirements of CCITT Recommendation G.165;
- Signaling tone generation and detection to properly support the calling process;
- Speech activity detector to turn the transmitter on and off in response to the local user's voice, which reduces the average uplink power usage by a factor of 0.4 to enhance capacity;
- Speech compression, using an accepted algorithm such as *adaptive differential PCM* (ADPCM), *residual excited linear predictive* (RELP), or *code excited linear predictive* (CELP).

Speech processing substantially reduces the information bandwidth and thus will seriously degrade a modulated signal from a conventional voiceband modem or fax machine. For these services, the speech processing is bypassed and the data is applied directly to the SCPC modem within the channel unit. A fax signal, when detected, is routed by the telephone interface to a separate fax interface unit where the audible fax signal is converted back into digital data. From there, it can be transferred to the downstream parts of the channel unit like data that has arrived over the data port.

The following is a brief and somewhat simplified explanation of how the fax type of call is handled between a communicating pair of SCPC channel units. The presence of a fax signal (instead of human speech) is detected in the fax interface unit. The speech processing function is then disabled and a data channel is established in its place through the data interface. Then, the fax interface on the sending end conducts a handshaking process with the fax interface at the distant receiving end. This includes the determination of the best data rate to be used for the fax transfer (a function normally conducted by standard Group 3 fax machines). At this point, the channel unit at the sending end engages the local fax machine to begin its transfer of the image. The two fax machines can then communicate in the normal manner as if connected over an analog telephone line.

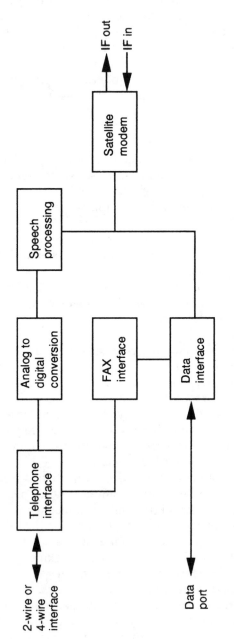

Figure 8.6 Typical elements and interface features of the SCPC channel unit.

A direct digital interface is provided when the efficient transfer of computer data is desired. The data interface must satisfy the requirements of a common standard such as RS-232 or V.35. In the case of the latter, the data interface provides bit scrambling to randomize data so that there are enough bit transitions for synchronization at the receiving end.

The satellite modem is another important element of the Earth station as it must achieve a high-quality link from end to end. The following are the functions of the modem:

- Differential coding of the data coming from the speech processor or data interface to set up the data for phase shift modulation by removing phase ambiguity over the satellite link;
- FEC coding to improve bit error rate performance over the satellite; options can be provided to allow various rates to be used, such as rate 1/2 and rate 3/4;
- Modulation, which is a series of functions including baseband filtering, modulation, channel frequency selection, carrier on/off control, and carrier power control;
- Demodulation, which is a series of functions including downconversion of the selected channel, demodulation, automatic gain control (leveling), automatic frequency control, carrier recovery, symbol timing recovery, and preamble detection; modems may either employ biphase PSK (BPSK) or QPSK.

The are a variety of potentially useful configurations for the remote FTS Earth station. The most basic has been called the jungle telephone, so named because it can provide a usable line in a jungle or literally anywhere else. The jungle telephone would be attractive where a subscriber needs a telephone line but none is available from the PSTN. Service is available through a GES to the national PSTN. Connections between jungle telephones are also possible and potentially of value. This would be useful at industrial sites or for emergency communications. In many ways, the jungle telephone resembles the mobile user terminal discussed in Chapter 9.

The multichannel Earth station configuration would supplement the PSTN by providing both local access to a village or industrial site or as an added means of trunking calls between toll exchanges. A typical configuration of a multiple-channel FTS Earth station is provided in Figure 8.7. This could be used as a higher capacity remote Earth station or, with enough equipment, as a GES in a major city. Indicated are a number of channel units, each of which contains all of the elements discussed previously. The outputs of the units are added in power and applied to the RF terminal, shown at the center of the figure. It interfaces with the indoor equipment at an IF frequency such as

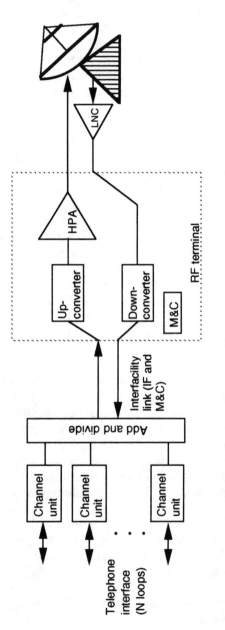

Figure 8.7 Interfaces and major elements of a multichannel FTS Earth station.

70 MHz or 140 MHz to provide access for the quantity of channel units. A twisted pair signal cable is often included with the IF cable between indoor equipment and the RF terminal to transfer *monitor and control* (M&C) signals.

The RF terminal can be designed for operation in any of a number of frequency bands, particularly C, X, Ku, and Ka bands. C and Ku are the most popular at the time of this writing simply due to the much greater availability of satellite capacity around the world. The other two bands would be introduced as needed and as the technology becomes more readily available. On the uplink side, the IF input to the RF terminal is translated as a group of channels to the particular transponder frequency within the up-converter. The output is on the right frequency but at too low a power level for the desired transponder operating point. Power is increased substantially by the *high-power amplifier* (HPA) at the end of the uplink equipment chain.

The types of HPAs that are found in FTS service are either the *solid-state power amplifier* (SSPA), which is constructed from high-power field effect transistors, and the *traveling wave tube amplifier* (TWTA). Typical power levels that are available on the market at the time of this writing are listed in Table 8.4. SSPAs under 10W have been in use in the RF head of the standard VSAT. The same type of amplifier and RF head can be used in an FTS remote station with a capacity of one or two channels. For higher capacity requirements, the medium-power SSPA is available. It consists of multiple low-power stages that have been paralleled to achieve an output of 20W. The table indicates that Ku-band power levels are lower than C band, which is a result of the difficulty of handling the power within the smaller dimensions of Ku-band FETs and associated amplifier circuitry.

For an FTS station with a high-channel capacity requirement, such as that used for a GES, the TWTA becomes necessary. This is the standard way to provide this level of power, but it brings with it higher cost and complexity in operation. The TWTA requires relatively high voltages and currents and so is

Table 8.4
Typical HPA Systems in Use for FTS Earth Stations

Type of Amplifier	Maximum Power		Comments
	C Band	Ku Band	
SSPA, low power	5W	2W	VSAT type
SSPA, medium power	35W	20W	
TWTA, medium power	150W	120W	Laboratory type
TWTA, high power	500W	350W	Video type
Klystron, high power	3 kW	2 kW	Video type

subject to a number of potential failure modes. Untrained people should not attempt to open or operate high-power equipment. In addition, it is absolutely critical that moisture not enter the high-power section of the outdoor unit. While an SSPA should have an indefinite lifetime, a TWT will wear out due to the limited lifetime of its cathode. It will normally be necessary to replace a TWT after two to four years of service. Earth stations with TWTAs usually have an active backup amplifier so that one is available in the event of failure or the need to perform routine or emergency maintenance. This is not shown in Figure 8.7.

On the receive side, the full downlink frequency range is amplified and translated to 1 GHz within the *low-noise converter* (LNC). The desired transponder is selected by the down-converter, and the spectrum of SCPC carriers is then applied to the interfacility link at IF (either 70 MHz or 140 MHz, depending on the type of equipment involved). The individual channel units within the indoor equipment can then select the assigned channels from within the IF range. Not shown in the figure are elements of the indoor unit that control the demand assignment function and control of the RF terminal from the indoor equipment.

8.3 PREASSIGNED POINT-TO-POINT LINKS

Satellite transmission has long been applied to FTS service with preassigned links that operate at the higher levels of the hierarchy in Figure 8.1. These are preassigned point-to-point links that carry several telephone channels through the concept of multiplexing. The level 3 links are between pairs of domestic toll exchanges, while level 2 links serve the international gateway. International links themselves are commonly provided over the INTELSAT system and are definitely candidates for FTS service.

The preassigned links first implemented over satellites employed *frequency division multiplexing* (FDM) with capacities ranging from 12 voice channels at the low end all the way up to the 1,800 channels on the highest capacity links found in service in the INTELSAT system. However, analog FDM, which is efficient on the basis of bandwidth per channel, has been replaced by TDM. This has raised the quality of satellite voice communication to be comparable to what the digital terrestrial network can deliver. INTELSAT's IDR carriers are now standard in the international system. Digital services like medium-speed data and video teleconferencing easily can be introduced and integrated with voice using multiplexers and bandwidth managers of various kinds. A private satellite service called *INTELSAT Business Service* (IBS) became popular in the late 1980s, joined by EUTELSAT's version called SMS.

The capacity per link is often composed of a T1 or E1 channel, although higher bandwidths have been used for time to time. For example, digital carriers

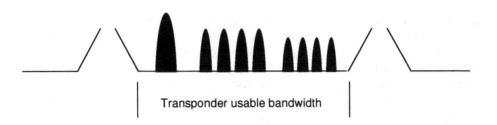

Transponder usable bandwidth

Figure 8.8 An example of a transponder that is fully loaded with MCPC transmissions.

with 6 Mbps or 8 Mbps can share the bandwidth and power of a common transponder through the *multiple-channel per carrier* (MCPC) technique. A graphic example of how MCPC carriers would be placed in a satellite transponder is provided in Figure 8.8. The large carrier would come from an uplinking Earth station that needs to transmit a relatively wide bandwidth, while the other two sizes match Earth stations with smaller traffic requirements. The carriers do not occupy all of the bandwidth because of the loss of capacity from providing sufficient output backoff. Even still, a properly loaded 36-MHz transponder can support up to about 24 Mbps of useful transmission in the multiple-carrier mode. The precise capacity will depend on the saturated EIRP of the satellite and the *G/T* of the Earth stations.

The approach to transponder loading is the same as SCPC, with the exception that a variety of carrier capacities occupies the available bandwidth. The power level of each carrier is adjusted to meet the needs of the particular link. Intermodulation distortion is controlled by allowing an appropriate amount of output backoff, typically in the range of 4 dB to 6 dB for a TWTA and 2 dB to 4 dB for an SSPA. The precise amount can be determined once the specific carrier arrangement and power amplifier characteristics are known. Fortunately, the answer in not particularly sensitive to the precise value of backoff, and it is often the practice to use a "round" number of 5 dB for the case of a TWTA. This means that a Ku-band transponder with a saturated EIRP of 46 dBw (single carrier) would be capable of delivering 41 dBw of total EIRP when transmitting a group of MCPC carriers such as that shown in Figure 8.8.

A permanently assigned FTS needs a good system of multiple access to keep the individual transmissions separate. Most FTS systems in operation today use either FDMA or TDMA; however, there is a decided interest in exploiting the interference rejection characteristics of CDMA. The Earth station equipment in the case of FDMA is perhaps the simplest because separation is maintained by permanently assigning different carrier frequencies (indicated in Figure 8.8). For TDMA to work properly, stations must transmit their information as synchronized bursts, as discussed in Chapter 7. The CDMA mode of operation does not require any coordination between Earth stations, which can transmit at the same time on the same frequency. However, the complications come in from the need to properly despread the desired carrier and synchronize to its timing. Also, the determination of the number of Earth stations that can share the same frequency channel or transponder is a complex matter, as we will discuss in Chapter 9. In the following subsections, we describe two topologies used in preassigned MCPC links.

8.3.1 Single-Destination Carriers

A single-destination carrier is an Earth station transmission that is intended for one other Earth station, representing half of a full-duplex link. The bandwidth is equal in both directions, as in a microwave radio path or fiber optic cable that connects two end users. The flow of information for such an MCPC link is shown in Figure 8.9. A simple TDM format is used so that the information transmitted by A is perfectly balanced by that from B. This provides full-duplex service and does not require any preprocessing within the Earth stations other than what is normally associated with the modems on each end of the satellite link. In the figure, we assume the first block of time is used for data transmission, the second for fax, the third for digital video teleconferencing, and the final block for telephony. The sequence repeats according to the frame format of the link.

8.3.2 Multidestination Carriers

The first generation of the INTELSAT system introduced full mesh connectivity using analog ground equipment that was available in the late 1960s. Up to that

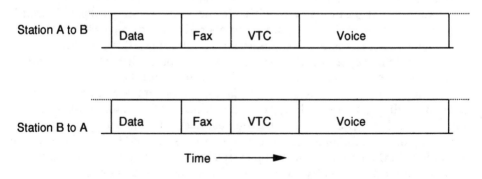

Figure 8.9 Transmission timelines for single-destination TDM carriers that connect two stations together in a point-to-point link.

time, the single destination carrier using FDMA was heavily applied to terrestrial microwave networks and was a natural choice. However, to apply it, every Earth station would have had to transmit a separate carrier for each destination. This would have resulted in a proliferation of transmission chains at the ground stations and a somewhat inefficient loading of the transponder. As a solution, the multidestination carrier was adopted to allow each station to transmit only one carrier that would contain the multiplexed traffic for all destinations.

An example with four Earth stations transmitting multidestination carriers is shown in Figure 8.10. Each station uses TDM to transmit blocks of information to the three other stations, achieving full mesh connectivity. The carriers are the same bandwidth and capacity, but blocks are unequal in size, in recognition of the fact that balanced information transfer is only required between pairs of stations. For example, station A transmits its half of a duplex link to station C within its second block. In return, station C uses its first block to transmit the complementary side of the link to station A. The remaining connectivities can be derived from the figure in the same manner. Note that there are two unused blocks at the ends of stations C's and D's time frames. This was not assigned to traffic and hence goes unused. It would have been possible to increase the size of the transfer between stations C and D only, since stations A and B already fully utilize their time frames. Alternatively, the carrier data rates of stations C and D could be adjusted downward and the blocks extended in time to fill out the unused time slots.

Transmission from:

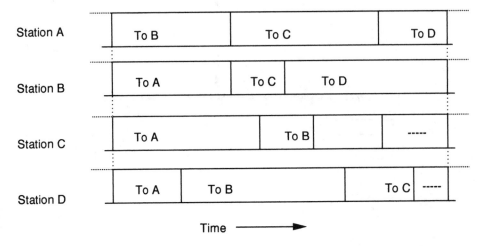

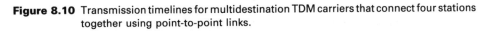

Figure 8.10 Transmission timelines for multidestination TDM carriers that connect four stations together using point-to-point links.

A simplified diagram of how the uplink and downlink equipment chains are arranged at the Earth station is shown if Figure 8.11. The top box in each station's equipment configuration is used to transmit the multidestination carrier to every other station. To make this possible, each station must have three receiver chains, as shown below the transmit chain. Two sets of direction arrows indicate how connectivity is achieved between station A and station B (the

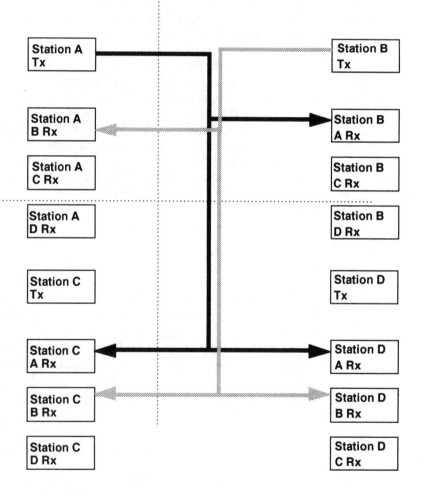

Figure 8.11 Equipment arrangement for multidestination carriers for a four-station network. Each station transmits one carrier and receives the transmissions from the three other stations. Connectivity between stations A and B is shown.

others are left out for clarity). Station A's carrier in the heavy black arrow set extends to B, C, and D; while station B's carrier in the shaded arrow set extends to A, C, and D. It looks complicated, but in actuality the arrangement is easy to operate since the transmissions are steady state and the equipment can be left online indefinitely.

One step at each station that is not shown in Figure 8.11 is the assembly of the three received downlink data streams into a single time frame. This is necessary to meet the terrestrial interface requirements to the telephone exchange or user-owned PBX. Station B, for example, must pull off its traffic blocks from the downlinks from stations A, C, and D and plug them into a time frame that corresponds to the exact format of its uplink. This bidirectional TDM link is then in proper format for use in standard digital exchanges.

The multiplexing equipment used for this purpose can be derived from standard TDM multiplexers or other bandwidth managers. A popular type is called *digital circuit multiplying equipment* (DCME), which is often found in INTELSAT Earth stations that employ the multidestination technique. Another advantage of DCME is that it can increase the number of telephone channels on a given transmitted carrier through *digital speech interpolation* (DSI) and speech compression. A multiplication by a factor of four is typical. DCME is available for both the single-destination and multidestination types of carriers.

Another important element of the transmission system is the modulator and demodulator. Most of these are packaged as pairs in the form of a satellite modem. A typical example of such a device can transmit at any data rate between 64 Kbps and 8.448 Mbps, allowing the bandwidth of the carrier to be adjusted to the traffic requirements. This is a good idea because satellite operators charge for transponder capacity based on the required bandwidth and power. The resulting loading of the transponder would be similar to that shown in Figure 8.8. The modem interfaces with the telephone multiplexer or telephone exchange using a standard such as V.35, T1, or E1 and with the uplink equipment at an IF frequency of 70 MHz or 140 MHz. The modulation format is either BPSK or QPSK, and a variety of FEC coding schemes are available to match the characteristics of the Earth stations in the network.

8.4 APPLICATION IN RURAL TELEPHONY

FTS has increased in popularity due to the rise in demand for good quality telephone service in the developing regions of the world. The techniques discussed previously make FTS a very versatile problem solver in remote areas where it is too expensive or difficult to extend the terrestrial PSTN. One big advantage of the satellite approach is that once satellite capacity is available in the particular country, the network can be installed in a few months or even one month. The network then services smaller cities, villages, and even individual locations with all of the capabilities of the modern digital PSTN. A well-designed FTS can enhance the quality of life and economic opportunity in these regions as many of the more advanced features of the intelligent public network become available to almost anyone.

An example of a complete FTS network implementation is shown in Figure 8.12. The network provides basic telephone, fax, and voice-band data service to villages and remote users with dedicated Earth stations. Connection to the PSTN is through a gateway Earth station at the capitol city (this may or may

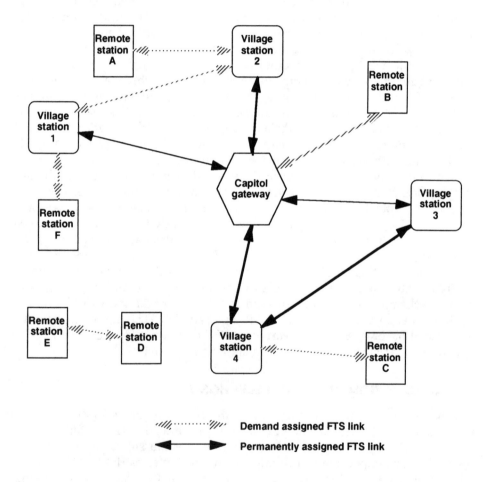

Figure 8.12 An example of a rural FTS network that employs both demand-assigned and permanently assigned links.

not be associated with the international gateway for the country in question). The heavy arrows represent permanently assigned trunk channels that support predictable traffic volumes between villages and the capitol. Villages 3 and 4 have enough traffic between them to require such a permanently assigned trunk. Any of the permanently assigned links can be implemented either with single-destination or multidestination carriers.

The real power of the rural network is achieved with demand-assigned links, indicated with the shaded arrows. These are established only when a telephone call is placed, allowing a station of any type to communicate with any other. Remote stations can call villages, the capitol, or even other remote stations. By providing connectivity on demand and without restriction, the FTS network creates a nationwide telephone infrastructure that does not depend on the degree of sophistication of the existing PSTN.

Satellite capacity for an FTS network can be drawn from a domestic C- or Ku-band satellite. Once Ka-band satellites are launched, they, too, can supply capacity for rural FTS networks. The capacity of a typical C-band transponder is in the range of 500 to 3,000 channels, depending on the voice channel data rate and the G/T of the Earth stations. In a typical network, a transponder could supply 1,000 simultaneous calls involving 500 pairs of subscribers. These could put rural callers in contact with people in cities who are connected through the PSTN to a gateway. Alternatively, they all could be in rural areas, using the satellite network as their only means of communication. Since individual subscribers do not use their telephones continuously during the busy hour, the same 1,000 channels would support a user community of between 50,000 and 100,000. Multiply the number of transponders, and the size of the user population can likewise be multiplied. One satellite with 24 transponders can therefore provide rural telephone service to over one million subscribers.

There is a natural extension of FTS from fixed communications to the area of providing mobility to users. Chapter 9 reviews the progress made in the introduction of mobile telephony using satellites that operate in GEO and non-GEO orbits. This is another exciting development in the application of satellites to commercial communications.

Reference

[1] Schwartz, M., *Telecommunication Networks*, Reading, MA: Addison-Wesley, 1987, p. 522.

Mobile Satellite Service (GEO and Non-GEO)

9

Satellite communications is a natural facility for serving users while they travel by the various means of transportation. These include ships on oceans, rivers, and lakes; commercial and private aircraft; land-based vehicles of various types; and individuals using portable and handheld devices. The MSS is an established user of spectrum at frequencies below about 3 GHz. This regime is preferred because of its greater ability to bend and reflect over and around obstacles. (The low end of the usable spectrum is probably 100 MHz, which is able to penetrate the ionosphere under all conditions.) Frequencies above 3 GHz, while more readily available, are easily blocked by natural and manmade obstacles and introduce practical difficulties when it comes to generating transmit power.

The year 1990 was a turning point for MSS, when Motorola introduced their concept of a *low Earth orbit* (LEO) satellite system capable of directly serving handheld terminals. When the idea first appeared, the belief within the established GEO satellite industry was that even the most advanced satellites of the day were not capable of supporting direct links to handheld phones. However, by 1994, several companies had devised schemes to provide mobile service to handheld phones, employing literally all altitudes ranging from Iridium's 800 km all the way up to 36,000 km at GEO. Beginning in 1998, there will be the prospect of handheld phones communicating directly with satellites. These systems will also see service for fixed telephone and temporary and emergency communications, which will have significant impact in areas and times of great need.

9.1 FOUNDATION OF THE MOBILE SATELLITE SERVICE

The role of satellites in mobile communications has expanded since the first MSS satellite, Marisat, in 1976. However, the basic connectivity has not changed even though a mobile terminal is handheld instead of mounted on a ship. As shown in Figure 9.1, a full-duplex link is provided so that the user can conduct

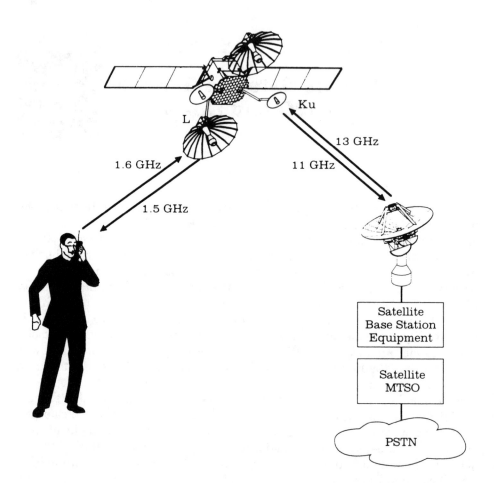

Figure 9.1 The mobile-to-fixed link, using L band to communicate with the mobile subscriber and a Ku-band feeder link to the gateway Earth station. The mobile subscriber can make and accept calls from the PSTN. A dual-mode capability would permit the mobile subscriber to use the same user terminal on the terrestrial cellular network as well as the MSS system.

a normal telephone conversation through the satellite and GES. In this example, the user terminal (UT) is in the form of a specialized handheld mobile telephone that can transmit and receive within the L-band part of the spectrum between 1.5 GHz and 1.6 GHz. This in itself is not a major accomplishment since there are existing terrestrial mobile telephone services called the *personal communications network* (PCN) and the *personal communications service* (PCS) that use the 1.8-GHz and 1.9-GHz bands, respectively.

The antenna of the UT provides very little gain because of the need to maintain small size and to provide a nearly omnidirectional pattern. The satellite in the figure, on the other hand, has a much larger antenna system than found in FSS systems in order to close the link to the user. In most of the modern MSS systems, the spacecraft antenna produces an array of small spot beams that cover the ground with a grid of cells that emulate a terrestrial cellular network. This concept will be reviewed in more detail later in the chapter.

The link between the satellite and the GES is called the MSS feeder link and can be implemented within a standard FSS allocation. The Ku band is illustrated in the example in Figure 9.1; in contrast, maritime mobile satellite systems employ C band for this purpose. The GES, representing a large fixed asset of the service provider, interfaces with the PSTN and has a large antenna and high uplink power capability. The feeder link can be engineered for adequate availability and thus minimizes its impact on the overall quality of service as rendered to mobile users. Within the GES is the RF terminal and associated baseband equipment much like its counterpart in FTS applications (reviewed in Chapter 8). In fact, the architecture of a mobile telephony MSS network draws heavily from its FTS counterpart. Figure 9.1 indicates that the Earth station includes or is directly connected to a *mobile telephone switching office* (MTSO) to service telephone calls that are routed over the satellite. The MTSO is a standard local exchange or cellular MTSO design that has been modified for satellite mobile telephone service. Through the MTSO, the satellite network delivers a range of telephone, data, and fax services to mobile users who can be literally thousands of kilometers away.

A state-of-the-art MSS system can provide other capabilities besides those given in Figure 9.1. The most striking is direct mobile-to-mobile calling that allows subscribers to talk to each other regardless of their location and situation. Terrestrial cellular networks do not see much demand for this type of connection, but the situation with regard to MSS is yet to be observed in practice. It will depend heavily on the mix of subscribers and their usage patterns. The quality of the terrestrial telephone network in different countries served by the system will also play heavily into the attractiveness of direct mobile-to-mobile calling.

Some systems will address this by connecting these calls through a common gateway, introducing the delay and degradation of a double hop. This may not be a concern in LEO and MEO systems, where the propagation delay is relatively low from the start. On the other hand, this type of call ties up double the GES resources of a standard mobile-to-PSTN call. The approach being taken in the next generation of GEO MSS systems employs a single-hop approach where the satellite acts as a direct connecting point. The propagation delay is comparable to the single-hop mobile-to-PSTN call, requiring that the network establish the connection on a call-by-call basis.

Satellite communication among moving Earth stations is very different from a technical perspective than the fixed and broadcasting services discussed previously in this book. The situation of a user moving relative to the satellite (or, in the case of non-GEO satellite systems, a satellite moving relative to the user) causes dynamic behavior in the link. This results in various forms of fading that cannot be predicted in the same way as in fixed situations with rain attenuation and ionospheric scintillation. On top this, the amount of link margin that is available to the mobile link is comparable to other satellite applications, subjecting the user to short or long outages, depending on the geometry and local terrain blockage that exist at any particularly moment.

Terrestrial cellular systems also undergo dynamic fading as the subscriber drives past buildings and trees, under overpasses, and into isolated locations where the base station simply cannot reach. This fading is very rapid in time because it is produced primarily by multipath propagation, where the strongest signal that is received may at times be a reflected signal off of one or more buildings. The various reflections add and subtract in random ways, with the instantaneous sum varying very rapidly. What tends to overcome the problem is the very high link margins provided (in excess of 30 dB) along with the availability of multiple signal sources from several base stations that could potentially reach the subscriber at any one time.

The propagation model for MSS depends on the situation within which the mobile user is found. Ideally, there should be a direct line-of-sight path between the user and the satellite. The service quality in this case, which is suggested in Figure 9.1, is ideal because the link can be engineered for no outage due to the mobile-to-satellite path. If the user or satellite is moving, then the link will experience periods of blockage when the user is "shadowed" from the satellite transmission. The MSS network would either (1) allow the inevitable dropouts in the data transfer, which would intermittently halt the conversation or information flow or (2) attempt to eliminate or reduce the dropouts through path diversity. The latter is very expensive because it requires that the number of satellites be increased by perhaps a factor of two.

In most cases, the mobile user will want to be connected with the PSTN, which is provided through a GES that employs FSS spectrum at C or Ku band. We must consider the fading environment on this side of the link in the same manner as with any FSS applications. The resulting mobile-to-GES service can be engineered to meet almost any availability requirement.

9.1.1 Radio Frequency Spectrum Availability

Radio spectrum has been allocated by the ITU to MSS just as it has to the other satellite and terrestrial radio services. The L- and S-bands spectrum currently in use or planned for application in the next decade is shown in Figure 9.2.

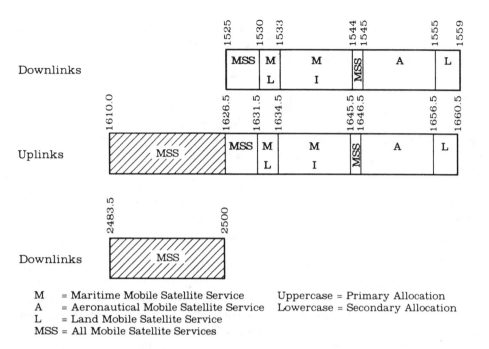

M   = Maritime Mobile Satellite Service   Uppercase = Primary Allocation
A   = Aeronautical Mobile Satellite Service   Lowercase = Secondary Allocation
L   = Land Mobile Satellite Service
MSS = All Mobile Satellite Services

Figure 9.2 Spectrum allocations for the MSS resulting from the World Radiocommunication Conference of 1992. The hatched portions represent a new allocation for expansion of the service through non-GEO satellites.

The unshaded downlink and uplink bands are already in use by Inmarsat and others, while the cross-hatched bands are reservations for the coming generation of non-GEO satellites that will be introduced within a few years of this writing. The impression that we take away is that the amount of available spectrum is substantially less than what is allocated to the FSS and BSS. This reflects the popularity of and competition for these frequencies. Much of this spectrum had to be taken away from fixed terrestrial services like short-haul microwave links and radar. Some countries, in fact, do not acknowledge that MSS has any particular right to these frequencies. This leads to coordination difficulties that any prospective MSS operator might need to address.

Notice from the footnotes to the table that the allocations have quite a few stipulations about its usage. Some portions give MSS a primary status, meaning that the nonprimary services, called secondary services, must give way. On the other hand, MSS is not even listed as secondary in the aeronautical bands, which is allocated to *aeronautical mobile satellite service* (AMSS) on a single-user basis. The AMSS is not in existence at this time, so the current practice is to use these bands until such time as the aeronautical people get their act together. This is allowed under a general provision of the ITU Radio Regulations, which permits the use of a band as long as harmful interference is not caused to anyone who is using the band in accordance with the official allocation.

The coordination of MSS satellite systems at L band is formidable as compared to both the FSS and BSS. Figure 9.3 indicates how ground antennas may or may not be able to separate the signals coming from satellites that operate on the same frequency. The satellites are spaced of the order of 3 deg apart, and their coverage areas overlap. In the typical FSS satellite system, users employ directive ground antennas that can discriminate among the satellites. The orbit spacing should be sufficient to allow the satellites to reuse the spectrum. For example, the antenna in the top illustration that is pointed at satellite A has a beam that is narrow enough to prevent the both uplink and downlink interference with satellite C. Satellite B, however, is within the uplink beam and, hence, can receive interference (but only if its footprint covers this particular Earth station).

Contrast this situation with the Earth station in the lower illustration, which in this case is using a much simpler antenna with a very broad beam. This is typical of the situation with a handheld or vehicular UT. (The reason for doing this is that the broad beam is not sensitive to the orientation of the antenna, either in azimuth or in elevation.) The closest orbit position that can reuse the spectrum is all the way over at C, which is effectively over the horizon from the user. MSS to this class of user is characterized by a complete lack of frequency reuse provided by orbit separation. The only reuse that we can count on would come from nonoverlapping satellite coverages, which is feasible as long as the spacecraft antennas on opposing satellites have excellent isolation properties. This can only be determined through the ITU coordination process, where the parties look in detail at the satellite designs, Earth station characteristics, and the types of carriers that are transmitted over the network.

9.1.2 MSS Link Design

The satellite links involved with MSS services are basically the same as their counterparts in FSS and BSS. The principle differences, addressed in a previous paragraph, are due to limited bandwidth available at L and S bands and the dynamic behavior of mobile link fading. To provide additional clarity, we include an example of a link budget in Table 9.1 for the forward link (from the GES to the UT) and return link (from the UT to the GES). Many MSS systems, such as the one illustrated here, assign Ku-to-L band to the forward link and L-to-Ku band to the return link. To maximize capacity and minimize uplink power from the UT, we employ speech compression, which reduces the basic transmission rate to 4.8 Kbps. The actual data rate on the link is 5.4 Kbps, which includes the extra bits to support supervisory signaling and forward error

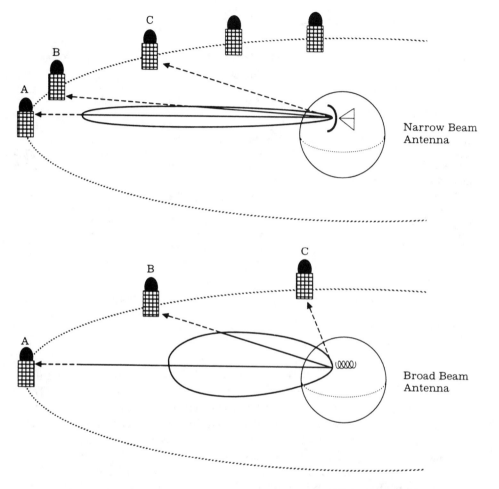

Figure 9.3 Earth station antennas allow the orbit and spectrum to be reused, provided that their beamwidth is sufficiently small. In the upper picture, this is the case because the main beam only includes satellites A and B, with C and the others not producing interference. The very broad beamwidth of the mobile-type antenna does not allow the orbit and spectrum to be reused unless the satellites cover different areas on the ground. Satellite C will not receive interference because it is beyond the local horizon of the user terminal.

Table 9.1
Simplified Link Budget for a GEO MSS Service to a Handheld User Terminal

Parameter	Forward Link		Return Link	
	Up	Down	Up	Down
4.8-Kbps voice SCPC bandwidth (kHz)	4.0	4.0	4.0	4.0
Frequency (GHz)	14.25	1.54	1.64	12.45
Transmit antenna	11m	12m	helix	2m
EIRP (dBW)	41.5	35.5	−4.5	5.0
Path loss (dB)	207.0	187.9	188.5	205.9
Atmospheric and other losses	0.4	0.2	0.2	0.4
Receive G/T	−4.0	−24.0	15.0	38.0
Receive C/T	−169.9	−176.6	−178.2	−163.3
Noise bandwidth [dB(Hz)]	36.0	36.0	36.0	36.0
Boltzmann's constant	−228.6	−228.6	−228.6	−228.6
C/N thermal (dB)	22.7	16.0	14.4	29.3
C/IM (dB)	30.0	22.0	50.0	24.0
C/I (dB)	24.0	20.0	23.0	22.0
C/N total (dB)	19.8	13.8	13.8	19.4
Link C/N (dB)		12.8		12.8
Required C/N (dB)		5.8		5.8
Margin (dB)		7.0		7.0

correction. The access mode is FDMA, and telephone channels are assigned to individual SCPC carriers that are voice activated. The detailed review of Table 9.1 provided in the next two paragraphs can be skipped by nontechnical readers.

In the forward direction, the GES transmits an EIRP per channel of 41.5 dBW at 14.25 GHz with an 11-m antenna. The uplink path loss to the satellite consists of 207.0 dB of free-space loss and 0.4 dB of atmospheric and other minor losses. The L-band downlink in the second column has an EIRP per corner of 35.5 dBw to reach a handheld UT. The "other" loss of 0.2 dB is from the polarization mismatch between the UT and the satellite. At L band, it is common practice to use circular polarization to make the link much less sensitive to Faraday rotation, which is more prominent at lower frequencies like L band.

The satellite provides a Ku-band footprint over a coverage area that contains all of the GESs in the network. In this example, this beam is regional in size and has a value of G/T equal to −4.0 dB/K. The uplink C/N (thermal) of 22.7 dB results from the uplink EIRP, path loss, G/T, and noise bandwidth.

The Ku-band uplink experiences interference from a variety of sources, including a cross-polarized uplink if it exists and other beams that operate on the same satellite. Interference from other satellites is very difficult to assess and so we need to treat it as a separate item during coordination (discussed in Chapter 10). The satellite repeater employs microwave power amplifiers that at L band are typically SSPAs. With multicarrier operation, there would be a contribution to the total noise from intermodulation distortion. Because the SSPA is loaded with a large quantity of small SCPC carriers, the intermodulation can be treated as noise through the use of a *C/IM* entry. Uplink *C/IM* is produced within the Ku-band GES because it also transmits multiple carriers. The value of 30 dB in the uplink is relatively high and so it has less of an impact than the downlink *C/IM* of 22 dB, produced within the SSPA. By combining each of these contributions, we obtain a total *C/N* for the forward link uplink of 19.8 dB.

Examining the GES-to-UT downlink, the operating frequency at L band is 1.54 GHz. The spacecraft antenna in this example is sufficiently large to produce adequate gain for the link to a handheld UT. The L-band EIRP per carrier of 35.5 dBW results from a transmit gain of the assumed 12-m antenna of 41.5 dBi combined with a transmit power of −16 dBW. Path loss at L band is less than at the Ku-band frequency by the term $20 \cdot \log(14.25/1.54)$, or 19.1 dB. The mobile phone in this example has a *G/T* of −24 dB/K, which is very low in comparison to even the smallest of dish antennas used in the BSS. The main reason for this is that the handheld UT uses a nondirectional antenna to allow the subscriber to be able to move around freely. The UT antenna is assumed to be a compact design that produces a pattern that is tolerant of the user's orientation. The gain of 2 dBi is an average transmit/receive value that would be achieved in the direction of the satellite with the UT properly held in the user's hand.

The Ku-band downlink of the return link is transmitted by a 2m area coverage spacecraft antenna, producing a *C/N* of 29.3 dB. Combining this with the indicated values of *C/I* and *C/IM*, produces a total downlink *C/N* of 19.4 dB. The last step in the analysis is to combine the total uplink and downlink *C/N* values, which gives 12.8 dB. The threshold *C/N* is determined from the vendor specification or, preferably, from laboratory measurements performed by the network implementer. In this example, the value of 5.8 dB considers the particular compression and coding system employed with the link operating an error rate of 0.01. The bottom line of Table 9.1 indicates an overall link margin of 7 dB, which is available to protect the link from shadowing losses as the UT moves from place to place.

The following comments highlight the factors in Table 9.1. In an overall sense, the forward and return links are balanced, meaning the resulting combined *C/N* values are the same. This is because of the dominant effect of the simple antenna on the handheld UT, which directly impacts the receive gain (forward link) and the transmit gain (return link). The uplink from the handheld UT is perhaps the most critical link in the system. The EIRP is limited by the low antenna gain of the omnidirectional type of antenna used with the UT. Transmit power to the antenna is constrained two important factors. The first is due to battery power, which must be held to a minimum to extend the duration of use between recharging. The UT is no good, after all, if the battery is dead after only one phone call. A general guideline is that the phone should have enough battery power for one full hour of talking (which includes dialing and holding time) and, alternatively, ten hours of standby time when the phone can potentially receive calls.

The second factor is the concern for a potential hazard to the human body from RF radiation. While this risk has not been quantified at the time of this writing, there are some specifications for power levels that are considered to be safe. The average power that is applied to the antenna should be under one-third watt, with the peak power allowed to be some factor times the average. This is not a recommendation by the author, and readers wishing to apply the standard should review the relevant documentation provided by the IEEE as well as the Federal Communications Commission of the United States. The resulting EIRP from the handheld UT is -4.5 dBW in the direction of the satellite. In reviewing the C/N thermal values in Table 9.1, it is clear from the value of 14.4 dB that the uplink dominates the overall C/N performance. A higher value is not practical because of the limited EIRP from the UT coupled with the challenge of producing a high-satellite G/T at L band. A C/IM of 50 dB results from the presumption that no intermodulation distortion emanates from the UT. The uplink C/N total equals the downlink value for the forward link, demonstrating the dominance of the L-band side of the service.

Because of its large antenna, the GES produces very strong values of uplink C/N in the forward direction and downlink C/N in the return (19.8 dB and 19.4 dB, respectively). Thus, the impact of the GES on the overall link is kept small. The Ku-band downlink to the GES, indicated in the right-hand column of Table 9.1, requires a large Earth station antenna communicating through an area coverage beam on the satellite. We see that intermodulation distortion and interference are much larger contributors to the overall link than the downlink thermal noise.

The significance of a link margin of 7 dB is that this link will work in a satisfactory manner for line-of-sight conditions. It can tolerate some additional loss from antenna misalignment, absorption by the body, and a limited amount of shadowing by trees and low buildings. The actual link will be exposed to a few additional losses and sources of degradation. For example, carrier levels from the UTs will vary over a range of measurement and adjustment. This means that some carriers will arrive at the satellite lower than the average, producing a loss of link margin. An allocation of 2 dB might be sufficient for this factor. Another such item is fading on the Ku-band links due to rain attenuation. In reality, the Ku-band GES can increase power during heavy rain through uplink power control.

Microwave link engineering is both a science and an art. In the case of MSS, there are several critical factors to consider, some of which cannot be accurately quantified either before or after the system goes into operation. What we are left with is a service that has great potential utility but which requires a certain degree of user cooperation.

9.1.3 Orbit Selection

MSS systems, by definition, involve mobile Earth stations, many of which do not employ directional antennas. The satellite involved in the link need not be stationary relative to the user, and thus the orbits may not be GEO. We mentioned the Motorola's Iridium system as a pioneering concept that introduced the handheld UT to a skeptical world. By the time of this writing, Motorola had been joined by Loral and its Globalstar system and I-CO Communications, a spin-off of Inmarsat, as proponents of non-GEO orbit constellations.

The general range of candidate orbits for use in a global MSS context is shown in Figure 9.4. The LEO constellation is exemplified by the Iridium system, which currently contains 66 satellites (the original FCC filing by Motorola showed 77 satellites, but they later reduced the number). The polar orbits of Iridium cause the satellites to provide the best coverage of the North and South Poles, with the least favorable operation occurring along the equator. The other leading LEO system, Globalstar, contains a reduced number, 48, made possible by tilting the orbits to focus the coverage away from the poles and toward the regions of interest to more users than to polar bears and penguins. At a higher altitude, the MEO or *intermediate circular orbit* (ICO) provides coverage comparable to Globalstar but with far fewer satellites. I-CO Communications takes its name from the orbits they employ.

At 36,000 km, the GEO approach to a global MSS system allows the service to be implemented with only three satellites (four are shown in the figure to allow for some backup capacity). The type of satellite to be employed is demonstrated in the previous discussion of the MSS link budget. By employing such a large reflector and high-power spacecraft, the GEO system still closes the link to a handheld UT of comparable design to competing designs at a lower altitude. We review some of these approaches in more detail in the following subsections. At the end of the chapter, we compare the systems on an economic basis using design and cost models developed by the author.

9.2 GEO MSS SYSTEMS

Essentially all of experience to date with commercial satellite communication has been at GEO. The overall simplicity of the arrangement and the ability to

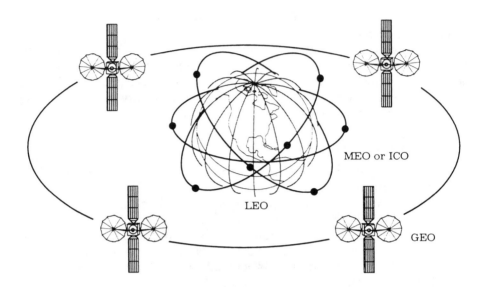

Figure 9.4 The three orbits that are applied to the MSS: geostationary (GEO), medium Earth orbit (MEO)—also referred to as intermediate circular orbit (ICO), and low Earth orbit (LEO). As the altitude of the orbit is decreased, the number of satellites required for continuous coverage increases.

use fixed antennas on the ground have allowed GEO to reach critical mass for the applications described in previous chapters. A number of GEO MSS networks are either in service or under construction at the time of this writing, demonstrating the technical viability of this approach.

However, some have raised the question about its attractiveness to mobile users, particularly when I-CO and the others enter the market. The major benefit of the lower orbits is reduced time delay for voice services. This factor is very important in terrestrial telephone networks, particularly with high-quality transmission as provided through fiber optic technology. As will be discussed later in this chapter, time delay is less of a factor in mobile communications.

We begin with a review of the established GEO systems for global, regional, and domestic service. The latter two types of systems demonstrate that GEO has the added benefit of allowing one country or a group of countries to implement and operate a system without having to enter the highly competitive and potentially more risky global market.

9.2.1 Inmarsat

The Inmarsat system evolved from an entrepreneurial start by COMSAT back in the late 1970s. COMSAT used its win of a U.S. Navy contract to launch

L-band MSS service for commercial vessels. The first satellites, called Marisats, were built by Hughes and at least one was still in operation at the time of this writing. Having established the INTELSAT consortium for international FSS, COMSAT went about organizing Inmarsat as an international MSS joint venture of governments and telecommunication operators. Inmarsat is headquartered in London, United Kingdom, where it manages and operates a global GEO MSS system that serves L-band UTs and C-band GESs. Only the satellites and network control equipment are owned by Inmarsat, which is the same approach taken by INTELSAT and EUTELSAT.

A progression of Inmarsat terminal types is provided in Figure 9.5, beginning in 1982 with the classic Standard A analog terminal for ship-board use. The antenna in this case is a parabolic dish about 1m in diameter, with a gimbal system to maintain pointing during ocean travel. Directional UT antennas were necessary to close the link to the low-power Marisat satellites, which were constrained by the small size of the spacecraft and the need to provide hemispheric coverage of ocean areas (recall that the oceans cover three-fourths of the Earth's surface). Digital communications were introduced first to improve service quality and then later, through the innovative Inmarsat M standard, as a means to reduce UT size, power, and cost. The Inmarsat M terminal, costing around $20K in 1996, introduced truly portable satellite voice service. Many who purchased such devices have found the quality to be very satisfactory and the cost per minute to be tolerable in light of the fact that worldwide telephone service is provided. Inmarsat P is the concept for handheld UTs, which has involved into the I-CO system.

The performance of Inmarsat GEO satellites has grown in power terms, culminating with the Inmarsat III series now being deployed. The spot beams shown in Figure 9.6 allow portable and vehicular terminals to access the satellite directly. These devices are similar to the first portable cellular phones, which were the size of a loaf of bread. In the future, Inmarsat may introduce GEO

1982	Inmarsat-A	Origninal voice/data terminal
1990	Inmarsat-aero	Aero voice and data
1991	Inmarsat-C	Briefcase data
1993	Inmarsat-M	Briefcase digital phone
1993	Inmarsat-B	Digital full service terminal
1994	Global paging	Pocket sized pagers
1995	Navigational services	Variety of specialized services
1998-2000	Inmarsat-P	Hand-held satellite phone (planning effort)

Figure 9.5 Inmarsat standards for services and user terminals.

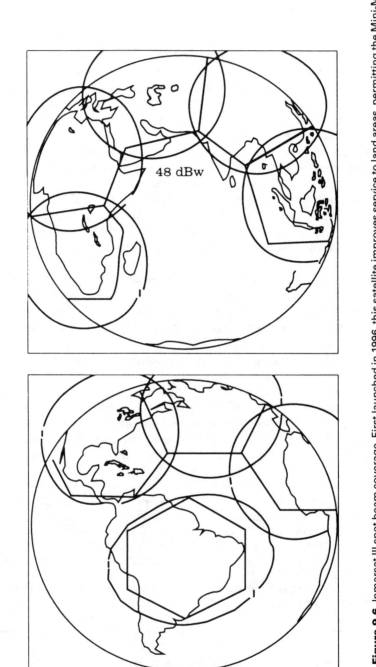

Figure 9.6 Inmarsat III spot beam coverage. First launched in 1996, this satellite improves service to land areas, permitting the Mini-M user terminal to be applied for the first time.

satellites capable of serving handheld UTs. The major thrust in personal UT services is through I-CO, which is discussed later in this chapter.

9.2.2 AMSC and Telesat Mobile

American Mobile Satellite Corporation (AMSC) is a publicly traded U.S. company corporation that was formed by the first applicants for domestic MSS in the United States. The FCC recognized that L-band spectrum cannot be reused with orbit spacing, so they insisted instead that the applicants form a joint venture company to which the single license was awarded. More recently, the FCC used auctions to make such decisions, which had the added benefit of producing substantial revenues for the U.S. government. Telesat Mobile, on the other hand, began as a subsidiary of Telesat Canada, the monopoly satellite operator in Canada. As such, they were awarded the only Canadian license for L band. The two companies entered into a joint purchase of a pair of satellites built in turn through cooperation between Hughes Space and Communications of the United States and Spar Canada. The systems are identical and offer SCPC voice and low-speed data services to moving vehicles and portable UTs. The limited satellite EIRP and G/T precludes the use of handheld units.

AMSC entered service at the end of 1995 after the launch of the first MSAT. Telesat Mobile used this MSAT until theirs was launched in 1996. The architecture of the network is such that calls must connect through a GES to the PSTN, allowing mobile users to make and accept calls from subscribers on the fixed terrestrial network. The network interoperates with the U.S. AMPS standard in the domestic cellular system. The vehicular UTs are therefore dual-mode AMPS and digital SCPC. The main market for service, determined by the technical capability of the network, is vehicular telephone service in areas not covered by cellular. If the subscriber is within the cellular coverage, then calls are placed terrestrially. The satellite link is employed only if terrestrial cellular service is out of range or restricted due to a lack of roaming.

The service offerings from AMSC at the time of this writing include the following.

- *Satellite Roaming Service:* This the foundation service that supports the use of dual-mode UTs. AMSC has entered into agreements with a substantial fraction of the AMPS cellular operators in the United States. A typical cellular subscriber can add SRS to the basic cellular service to allow extension over the satellite when terrestrial cellular is not available. The call records are collected by AMSC and then transferred to the subscriber's home cellular operator who in turn adds appropriate changes to the regular cellular bill. This works similarly to how long-distance service might be included.

- *Satellite Telephone Service:* The user subscribes to the service either through a nationwide reseller or directly with AMSC. Dual-mode UTs need not be used because roaming between terrestrial and satellite communication is not provided. Instead, the STS subscriber employs a single-mode UT to gain access to the satellite link. Billing is either direct from AMSC or through the reseller.
- *Fleet Communications:* AMSC entered into the MSS business long before their first satellite was launched. The Fleet Communications service is based on Inmarsat's Standard C low-data rate MSS system. A data communication UT is installed in the vehicle (usually a truck) so that the driver can remain in contact with the dispatch center of the associated transportation company. This affords the driver a reliable data link for routine and emergency purposes. For the transportation company, they will know where each vehicle is located through the use of position determination equipment associated with the UT. The most popular type of position location system is the *global position satellite* (GPS) system. As part of the service, AMSC provides a link from their hub in Reston, VA, to the headquarters of the transportation company, along with the PC software needed to track, communicate with, and display vehicles.

The satellites of AMSC and by the Canadian MSS operator, Telesat Mobile, are identical in every respect and support the services discussed in the previous paragraphs [1]. As shown in Figure 9.7, the satellite employs dual 5m L-band antennas, one for transmit and one for receive. A small Ku-band antenna provides feeder links to the gateway Earth stations. The antenna coverage pattern in Figure 9.8 indicates how four beams provide coverage of the 48 continental states and Canada. The eastern- and western-most beams reuse the same segment of the band, while the two internal beams employ nonoverlapping band segments to preclude self-interference. This reuse, while relatively modest, was an innovation in the MSAT design, allowing the satellite to achieve a capacity of 2,000 voice channels. Other beams are pointed at Alaska, Mexico, the Caribbean, and Hawaii.

Figure 9.9 shows a simplified block diagram of the L/Ku-band repeater subsystem onboard the MSAT satellite. At the top is the forward direction (from Ku-to-L band), and the return is shown at the bottom (from L-to-Ku band). This analog repeater provides a bent-pipe transfer of uplink to downlink. There is, however, a significant degree of sophistication within the repeater. The received L-band spectrum is converted down to an IF within the satellite to permit further filtering and then converted back up to Ku band for transmission to the GES. This filtering of smaller segments is accomplished within the forward and reverse switch matrices, shown at the center of each chain of equipment. Using *surface acoustic wave* (SAW) filters, a narrow piece of spectrum can be

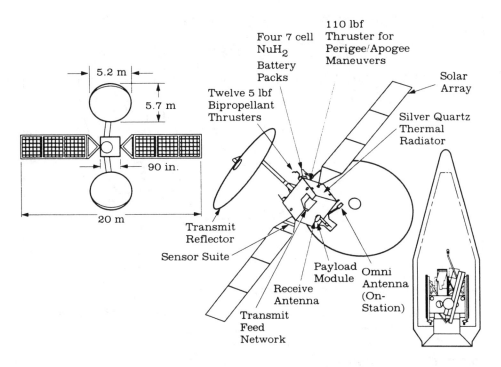

Figure 9.7 The American Mobile Satellite Corp. spacecraft, featuring two 5.5m semiflexible reflectors and an analog bent-pipe mobile-to-fixed repeater (courtesy of Hughes Space and Communications).

passed or blocked under ground command. This prevents the retransmission of interference that would otherwise waste downlink EIRP and interfere with operations. The SAW filters can be selected by ground command by changing the translation frequencies within the Ku-band and L-band frequency converters.

The traffic flows from UT to GES and from GES to UT, thus connecting mobile subscribers to the PSTN. As mentioned previously, all calls over AMSC's satellite land at their GES in Reston. From there, the calls are applied to one of the long-distance networks within the United States, with international direct distance dialing connectivity provided. With only one GES, the network is simple to operate and low in cost. In time, AMSC will add a second GES on the west coast. One conventional Ku-Ku transponder onboard the satellite provides coordination and signaling between the GESs, which must interoperate around the coverage region. Some of the bandwidth can also be used for routine communications between their facilities to save on external telecommunication cost.

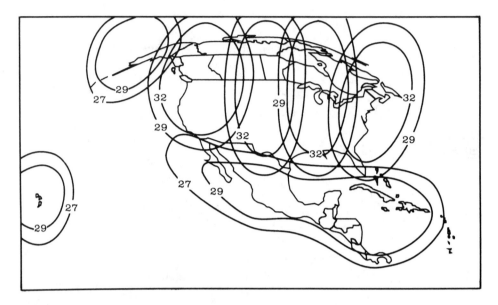

Figure 9.8 Typical AMSC L-band antenna gain, in dBi (courtesy of Hughes Space and Communications).

9.2.3 Optus B MobileSat

Australia was the first nation to introduce LMSS services through the MobileSat system. Implemented and operated by Optus Communications, the MobileSat service employs the L-band repeater onboard the Optus B satellites [2]. With a relatively small antenna system and low amplifier power, Optus B can service up to 500 simultaneous subscribers per satellite. Figure 9.10 illustrates the single beam coverage of Optus B. Users with vehicular or portable terminals can obtain mobile telephone service from anywhere in Australia. Currently, terrestrial cellular is restricted to the larger metropolitan areas and corridors providing MobileSat with a captive market in the Australian outback.

Figure 9.11 contains a plot of elevation angles from the ground toward the Optus B satellite. The ground elevation angle is one of the most important parameters for LMSS service since it defines which users can reasonably expect to have a direct line-of-sight. Low elevation angles more likely subject the user to blockage from local terrain. For example, in the area around Sydney, the elevation angel is approximately 50 deg, allowing the mobile antenna to clear many obstacles.

The supporting ground segment for MobileSat is very similar to that of AMSC, with the exception that a pair of GESs provide a fully redundant network. In addition to reliability, the scheme also saves on long-distance costs since

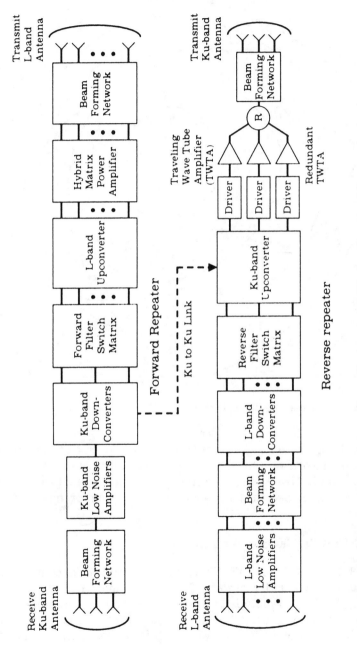

Figure 9.9 The bent-pipe repeater found in the AMSC satellite (courtesy of Hughes Space and Communications).

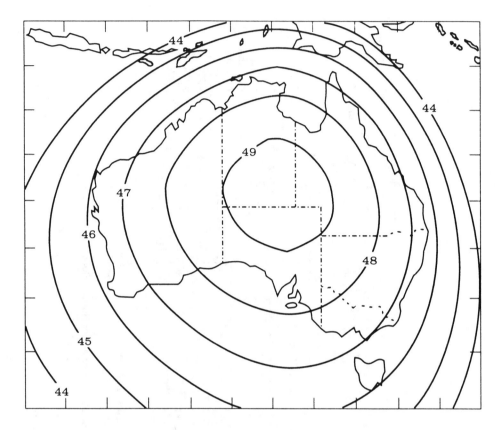

Figure 9.10 Optus B L-band coverage (EIRP in dBW).

calls can be directed to the GES closest to the PSTN end of the connection. The hardware and software for the MobileSat ground segment were provided by NEC Corp. and CSC of Australia, respectively. User terminals were supplied by NEC and Westinghouse, the latter also being a supplier for the AMSC system.

9.2.4 GEO Mobile Handheld Systems

In June 1993, Hughes Communications, Inc., introduced the concept of hand-held service from a GEO MSS system. Originally named *Asia Mobile Telecommunications* (AMT), HCI provided the encouragement for a number of systems that began construction during the 1995–1996 period, including *Asia Cellular Satellite* (ACeS) and *Asia Pacific Mobile Telecommunications Satellite Pte* (APMT). Hughes Space and Communications, a sister company to HCI, supplied

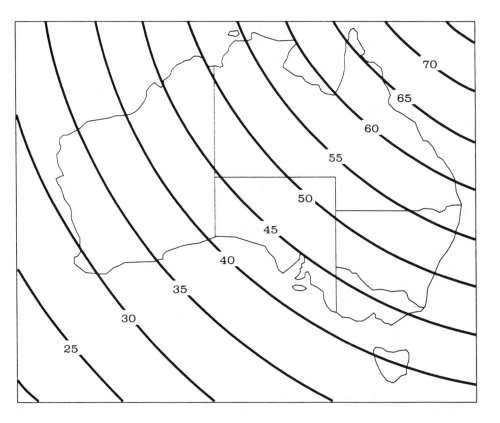

Figure 9.11 Optus B ground elevation angles.

APMT; while Lockheed-Martin supplied ACeS. The satellites used in all of these systems satisfy the requirements of the link budget provided in Table 9.1, which means that they all employ 12m antennas and very high power downlinks. The frequency bands employed are based on the L-band allocation to GEO systems in Figure 9.2; the feeder link for APMT is at Ku band while that for ACeS is at C band.

A brief summary of the characteristics common to the GEO MSS systems in production at the time of this writing is provided in Table 9.2. The actual specifications of one of these systems are too detailed and complex to include here. Only a brief selection of top-level capabilities is provided in the table. Still, the current generation of GEO MSS spacecraft has remarkable capabilities to provide the standard range of terrestrial digital cellular services. The systems are all based on the GSM standard, which is fast becoming the primary global cellular standard. A dual-mode GSM/MSS phone can roam between countries on either a terrestrial or satellite basis.

Table 9.2
Technical Summary for a Typical GEO MSS System Capable of
Supporting Handheld User Terminals

Characteristic	Value	Units or Comments
Frequency bands		
Downlink	1525 to 1559	MHz (GEO MSS allocation)
Uplink	13.75 to 14.50	GHz (alternative)
	5.925 to 6.425	GHz (alternative)
Connectivities	Mobile to gateway	Connection to PSTN
	Mobile to mobile	Single hop, on demand
Services provided	Telephone	
	Fax	
	Circuit-switched data	
	Packet-switched data	
Satellite EIRP	72	dBW (aggregate)
Satellite G/T	15	dB/K
Polarization	Circular	
Stationkeeping	±0.1 deg	Inclined orbit, up to ±6 deg
Channel capacity	16,000	Channels, mobile to PSTN
	10,000	Channels, handheld UTs
Call setup time	<6	Seconds, to domestic PSTN

The capacities listed in Table 9.2 represent the maximum number of simultaneous telephone calls that can be conducted over a single satellite. It is reasonable to ask what the number of subscribers is that goes along with this circuit capacity. A regional GEO MSS covers several time zones, allowing each channel to support a total of approximately 100 subscribers. Each subscriber, in turn, might use the service about 60 minutes, on average, per month. The satellite can therefore serve a subscriber population in the neighborhood of 1 to 2 million.

A critical part of the design of this generation of GEO MSS satellite is the technique for routing channels and calls between the beams (which number in the hundreds). Improving on the analog processing concept of MSAT, these advanced satellites take the next step of performing the operation in digital form using an onboard digital processor such as that suggested in Figure 3.8. The L-band spectrum in each beam is translated to IF as in the analog approach and converted to a digital representation in an analog to digital converter. From this point, the digitized information can be filtered, routed, and, if appropriate, demodulated down to individual bits. After the necessary operations are performed, the selected information is routed to the appropriate transmission channel where it can be converted back to analog form. The resulting band of carriers is translated to either Ku band (for mobile to gateway service) or again to L band (mobile to mobile service) and transmitted through the appropriate amplifier and antenna feed. The specific way the processing, routing, and multiplexing is accomplished depends on the approach taken by the supplier and is currently considered to be proprietary.

The evaluation of this type of system against competitors that use non-GEO orbits is provided at the end of this chapter. The next section discusses the services that an advanced MSS system are capable of supplying. These are based on the offerings of the first GEO MSS systems under construction at the time of this writing.

9.3 ADVANCED MSS SERVICES

The typical MSS system is capable of delivering a comprehensive package of advanced telephone services, similar to the offerings of GSM and other digital telephone systems. This means that a subscriber with an appropriate UT can engage in a variety of activities from anywhere in the coverage area. These services mirror the intelligent network, including narrowband ISDN, affording users the freedom to conduct business from remote locations just as if they were in their home offices.

9.3.1 Mobile Telephone and Data Services

The foundation of any MSS system is the provision of mobile telephone service that is capable of connecting calls to the PSTN. In Chapter 8, we reviewed how a satellite network can provide telephone services among fixed locations on the ground. The concept behind mobile telephony is to carry out the same function but to allow the added dimension of mobility. It should be no surprise that much of the same architecture and technology that was proven in FTS networks has been taken over to the MSS field. Furthermore, terrestrial cellular telephone standards like AMPS, TACS, and GSM have achieved total acceptance by the user community. This is because the subscriber is allowed to conduct business over the telephone without being tied to the wireline network. Satellite mobile telephone achieves the same result but of a much broader area.

In the area of circuit-switched services, voice calls and fax calls are the dominant applications. A voice call provides a point-to-point connection between mobile user and either a PSTN subscriber or, if provided, another mobile user. Subscribers must be able to originate calls in either direction. The network must be able to locate a mobile subscriber, ring the UT, and establish the communication path. To make this happen, the satellite broadcasts an alerting channel throughout the coverage region. The UT, when first activated, detects this channel and responds by transmitting a message that indicates it is prepared to receive calls. The MSS network then goes through an authenticating process before allowing service. From that point, a GES can direct calls to this particular UT when they enter the network from any other direction or location. The origination of calls from the UT is very straightforward, since it is already in contact with a GES.

The quality of voice calls is determined by the type of digital compression that is employed as well as the local fading environment. All MSS systems that support handheld UTs use highly compressed speech to get the data rate below 5 Kbps. Speech quality is rated as good (numerically, 3 out of 5), meaning that the subscribers consider it to be significantly poorer than modern PSTN services using fiber optics and uncompressed speech and only slightly worse than terrestrial digital cellular. A rating of good means that subscribers will gladly make telephone calls over the satellite network when nothing else is available. Old technology like long-distance wireline connections and high-frequency radio would receive a lower rating still. Additional information on compression standards and how they are rated is presented later in this chapter.

The intelligent nature of the MSS network will allow some attractive voice services to be introduced quickly and efficiently without added cost. Voice mail (also called call answering) is gaining in popularity on cellular networks. It is provided through the mobile switch that is associated with the GES. Each subscriber is given a voice mailbox to retain all messages. A call will be directed to the voice mailbox if the subscriber's phone is off, out of range, or engaged in a call. The function is exactly the same as voice mail over a PBX or local telephone network. Voicemail is potentially very valuable to a mobile subscriber, allowing access to the mailbox at all times from anywhere within the coverage area of the satellite.

Other services through the GES include call forwarding, call blocking or baring (to filter or block calls based on the caller's number), caller identification (subject to local regulatory rulings), call holding (e.g., putting a caller on hold to allow another call to be taken or placed), and call waiting. While any and all of these are relatively easy to implement, there may be circumstances where the network operator will not make them available to subscribers. There are also some advanced features that soon will be added to the ISDN and GSM standards. Once this is accomplished, MSS systems can introduce them as well.

Data and fax calls represent an important and growing application of mobile telephone networks. A satellite MSS service must be capable of supporting these needs in order to compete effectively. The speech processing and compression must be bypassed so that the data is applied directly to the UT transceiver subsystem, a technique already applied in FTS networks (Figure 8.6). At this point of interface, the channel unit can apply forward error correction to the unmodulated data. The speed of transmission will depend on the operating mode of the UT. In one system, a fixed data rate of 2,400 bps is available to data and fax. However, this speed is slower than that which is available over the PSTN, so newer MSS systems are pushing to provide 9600 bps to meet user needs for reasonably fast fax transmission and computer network access.

The appearance of the first of the little LEO systems with their promise of cheap data packet transmission is now pushing the satellite telephony systems to introduce comparable capabilities. This is relatively easy because voice and real-time data transmission are much more challenging that intermittent data transfer. Two possible schemes exist, namely, packet data networking, using the X.25 or IP protocols, and short messaging, which works much like e-mail. Packet data networking was developed to provide a reliable and low-cost substitute for dial-up telephone connections. The X.25 protocol is an international standard now supported by public data networks in nearly every country. Also, all computer manufacturers support X.25 for full-time and occasional access. The subscriber would connect a terminal or PC to the UT on one end, and the server or host computer would be connected to a GES using a backhaul circuit. This mirrors the VSAT star network discussed in Chapter 7.

Short messaging is a direct competitor to the offerings of little LEO systems. As a one-way receive only service, it effectively implements satellite-based paging. A connection between sender and recipient is not required because the network will forward the message. The time to accomplish this will depend on the quantity of messages being processed at the same time and the data transfer rates along the information path. While not as effective as terrestrial paging in terms of its ability to penetrate buildings, satellite message broadcasting would be available over an extremely wide area where paging is not available commercially. A two-way messaging capability is provided by the UT using a burst mode of transmission.

9.3.2 Handheld User Terminals

Voice services can be provided through a number of UT configurations. Expectations are that the most popular will be the handheld variety, which mirrors their success in the cellular market. In many developing regions, such as Asia and Latin America, the handheld or portable UT is a necessity because people spend less time in their automobiles than in North America and Europe. An exception might be in crowded cities like Bangkok and Rio de Janeiro; however, these are not markets for MSS services simply because they are already well served by terrestrial cellular systems. Most handheld phones will have dual-mode capability, where the terrestrial standard would match what is available locally. It appears that GSM will be the most effective terrestrial mode, which is advantageous in terms of sharing elements within the UT to reduce its size, complexity, and weight.

There are a few issues regarding the use of handheld UTs in a GEO MSS system. Because the link requires a near line-of-sight path to the satellite, the user must employ the phone either outside in the open or, if within a building, at a window that faces toward the satellite position. Users who are accustomed to cellular service where it is possible to make and receive calls from within a building and on the outside without a direct path to a base station may find MSS to be more constraining. The end result is that any MSS system providing service to handheld phones requires a certain degree of user cooperation. By this we mean that the user must be aware of the unique constraints of MSS and act accordingly. For best performance, the user should remain stationary after acquiring the satellite. Since the path to the satellite is stable under this condition, the user can continue to talk for an indefinite period, limited only by the available battery power. The situation for non-GEO MSS links is less favorable because the path changes even if the user is stationary. Operation from within a moving vehicle, a popular mode in cellular systems, is not recommended because of the difficulty of obtaining and maintaining the line-of-sight path. For vehicles in motion, the preferred approach is to use an external antenna and a *docking set* that transfers signals between the antenna and the handheld phone.

9.3.3 Vehicular Terminals

The vehicular UT will normally consist of a standard handheld phone and the docking set mentioned in the last paragraph. This allows the unit to be powered by the vehicle battery and the antenna to be attached to the exterior of the vehicle. RF output power will be limited to what the handheld UT can produce. The net EIRP and G/T will not be appreciably different from the handheld by itself because the loss in the cable will be compensated by the improved gain of the exterior-mounted antenna.

There will be applications in commercial transportation involving trucks and busses where the docking set is inappropriate. Instead, an installed transceiver and improved antenna will be the likely approach. Within the vehicle cab, the user will employ a mobile phone handset of the type currently in use in cellular vehicular installations. A handsfree capability will be included, as it is unlawful in many countries to hold a handset while driving. The transceiver, with its power amplifier, will be installed in the trunk or boot of the vehicle, where it is out of the way and in close proximity to the antenna to minimize RF losses. A control and signal cable connects from the handset to the transceiver.

Both AMSC and MobileSat began their services with this type of installation. The service quality from such an installation is enhanced by a more stable link and features like handsfree operation. Another benefit of the permanent approach is that other facilities can be added, such as fax interface, data communication adapter, position location, and imaging. The vehicular terminal could cost significantly more than the consumer-oriented handheld UT; however, the user has access to low airtime charges from the GEO MSS service provider.

9.3.4 Fixed Telephony User Terminals

We introduced in Chapter 8 the concept of using an FSS satellite as a networking resource for thin-route telephony services. The types of FTS terminals discussed previously are probably too expensive to be used for home services; they likewise may be out of the reach of small businesses wishing to use them as part of a private network. The associated FSS space segment costs also tend to favor bulk transmission of telephone channels as opposed to selling minutes of transponder time to individual users. As a consequence, FTS has not been extended to the general public in the same manner as DTH or even data broadcasting. This may change one day, but as of the time of this writing, only MSS systems offer the potential of direct-to-user telephony service.

To address the individual user in need of basic telephone service, a GEO MSS system has a number of attractive features. First, and most important, the basic handheld UT design must be produced to sell at a price that is affordable to middle-income individuals in developed economies. Airtime is offered on a per-minute basis, and access to the PSTN is seamlessly provided through the facilities of a GES. A fixed telephony installation only requires an antenna aligned with satellite and the necessary elements of the UT. The subscriber could employ a four-wire telephone that interfaces with the UT. The entire installation would cost only slightly more than the handheld unit. The antenna, about half the size of a UHF TV antenna, would be attached to the exterior of the building and pointed in the direction of the satellite. A low-noise L-band preamplifier on the antenna provides gain to overcome the loss of the cable that connects to the actual UT/telephone. Transmit RF output from the UT would, when combined with the enhanced gain of the antenna, be sufficient to deliver adequate EIRP.

The quality of service for this type of FTS installation would be superior to either the handheld or vehicular UTs. The reason for this is that, with constant alignment of the antenna to the satellite, the link will rarely fade below threshold. Voice quality will be limited only by the selected digital compression system. The UT can support fax and circuit-switched data through the appropriate access port.

The attractiveness of this approach to FTS results from the availability of a low-cost UT. Satellite air time would be charged based on the quantity of users and the amount of power and bandwidth per call. With the improved gain of the FTS antenna installation, it is likely that airtime can be sold to such subscribers for about half of what a handheld user would experience. This might still be expensive in comparison to basic PSTN telephone service within a city but would be competitive with what is charged for long-distance service.

A special application of FTS via MSS would be to install a shared terminal in a village or industrial site. Individuals would gain access to the UT through a manual switchboard or a small digital exchange. The cost of installing this type of UT is substantially less than that of extending the PSTN. Therefore, it might be effective to subsidize the price of making calls in order to reduce capital commitments. PSTN service can be extended terrestrially once the demand exceeds a particular level. Commercial installations would use a PBX to permit users to speak to each other within a facility (e.g., factory, construction site, and petroleum production facility) and to access the FTS terminal to make long-distance calls.

Aeronautical and maritime UTs offer special opportunities to improve communications to passengers and crew. In this case, the cost of the UT is substantially more than that of the handheld or vehicular models. This is because of the complexity of installation of the antenna and cabling system within the structure and electrical system of the aircraft or vessel. When one examines the amount of calling that such an application can generate, it always turns out to be a relatively small usage as compared to land-based demand. However, the operation of an aeronautical or maritime service can generate some nice revenues that justify the incremental investment.

9.4 NON-GEO MSS SYSTEMS

Non-GEO orbits, as shown in Figure 9.12 for the Iridium system, give up the synchronization of satellite revolution to Earth rotation in exchange for the advantages of being closer to the Earth. The advocates of the non-GEO MSS systems emphasize the benefits of reduced propagation delay, which certainly is advantageous for interactive voice service. Reduced path loss might, of itself, be useful in reducing demands on the UTs; however, there are several inefficiencies in system application for LEO and MEO as opposed GEO systems. Real-time services like telephone and circuit-switched data require that at least one satellite be in view between the two ends of the link at any given time. Developers of non-GEO systems must deploy significantly more satellites in order to maintain continuous coverage of the user. The consequence is that all of the non-GEO MSS systems under development are expensive to implement and must be targeted to a global market.

The non-GEO MSS systems that are intended for real-time applications in the global mobile telephone market are collectively referred to as the Big LEOs. This is a bit of a misnomer because two of the contenders, I-CO and Odyssey, are MEO systems. By way of contrast, the little-LEO systems are smaller in terms of the size and number of satellites, concentrating on market niches in applications such as low-speed two-way burst data messaging, radio broadcasting, and positioning. The economic viability of the latter has never been demonstrated. A company called Orbcom began launching little LEO satellites in 1995 and initiated commercial service in February 1996, by offering two-way data messaging to inexpensive handheld radio transceivers.

In the following paragraphs, we review the basic strategies available and the leading non-GEO systems under development. The expectation is that one or perhaps two of these systems will be in operation by the year 2000. There are several non-GEO satellite networks for MSS that have been proposed. The leading contenders in the opinion of the author are Motorola's Iridium, Loral's Globalstar, TRW's Odyssey, and Inmarsat's Project 21, now spun-off as I-CO Communications. Iridium is the only system proposed by a true cellular company, while the others are from traditional satellite communications companies. Motorola, Loral, and I-CO are the leaders in terms of organization, financing, technology development, and marketing. The U.S. FCC has made frequency assignments to Iridium, Globalstar, and Odyssey, allowing these organizations to proceed with production of the satellites and eventually to offer service in the United States and its possessions. Operation in other countries will depend on obtaining local regulatory approval, which will require a considerable effort on the part of these organizations. Another LEO system called Teledesic shares many of the same technical and business challenges but is not discussed here because it does not provide mobile services.

9.4.1 Iridium

The LEO approach, adopted by Motorola, locates the satellites in closest proximity to Earth, typically 900 to 1,200 miles with an orbit period of about 90 min. This provides the minimum time delay and propagation loss but raises the number of satellites for continuous service.

Importantly, Iridium represents the first proposal to offer ubiquitous service from space to handheld phones. The system, however, is not without its technical and financial challenges. Hand-off of calls is a necessity because a given satellite is only in view between the two ends of the link for a matter of few minutes. The fact that this visibility is further reduced by local terrain blockage forces Iridium and other LEO systems to add extra satellites to the constellation. The Iridium system also provides total flexibility for the user by relaying calls from satellite to satellite through a network of intersatellite links.

Except for control and tracking stations, the network is independent of the terrestrial infrastructure, allowing a user in, say, the Gobi Desert to speak directly to another user in the Sahara Desert.

The basic orbital arrangement of Iridium is shown in Figure 9.12, and some of the basic characteristics of the system are presented in Table 9.3. The constellation consists of six polar orbital planes (60 deg apart), with 11 satellites in each. The benefit of this architecture is that is assures 100% coverage of the globe, including the poles. In fact, the polar coverage is better than that over any other region. Satellites are constantly in motion relative to the Earth and

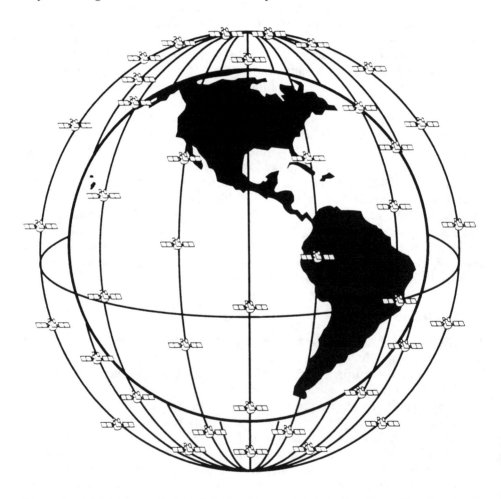

Figure 9.12 Orbital constellation of the Iridium network, including 66 satellites in 11 polar orbital planes.

Table 9.3
Summary of Key Characteristics of the Iridium System and Satellites

Characteristic	*Value or Comments*
Orbit altitude	900 km
Geometry	Polar orbits
Number of orbits	6
Satellites per orbit	11
Total number of satellites	66 plus spares
Number of beams per satellite	48 for Earth to space at L band
Intersatellite links	Provided at Ka band
Repeater design	Onboard processing of packets
Multiple access	TDMA
Satellite lifetime	Up to six years

are interconnected through a system of Ka-band intersatellite links. This raises the complexity of the network very substantially. Furthermore, each satellite has 48 spot beams generated by a phased array antenna. The links to the mobiles are at L band, using the allocation from 1610 MHz to 1626.5 MHz, identified by a cross-hatched area in Figure 9.2. Feeder and intersatellite links operate at Ka band, ranging from 19.3 GHz to 19.6 GHz.

The Iridium network, illustrated in Figure 9.13, will use these satellites to connect directly to users anywhere in the world, without passing through ground facilities. Alternatively, the network includes GESs that support calls to the PSTN and thereby to fixed wireline and cellular subscribers. Each beam covers a small area that, of necessity, moves across the surface of the Earth. Telephone calls must be handed off from beam to beam and satellite to satellite, whether the user is in motion or is stationary. Someone standing outside in the clear can expect continuous service, as the packetized speech is routed seamlessly from satellite to satellite. The user shown in the lower right is within a home and standing by an exterior window. During the time that a satellite is visible, the call can be maintained. However, the call may be dropped when the satellite passes out of range. The GESs all use Ka-band tracking antennas to follow the satellites as they orbit the Earth.

Motorola is the prime contractor for the entire system but has spun off the business to a company called, not surprisingly, Iridium, Inc. Their approach in establishing the business is to sell shares to other hardware suppliers, service providers, and governments around the world. This would encourage these entities to provide the elements and relationships that are needed to implement the system and arrange for service within the various countries. For example, at an early point in the program, they entered into partnership with Lockheed-

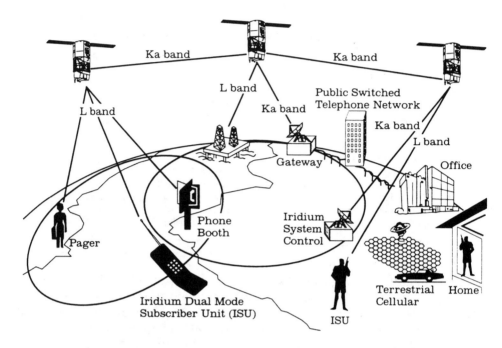

Figure 9.13 The service potential of the Iridium network.

Martin for the construction of the satellites. Many of their national service providers are existing cellular joint ventures between Motorola and local companies who have facilitated the development of the business. At the time of this writing, Iridium had obtained some but not all of the required financing to complete their constellation and ground network.

9.4.2 Globalstar System

Motorola picked up some early imitators, the most ambitious of which was Loral Corporation's Globalstar. By simplifying the system and introducing some of its own innovations, Globalstar represents a strong competitor to Iridium. Loral Corp. was since acquired by Lockheed-Martin, but the Globalstar activity has been retained by the investors and its primary manufacturer, Loral Space and Communications. Having acquired a significant fraction of the necessary capital, Globalstar is in as good a position as Iridium to proceed to a trial system. A further boost came when the FCC issued operating licenses to Iridium and Globalstar, and the ITU allocated the necessary feeder link frequencies at Ka band.

A summary of Globalstar technical characteristics is provided in Table 9.4. In comparison to Iridium, Globalstar is potentially less risky and expensive to implement. The higher altitude reduces the required number of satellites, and the inclined orbit raises the operating elevation angles. Coverage of the poles is not provided, while coverage of populated areas is enhanced. However, Globalstar like Iridium provides too much coverage of the ocean regions, resulting in less effective capacity over land areas that can be obtained with either a MEO or GEO strategy.

Globalstar has rested its technical foundation on two principles. Through a relationship with QUALCOMM, Inc., Globalstar will provide service using CDMA. This is drawn from QUALCOMM's work in the CDMA terrestrial cellular standard. Originally advertised as having a 20 times capacity advantage over analog cellular, CDMA has since fallen somewhat in its anticipated benefits. The matter of its superiority over TDMA as well has been questioned [3]. In any case, CDMA will represent an exciting alternative for implementing a new capability of this type. The alternative possibility of the Globalstar system supporting the GSM standard and TDMA has been discussed as well [4].

Globalstar has public stockholders as well as private investors [5]. The stock of one of its affiliated companies is listed on the U.S. NASDAQ exchange. Among the telecommunications service providers who back the venture are:

- AirTouch, a leading U.S. cellular company doing business primarily in California, was spun off from Pacific Telesis;
- Dacom, one of the two Korean cellular operators;
- France Telecom, the French PTT;

Table 9.4
Summary of Key Characteristics of the Globalstar System and Satellites

Characteristic	Value or Comments
Orbit altitude	1389 km
Geometry	Inclined 52 deg
Number of orbits	8
Satellites per orbit	6
Coverage	Latitudes up to 70 deg
Total number of satellites	48 including spares
Number of beams per satellite	48 for Earth to space at L band
Intersatellite links	None
Repeater design	Analog "bent pipe"
Multiple access	CDMA
Satellite lifetime	Up to 7 1/2 years

- Vodaphone, a very strong British cellular operator with joint ventures in other countries such as Australia, Sweden, and South Africa.

Globalstar has chosen to divide up the world among these service providers, giving them their own exclusive franchises. This may or may not prove to be an advantage to this venture since not many of the potentially largest markets are represented among these providers.
Equipment providers who have invested in Globalstar include:

- Alcatel of France;
- Alenia of Italy;
- Daimler Aerospace of Germany;
- Hyundai of the Republic of Korea;
- Loral Space and Communications of the United States;
- Qualcomm of the United States.

From a technical standpoint, the satellite uses a simple bent-pipe repeater rather than a digital processor such as that being flown on I-CO and APMT. A multibeam antenna with six spot beams provides higher gain and a degree of frequency reuse. Globalstar intends to launch from 4 to 12 satellites at one time. The first-generation satellite constellation, comprised of 48 satellites, is designed to operate at full performance for a minimum of 7 1/2 years. All transmissions are relayed through GESs that track the satellites as they pass by.

9.4.3 I-CO Communications

The last non-GEO system to review is the I-CO system under development by an organization that was spun off from Inmarsat. The Inmarsat Council, the governing body of Inmarsat, established I-CO Communications Ltd. in the United Kingdom to implement and operate the project that previously had the working title of Inmarsat P. With a strong link to Inmarsat, I-CO nevertheless is a completely commercial company that became fully funded by its founders. The list, consisting of 34 organizations representing an equal number of countries, is presented in Table 9.5. With 37 supporting founders representing as many countries, I-CO is in the strongest business and financial position of the non-GEO systems to reach the operation stage and, ultimately, success in the market. It has the backing of the largest MSS operator as well as many of the leading telecommunication operators and satellite service providers of the world.
Inmarsat began its evaluation of how to provide handheld service by making a thorough comparison of LEO, MEO, and GEO constellations. Their

Table 9.5
The Founders of I-CO Communications Ltd.,
the Commercial Implementer of the Inmarsat P System

Company	Country Represented
Beijing Marine Communication and Navigation Company	People's Republic of China
Bureau of Maritime Affairs	Liberia
Companhia Portuguesa Radio Marconi SA	Portugal
COMSAT Argentina SA (a wholly owned subsidiary of COMSAT (USA)	Argentina
COMSAT Corp.	United States of America
CS Communications Company Ltd. (a joint venture of the Communications Authority of Thailand and the Shinawatra group)	Thailand
Cyprus Telecommunications Authority	Republic of Cyprus
DeTeMobil (a wholly owned subsidiary of Deutsche Telekom)	Germany
Emirates Telecommunications Corp.	United Arab Emirates
Empresa Brasileira de Telecomunicaoes SA	Brazil
Empresa Nacional de Telecomunicaciones	Colombia
Empresa Nacional de Telecomunicaciones de Cuba	Cuba
Hellenic Telecommunications Organization	Greece
International Communications Organization of Cameroon	Cameroon
Kokusai Denshin Denwa Co. Ltd.	Japan
Korea Telecommunications Consortium (with Samsung Electronics and Shin Segi Cellular Communications)	Republic of Korea
Kuwait Mobile Telecommunications Authority	Kuwait
Ministry of Communications	Israel
Ministry of Post and Telecommunications	Lebanon
Ministry of Posts, Telegraphs, and Telephones and Saudi Public Investment Fund	Saudi Arabia
Ministry of Posts, Telegraphs, and Telephones	Oman
Morsviazsputnik	Russia
Philippine Communications Satellite Corp.	Philippines
PT Indosat	Indonesia
PTT Telecom BV	The Netherlands
Singapore Telecommunications Ltd.	Republic of Singapore
Societe Nationale des Telecommunications du Senegal	Senegal
Swiss Telecom PTT	Switzerland
Telecom Finland Ltd.	Finland
Telecomunicaciones de Mexico	Mexico
Telefonica de Espana, SA	Spain
Telekom Malaysia Berhad	Malaysia
Telekomunicacja Polska SA	Poland
Telemalta Corp.	Malta
Telkom SA Ltd.	South Africa
Telstra	Australia
Videsh Sanchar Nigan Ltd.	India

first decision was to drop the LEO approach in favor of either MEO or GEO. They based their decision on the cost and risk associated with each approach. Then, after working closely with Hughes Space and Communications as well as other suppliers of satellite and ground equipment, they made their final selection of the MEO.

The use of the MEO, illustrated in Figure 9.4, brings with it a greatly reduced quantity of satellites that operate at higher altitudes. The lengthened propagation delay and increased path loss must be accounted for in the network, but the satellite moves much slower and would normally be visible for the duration of a typical telephone call. Elevation angles also are favorable for the majority of the time. The global nature of the network fits nicely with the support of its founders. Coverage of the ocean regions, while not as profitable as for the land, is consistent with the Inmarsat charter to serve vessels of various types.

The twelve satellites that are needed to meet the service needs are constructed by Hughes Space and Communications. The HS-601 spacecraft was modified for the MEO mission, and the repeater is based on digital onboard processing. The frequencies to be used for the links to UTs are at the upper end of L band, being taken from a separate allocation by the ITU that is essentially dedicated to Inmarsat services. Feeder links operate at C band, which is the Inmarsat standard. The system employs TDMA, and all calls are between the mobile user and the GES for access to the global PSTN. Direct mobile-to-mobile calling requires a double hop, which should still be acceptable due to the shortened propagation delay relative to a GEO MSS system.

I-CO Communications anticipates that most users will want to employ handheld UTs. For this reason, they have opened up the specifications to industry to draw as many sources of UTs as possible. This will produce competition in the market from companies like Ericsson, Nokia, and HNS, who wish to expand sales of dual-mode phones. Like APMT discussed in the section on GEO MSS, I-CO uses GSM as its foundation. However, the dual-mode phones that support GSM and I-CO would not function on the APMT system, and vice versa.

9.5 APPLICATION OF MULTIPLE ACCESS TECHNIQUES

The selection of the multiple access method is always an important technical and operational decision because it can influence critical parameters like service availability, system capacity, and network connectivity. From a financial standpoint, capacity is very critical since it directly determines the quantity of users that a fixed investment in space can support. MSS systems generally have a limited amount of available spectrum and are subject to the complicating factor of mobile link fading.

The standard choices in increasing order of bandwidth per carrier are FDMA, TDMA, and CDMA. FDMA signals are assigned to individual SCPC frequencies and appear and disappear as users use their respective terminal devices. In TDMA, there are fewer transmissions (carriers) visible, but each is the result of several users time sharing the same channel. Finally, in CDMA, each single channel of communication is expanded by a digital code to occupy a much wider bandwidth. Because the code is noise-like, users can transmit one on top of the other, as will be explained below. All three access methods are at least 30-years old in terms of actual use in radio communication systems. There is really nothing new under the sun here; rather, it is a case of how effective the scheme performs under specific conditions (and these vary greatly) and how the associated hardware can be implemented within, say, portable and handheld phones.

9.5.1 Applying FDMA to MSS Service

The first multiple access scheme to be applied to satellite communication and the one with the most established track record (including in terrestrial cellular) is FDMA. An example of spectrum utilization along with the three key benefits are presented in Figure 9.14. The simplicity of FDMA lies in the fact that each user transmission is separated from every other by using a different frequency. A user terminal can receive a selected signal by simply tuning to the associated frequency in the same manner that you tune a television or digital FM stereo radio. The frequencies are assigned to users from a pool that is under the control of a central management resource that is also in direct contact with all users through a common signaling channel. This same principle has been applied to satellite telephone networks for the past 25 years and makes good use of the common relay point afforded by the satellite itself.

6 to 30 kHz

- Flexible
- Widely used in satellite and terrestrial networks
- Low complexity

Figure 9.14 Application of FDMA for MSS networks. FDMA is applied in the Inmarsat M, Optus, and AMSC systems, using digital speech compression.

An MSS FDMA network and its FDMA terrestrial cellular counterpart can multiply the available channel space by reusing frequency channels across geography. The available frequency bandwidth (typically in the range of 10 MHz to 30 MHz) is divided equally among the seven cells in the pattern so that adjacent cells do not use the same frequencies. Terrestrial cells are defined by the radio transmission range of cell site towers, while MSS cells are created by contiguous spot beams from the spacecraft antenna. This is illustrated in Figure 9.15 for an array of 14 spot beams and a seven-segment division of the usable spectrum. Dividing the total number of beams by the number of segments, we find that the effective frequency reuse is 100% in this example (e.g., 14/7 = 2). The seven-cell reuse pattern requires that there be a gap of two cells between any cells that use the same band segment. There is evident in Figure 9.15 in that segment 1, for example, is used in the center of the pattern and not again until the lower right-hand corner.

This kind of frequency reuse arrangement effectively overcomes one of the most serious problems with FDMA, namely, that you cannot allow two

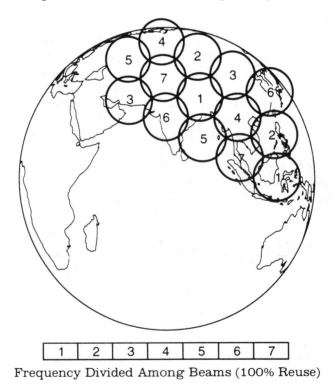

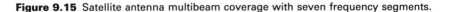

Frequency Divided Among Beams (100% Reuse)

Figure 9.15 Satellite antenna multibeam coverage with seven frequency segments.

channels to lie on top on one another. TDMA, which will be discussed in the next section, must follow this rule as well and therefore will employ the same type of reuse strategy. We will see that in CDMA, this rule may be overridden at the expense of appearing to waste bandwidth.

FDMA can be practiced with analog or digital modulation because the individual carriers are completely separate from one another. The flexibility of channel assignment also means that different modulation methods can coexist within the same piece of spectrum. This provides an easy way to integrate nonhomogeneous networks and to provide for a smooth transition from analog to digital.

Frequency modulation (FM) is the preferred analog modulation technique because it can operate in a narrow bandwidth in the presence of various kinds of noise and interference. It is in use in terrestrial cellular systems (e.g., AMPS in North America and TACS and NMT in Europe), and it continues to have a following on satellites as well. Many of the early MSS and FTS networks were implemented using companded FM, which provides a boost in audio quality at minimum bandwidth and power. The trend is now toward digital modulation and so it is likely that FM will not find much application in MSS systems outside of Inmarsat (e.g., the nearly obsolete standard A).

Digital SCPC with demand assignment was introduced before in the early 1970s in the INTELSAT system to provide full connectivity with a common satellite. Voice and data service at high quality was afforded by a channel rate of 64 Kbps. This eased the problem of providing international telephone service among all of the countries of a region, which would have required literally thousands of permanently assigned links. The early networks did not pursue high degrees of digital compression simply because of the limitations of microchip technology available at the time. Today, digital compression allows the data rate to be reduced to about 5 Kbps, which greatly benefits the link compared to analog FM. The channel data rate, including overhead and FEC, would be approximately 6 Kbps. Using QPSK, the bandwidth of a typical SCPC carrier using speech compression and FEC is under 4 kHz.

The ultranarrow bandwidth of compressed digital SCPC is a potential source of concern when it comes to assuring adequate frequency stability. The guardband between carriers must be held to a small number, perhaps only 500 Hz, to achieve good bandwidth utilization. The necessary stability can only be obtained with a system of frequency tracking within the Earth stations and the use of very stable local oscillators onboard the satellite. Another source of error is Doppler shift due to relative motion between the satellite and the various terminals on the ground. A GEO system experiences low Doppler rates and therefore can employ FDMA without much loss of bandwidth. For LEO systems, this is a serious problem, and for that reason FDMA has not been adopted. A MEO system could be a possible candidate for FDMA, provided that the relative velocities are within certain bounds. Implementers of LEO and MEO systems have chosen TDMA or CDMA, which are of sufficient bandwidth not to suffer substantial loss of bandwidth due to Doppler.

In terms of hub utilization, FDMA requires a separate channel transceiver per active mobile user. Therefore, the hub must be equipped with a number of such transceivers to support the peak expected voice calling demand. Channel transceivers are tuned to appropriate frequencies under control of the demand assignment system. Other parts of the hub, notably the antenna, transmitter, switching equipment, and support systems, can be shared.

9.5.2 TDMA

TDMA is inherently digital because information must be stored and subsequently transmitted in the form of a burst at high speed. It was developed for INTELSAT service back in the late 1960s to efficiently use a transponder at a burst rate of 60 Mbps. This is one thousand times faster than we need apply to MSS. Figure 9.16 indicates how spectrum is occupied by TDMA carriers, each of which supports several users. For example, a channel that is shared by eight users would operate at eight times the data rate and bandwidth of a single FDMA user. This amounts to about 48 Kbps, assuming 6 Kbps per user. The channel bandwidth would be approximately 30 kHz. TDMA is currently in use in VSAT satellite networks and is the core of the GSM digital cellular standard. Likewise, the currently operating North American digital cellular standard, IS-54, employs TDMA. Motorola has further advanced TDMA by adopting it for their Iridium project.

TDMA systems demand precise timing among users that share the same channel. The necessary synchronization is accomplished with circuitry at the

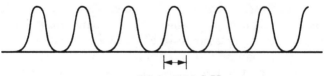

30 to 500 kHz

- Inherently digital
- Less hub equipment
- Supports frequency hopping

Figure 9.16 Application of TDMA for MSS networks. TDMA is applied in the GEO mobile MSS networks that support handheld user terminals. It was also adopted by Iridium and I-CO Communications.

hub that locks to the individual transmission from the UTs. In reality, there are several TDMA channels in operation at the same time, each on a different frequency. This reduces the required power per carrier and simplifies the design of the receiver. Multiple carriers therefore interpose a form of FDMA on top of TDMA. Due to the wider bandwidth, TDMA is more tolerant of frequency errors and Doppler shift. However, Doppler frequency shift has a smaller but significant effect on frequency error, and the necessary guardband will produce some bandwidth loss. The multiple carrier approach provides the opportunity to improve channel utilization through frequency hopping. A UT can jump between carrier frequencies to grab unused timeslots. This also facilitates seamless hand-off in non-GEO networks.

9.5.3 CDMA

CDMA approaches multiple access in a unique way by transmitting signals literally one on top of another. A basic introduction to CDMA is provided in Chapter 2. Such direct jamming would render FDMA and TDMA useless, but for CDMA this kind of interference is controlled because signals are uncorrelated. This is achieved either by using a different code for each transmission or by offsetting the transmissions in time. The code is in the form of a random bit sequence that is multiplied by the digitized data. The speed of this random bit stream, called the *chip rate*, is hundreds or thousands of times faster than the original data, thus expanding the signal bandwidth by the same ratio. The bandwidth-expanding property of CDMA has caused it to be called spread spectrum modulation. (Another CDMA approach called *frequency hopping* takes an FDMA carrier and moves it randomly across a much wider frequency band.) The information can only be recovered by multiplying the incoming CDMA signal by the original high-speed chip stream (a process called *correlation detection*). If the two streams are precisely synchronized, the user data literally pops out. Otherwise, the output is indistinguishable from noise.

CDMA is actually older than TDMA in terms of the art and practice of radio communication. Much of the research and application engineering work on CDMA has been shrouded in secrecy since World War II. This is because CDMA has some particular benefits in military communications and air defense. For example, a signal can be spread so much that it disappears in the ambient noise present in radio receivers. Another "military" benefit is that a narrowband jammer that is present within the band of the CDMA signal is spread out when mixed with the chip rate in the receiver. In contrast, the spread CDMA signal shrinks back into its original form in the same action so that the jammer has practically no effect. The other side of the coin is that if, say, 100 jammers were within the bandwidth of the CDMA signal, the interference in the above example potentially could certainly be harmful.

The following is a typical example. The user data is at 6 Kbps, while the random bit stream (and hence the CDMA signal) is 1 mega chip per second. Here, the ratio of chip rate to data rate is approximately 167, which is also the ratio by which the bandwidth is expanded. This factor also determines how well the correlation receiver will reject the jamming and interference. The spreading of the carrier reduces both its tendency to cause interference (because the power spectral density is reduced) and its sensitivity to interference from any signal, which is narrower than the spread spectrum signal. A narrowband interference within the bandwidth of the CDMA signal would have to be approximately 167 times as powerful to produce the same effect as two equal FDMA or TDMA signals interfering with each other.

Bandwidth utilization as well as some of the applicable characteristics of CDMA are presented in Figure 9.17. The narrowband form of CDMA occupies a channel bandwidth of 1.25 MHz, similar to the example in the previous paragraph. This is the approach taken by Qualcomm in the Globalstar system. Multiple CDMA channels can be assigned across the allocated spectrum, which allows the network to use frequency reuse through beam isolation techniques. Alternatively, the network might assign the same channels to adjacent beams and use the inherent isolation afforded by CDMA itself. Wideband CDMA, shown on the second line of Figure 9.17, is based on allowing the carrier to occupy the entire allocated spectrum. This might amount to 10 MHz, for example. All signals would have to be transmitted on top of each other.

There are some important issues with regard to the performance of CDMA in a fully loaded system. If the CDMA carriers within the satellite repeater are all equal in power, then it is very easy to calculate the capacity. However, power levels are not equal, and it is very difficult to maintain them within tight

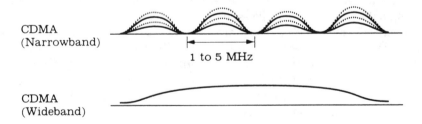

CDMA
(Narrowband)

1 to 5 MHz

CDMA
(Wideband)

- High signal isolation
- Reduced power spectral density
- Critical need for power control

Figure 9.17 Application of CDMA for MSS networks. Both the narrowband and the wideband cases are illustrated. The Globalstar and Odyssey non-GEO MSS networks plan to use narrowband CDMA.

bounds. As a user moves from position to position, the power reaching the satellite can vary over a wide range. The system will use automatic power control to compensate for variations in path and shadowing loss, but this cannot be perfect. This is the defining factor for capacity, which can be degraded by a factor of two or more. Another consideration is the degree to which CDMA networks can share the same bandwidth. While cohabitation was advanced by some proponents of non-GEO systems, the FCC ultimately decided that the three successful CDMA applicants were to be given their own nonshared band segments.

9.5.4 Comparison of FDMA, TDMA, and CDMA

A summary of the comparative properties of FDMA, TDMA, and CDMA is provided in Table 9.6. The properties in which we are interested concern the use and reuse of frequency spectrum and the ultimate capacity that the multiple access technique can deliver. We see that FDMA is the most straightforward in terms of how it is applied to MSS and the performance that results. It is the

Table 9.6
A Comparison of the System Characteristics of Multiple Access Techniques

Characteristic	FDMA	TDMA	CDMA
Bandwidth utilization	SCPC	MCPC, partial allocation	SCPC, partial or full allocation
Interference rejection	Limited	Limited, with frequency hopping	Can suppress interference, up to noise limit
Frequency reuse	Requires 7-cell pattern	Requires 7-cell pattern	Zero to 4 cell reuse possible, up to noise limit
Intermodulation effects	Most sensitive (most backoff required)	Less sensitive (less backoff required)	Least sensitive (least backoff required)
Doppler frequency shift	Bandwidth limiting	Burst-time limiting	Removed by receiver
Spectrum flexibility	Uses least bandwidth per carrier	Moderate bandwidth use per carrier	Largest demand for contiguous segment
Capacity	Basic capacity available	Can provide capacity improvement through hopping	Capacity indeterminate due to loading unknowns

baseline for comparison. TDMA has a few benefits from its ability to hop among carrier frequencies. This feature may or may not be implemented in a particular system. Furthermore, its wider bandwidth is more tolerant of frequency error and Doppler shift. Finally, CDMA offers superior interference rejection properties. In exchange, the performance of CDMA in a fully loaded MSS system has not been determined on a practical basis at the time of this writing. Each technique can provide a viable access methodology for terrestrial cellular and satellite mobile systems. There are many tradeoffs in the selection, but there does not appear to be an overriding benefit to one with respect to the others.

9.6 NETWORK ACCESS CONTROL AND SECURITY

National, regional, or global MSS offers many options for subscriber services, including making calls within and between countries. A majority of calls will no doubt be placed within the borders of a country, between a mobile subscriber and a fixed subscriber connected through the PSTN. Because the network provides PSTN access, MSS can become an extension of the national or regional infrastructure. Whether the mobile subscriber is truly mobile or not is immaterial to the network. However, this enhances the value of the service.

9.6.1 Subscriber Access and Connectivity

The various categories of calls that subscribers can make over a regional or global MSS system are indicated in Figure 9.18. The most basic and common connection is that of a UT to a GES within the same country, indicated as the top double-headed arrow. UT1 can place and receive calls from any PSTN subscriber on the domestic PSTN. It is also possible for the terrestrial side of the call to connect over the international PSTN infrastructure to a subscriber anywhere in the world. The second most important connectivity is between mobile subscribers within the same country. As the second solid arrow indicates, UT1 and UT2 are located in the same country and may call each other as if they were fixed line subscribers. The quality of the call will depend on whether the network provides single-hop or double-hop calling (e.g., the call setup time and total time delay).

The remaining two broken arrows indicate categories of calls that may be technically feasible but will depend on the regulatory environment. This would allow a mobile subscriber in country A to place a call to a PSTN subscriber in country B, permitting the call to bypass the national PSTN gateway. The call from UT1 would be landed at the GES in country B. A similar type of scenario might allow two UTs in different countries to connect to each other without passing through GESs or international gateways. This represents true borderless communication, which is on the horizon.

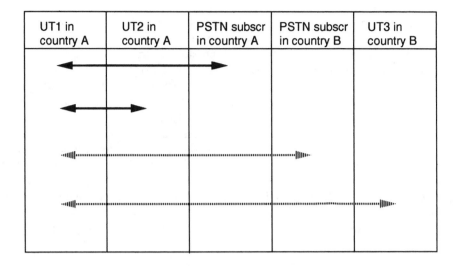

UT1 in country A	UT2 in country A	PSTN subscr in country A	PSTN subscr in country B	UT3 in country B

Figure 9.18 Potential connectivities in a regional or global MSS system—mobile UT to PSTN and mobile UT to UT. These connectivities may extend within the same country to provide domestic roaming and fixed telephony services and between countries for international roaming and emergency communications.

The MSS network can be managed as a single, unified entity, much like a cellular system within a city. This is the most straightforward approach and provides for the greatest degree of control and ease of management. In a domestic network, this is obviously the way to do things. There might be one or two gateways, such as provided in the AMSC and MobileSat systems, respectively. The other approach is that taken by I-CO and some of the regional MSS operators. This is to distribute the control and management of the network among the national operators. They in turn have their own respective GESs that control and manage the domestic subscriber base. Calls that traverse borders must be coordinated among the respective operators, who may or may not wish this to occur in practice. There still might be a need for a higher level of network control and coordination. For example, the operator of the space segment portion of the network will need to allocate capacity among the "members" of the system as well as arrange for charging for their relative usage. Also, calls that cross borders may also need coordination at the system level.

9.6.2 Network Security

Another important requirement is security of communications and network operation. Since this is a radio communications system, it is a relatively simple matter to monitor a given transmission in the downlink from the satellite. Subscribers on analog cellular networks should be aware that their conversations can easily be monitored. The only protection in this case is that the origin of the speaker is not directly known to the eavesdropper. With the advent of digital speech processing, it becomes much more difficult to overhear conversations with simple receiving equipment. The security is further enhanced with the introduction of encryption either on an end-to-end basis or simply over the radio path. The latter approach is taken in most digital MSS networks.

Advances in network security are available to the developers of MSS systems. In particular, security plays a major role in the architecture of the GSM standard. This is a good starting point for its inclusion in a compatible MSS system such as the regional and global networks under development at the time of this writing. The two important provisions of security are authentication and encryption, both of which are reviewed in the following paragraphs.

9.6.2.1 Authentication of UTs

Authentication is the process where the network verifies that a UT is legitimate and therefore authorized for access to its resources. This can be a very complicated process since the transmissions to and from the UT can be monitored. If the authentication codes and processes can be duplicated, then an unauthorized user can access network services without permission and therefore without paying. This can be countered with some form of password and a system of encryption keys to block monitoring of password transmission. If encryption is not used, then it would be possible to monitor the password when it is sent over the network and then simply replay it by the unauthorized user to gain access.

A very effective type of password and key system is called public key encryption. Now popularized as the RSA algorithm, from RSA Data Security of Redwood City, CA, public key encryption is a readily available but very strong technique for protecting passwords from theft and abuse. Each subscriber device has a private key that is contained in a chip called the *subscriber identity module* (SIM). These SIM chips or cards are common to users of GSM phones. The private key is used to encrypt an authorization request from the UT, resulting in access to network services. Decryption of the authentication is accomplished in the appropriate GES using a complementary public key. Since the private key portion of the encryption scheme is maintained secret, the authorization center will know that only the authorized SIM card could have been used to encrypt the message.

In the case of a regional or global system, the subscriber's authorization is granted by its home service provider, which could be located in a different country. When the subscriber roams to another country or service area, the operator in that area retransmits the authorization request over the terrestrial network to the home system. There, the authorization is verified and access granted for roaming services.

9.6.2.2 Encryption of MSS Transmissions

Privacy of communication can be provided through the familiar technique of encryption. It would be possible to use the public key approach previously mentioned, but it is generally accepted that the private key approach is more efficient. The important point is that it is very difficult to listen in on the communication once it has been encrypted using a private key system. Theory states that with enough computation resources it would be possible to obtain the information. From a practical standpoint, the information involved is simply not worth the effort.

A popular private key system is the DES that was developed for commercial encryption applications by IBM and the U.S. *National Institute of Standards and Technology* (NIST). Other private key techniques are available from commercial suppliers and over the Internet for free. In private key encryption, the same key is used to encrypt and decrypt the information. The basic approach is for the UT to select a random number as the encryption key. This itself is encrypted using the public key technique and sent over the link to the GES. From there, the randomly selected private key is decrypted and used for subsequent decryption of the communication.

Breaking the authentication or encryption system is generally beyond the means of the majority of potential abusers. However, it is worthwhile to find a way to duplicate or clone the SIM card and thereby be able to obtain service for free. SIM card technology is somewhat difficult to duplicate but not impossible. There were reports of the reverse engineering of the SIM card used for DBS service in Europe. The supplier of these bogus cards was allowed to continue in business under a gap in the regulations of the particular country. Such compromise of the GSM security system has not been made public but would be a source of concern if it occurred. Fortunately for MSS, the system is under direct control of the service provider and so it would be possible to shut down a user once an abuse is detected.

9.7 DIGITAL SPEECH COMPRESSION

Voice coding and compression is a process whereby speech is converted into a digital bit stream and compressed to reduce the amount of bandwidth needed for transmission. It is a technique that has literally dozens of alternative implementations, some of which we will review later. The goal is that the number of bits per second is cut substantially; yet the listener can hardly tell, if at all, that there has been an alteration in the signal. The reality for highly compressed speech is that what is actually recovered at the receiver has been changed to some extent. Compression of this type can increase the effective capacity of an MSS or cellular network by a factor of two to five. Interestingly, much of the improvement claimed by some proponents of CDMA and TDMA is actually the result of coding and compression and not from the multiple access method at all.

There is a wide range of coding and compression systems that have evolved over the decades. As digital technology has improved on the microchip side, so have the algorithms that provide the theoretical basis for efficient and high-performing compression systems. Table 9.7 provides a summary of the progres-

Table 9.7
Digital Speech Coding Systems and Standards in Use Since 1972

Data Rate (Kbps)	System in Which Applied	Compression System or Standard	Year First Applied
64	PSTN (foundation standard)	Pulse code modulation (PCM)	1972 (and before)
32	PSTN (second generation)	Adaptive differential PCM (ADPCM)	1984
16	Inmarsat Standard B	Adaptive predictive coding (APC)	1985
16	VSAT networks (HNS)	Residual excited linear predictive (RELP)	1988
13	Global system for mobile communications (GSM)	Regular-pulse excitation long-term prediction (RPE-LTP)	1991
9.6	Skyphone (aeronautical)	Multipulse linear predictive coding (MPLPC)	1990
8	North American Digital AMPS (IS-54)	Vector sum excited linear predictive (VSELP)	1992
6	Half-rate GSM	Code excited linear predictive (CELP)	1994
4.8	Inmarsat M, AMSC	Improved multiband excitation (IMBE)	1993
4.2	Inmarsat mini-M	Advanced multiband excitation (AMBE)	1996
2.4	U.S. Government Federal Standard	Linear predictive coding (LPC-10)	1977

sion of coding standards, beginning in 1972 when PCM became firmly established as the 64-Kbps standard for the PSTN worldwide [6]. Low delay coder/decoders (codecs) were introduced over the years, bringing the rate down to 16 Kbps without seriously affecting the perceived speech quality. Breaking the 12-Kbps barrier occurred with the Skyphone and North American digital cellular systems, which became available in the early 1990s. While the commercial systems pushed the envelope further to the 4.8-Kbps level by 1995, the U.S. government had long ago been using high degrees of compression down to 2.4 Kbps as far back as 1977. The quality of these vocoder systems was far from satisfactory for commercial service because of the "robotic" nature of the speech sound. No network or cellular quality codec was on the market at 2.4 Kbps at the time of this writing.

Human speech is a remarkable communications facility. Often, our communication involves more than just words and phrases, that is, content that can be transferred more quickly in text form. We also communicate our respective personalities and feelings in the tone, velocity, and articulation of what we say. A standard telephone channel, which occupies 3,000 Hz of bandwidth and requires 64 Kbps of data rate for transmission with good fidelity, will convey essentially 100% of this information without degradation. Specialists in this field refer to this level of service as network quality or toll quality. Essentially 100% of users would rate this quality as good to excellent.

Voice quality is measured using a scale of 1 to 5, as follows: 1 = bad, 2 = poor, 3 = fair, 4 = good, and 5 = excellent. This scale, called the Opinion Score, is published by ITU-T in their P-series of recommendations [7]. The opinion score technique is widely accepted as the laboratory technique to evaluate subjective quality of wireline and radio telephone systems. In a typical test, a group of human subjects is allowed to listen to the quality of speech that has passed through the trial link. Depending on the nature of the test, it could take anywhere from 20 to 100 subjects to get a meaningful result. The *Mean Opinion Score* (MOS) is obtained by averaging the numerical results for the subjects who have been exposed to the same test conditions. This is a rather complex test because there are potentially so many variables. In the simplest procedure, a recording is made of the output of the channel from the speech of a talker. The subjects are then allowed to listen and rate the speech quality according to the five-point scale. In a more accurate simulation, the channel is also subjected to impairments such as bit errors and interruptions. A very realistic simulation results when there are two subjects who talk to each other over a full duplex link. Various impairments such as fading and time delay can be introduced to test their relative impact on the MOS score.

A performance comparison of various voice coding and compression systems is presented in Figure 9.19. Speech quality, as measured by the MOS score for listening only, is plotted against the output speech bit rate. The curves

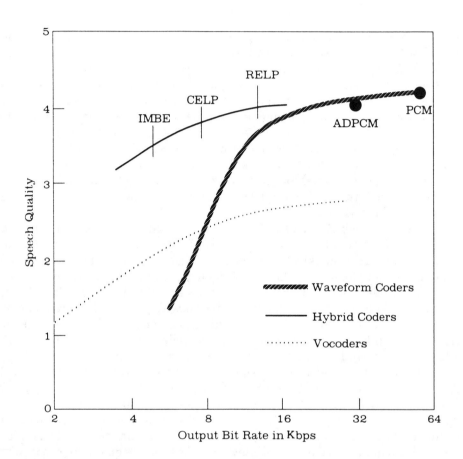

Figure 9.19 A comparison of the performance and quality of various digital speech compression systems. Three categories of compression systems are presented, namely, waveform coders, hybrid coders, and vocoders. Most MSS networks employ hybrid coders because of their ability to maintain acceptable quality at bit rates down to about 4 Kbps.

represent three different classes of speech coders. These categories are waveform codecs, vocoders, and hybrid systems. The most popular form of coding and the least effective in terms of compression is the waveform coder. Included is the familiar PCM coding scheme, which is the standard for all telephone networks. The standard 64-Kbps rate of PCM can be halved by employing ADPCM, which has also been adopted as a world standard. ADPCM is indistinguishable from PCM for voice communication, as indicated by the flat portion of the upper curve.

The compression techniques indicated by the heavy solid and dotted curves require a considerable degree of complexity within the coder and decoder. As indicated in Table 9.7, there are a variety of techniques with their associated names and acronyms. A more comprehensive discussion and comparison of these techniques can be found in [6] and [8]. To achieve a data rate of 12 Kbps or less, one needs to filter out nonspeech information and compress the rest, which is the approach taken in the vocoder and hybrid systems as well. A vocoder actually substitutes synthesized speech elements for those of the sender. Hybrid systems do a little of both, with the intent of producing lifelike speech (which tends to increase the data rate) while compressing much of the data through some of the techniques found in vocoders.

The two most popular speech coders use linear prediction, which requires that a segment of digitized speech be read into a memory so that analysis can be performed on it. What results is a compressed datagram that can be forwarded to the distant end. These work very well in practice but increases the time delay due to the requirement to store a segment of speech at the sending and receiving ends. In comparison, the feed-forward technique found in RELP tends to have much less time delay, but at the expense of delivering less compression. Currently, coders that operate around 4.8 Kbps impose the memory requirement, while those that operate at 16 Kbps or more do not. Between these limits, we incur some delay. Another consideration is that the greater the complexity of the coder, the more physical circuitry and electrical power that must be accommodated. This tends to work against allowing a very compact handheld instrument but is not a problem for portable and vehicular units.

Hybrid coders are clearly the best choice for MSS applications at the time of this writing. At data rates around 4.8 Kbps, they deliver communication quality at an MOS score in the range of 3.2 to 3.6. However, anyone who is considering the development of a new MSS system for telephone service must consider the selection of the speech codec very carefully. The technologies listed toward the bottom of Table 9.7 will be obsolete within a few years of publication of this book because of the rapid pace of advancement in digital electronics and compression technology.

9.8 ECONOMIC COMPARISON OF MSS SYSTEMS

Selection of the orbit geometry and network architecture are major decisions for developers of MSS systems. In this section, we consider the tradeoffs through which such developers must go to arrive at a cost-effective system. Any network, no matter how technically advanced, must satisfy investors who demand an adequate rate of return and subsequently survive in a competitive environment. Both of these are difficult challenges, particularly when you consider how many technical features must be proven before MSS to handheld phones will become a reality.

9.8.1 Methodology of the Comparison

We evaluate the economics of MSS in the context of a global network. Also, we assume that any such network must meet a reasonable financial return requirement, here taken to be 15%. Depending on the type of constellation, the network will be designed and constructed differently. The end result will be the same in that mobile subscribers with handheld phones will be served. We have made a number of simplifying assumptions to allow comparison on an apples-to-apples basis. The orbit altitudes range from LEO to MEO, providing a rough approximation to what has been proposed by Motorola, Loral, and Inmarsat. For comparison purposes, we also consider a hypothetical GEO that is implemented on a near global basis with three satellites. In actuality, GEO satellites can be introduced one by one, in response to specific demands. This is something that is beyond the range of possibility for non-GEO systems.

9.8.2 Required Quantity of Satellites

The single most important factor in the cost of the network is the number of satellites required to complete the constellation. Figure 9.20 presents the number of satellites required to complete a global network as a function of the altitude, in statute miles. This curve is based on geometry alone and demonstrates how the required number of satellites drops rapidly from approximately

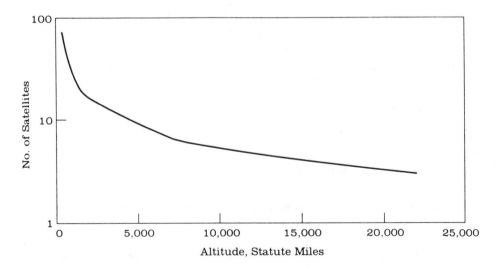

Figure 9.20 Theoretical number of satellites required for continuous coverage, shown as a function of altitude in statute miles. Note the logarithmic scale.

70 at 500 miles to 8 at 7,000 miles, leveling off at three satellites at GEO altitude of 23,000 miles.

LEO systems must all employ lots of satellites. However, the number of satellites required drops rapidly as the altitude is increased. MEO/ICO systems can make due with only 6 to 8 satellites. The lowest number is achieved at GEO, which totals only 3. In this case, GEO does not provide coverage of unpopulated polar regions. The lower altitude non-GEO systems have a significant fraction of their satellites wasting power and time over ocean regions where few subscribers are located. It has been estimated that the typical LEO constellation only delivers an effective capacity over land of about 20%. A MEO or GEO system can theoretically provide 100% of capacity to land-based subscribers.

A factor that we have to keep in mind is that as the altitude is raised, the power required to satisfy the link budget also increases. There is a direct increase in path loss of about 30 dB over this range of altitude. A countering factor exists in the manner in which that satellite covers the useful land area. As the altitude is increased, the beamwidth that illuminates a particular number of square kilometers of land decreases, which provides for a corresponding increase in antenna gain. The gain increase from LEO to GEO is approximately 20 dB. Taking these two factors into account, we see that the net disadvantage going from LEO to GEO is only 10 dB, maximum. The GEO satellites are therefore somewhat larger and more expensive than lower altitude MSS satellites due to the need to carry a larger antenna and prime power system (e.g., solar arrays and batteries). Their capacity must also be greater to compensate for their reduced number (however, capacity is not wasted over the open oceans). As we shall see, however, these factors are more than compensated by the fact that there are so many fewer of them.

9.8.3 Launching of the MSS Constellation

Launch services are needed to get the satellites into orbit. LEO systems have the advantage that substantially less energy is needed per satellite and it is possible to simultaneously launch several satellites on the same vehicle. For GEO, it is possible to launch at most two satellites at the same time. MEO systems lay somewhere in between. When examining the total cost of placing the constellation into orbit, the advantage that LEOs have with regard to multiple launch with the fewest number of stages is washed away by the need to put so many satellites in orbit. The results of a comprehensive analysis are provided later in this section.

9.8.4 Ground Segment (Infrastructure)

The ground segment that provides the gateways and network control is different for each of these networks. GEO systems tend to be the simplest to operate for two reasons. First, the satellites are more or less stationary and do not require much attention. Second, there are fewer satellites. A LEO system will be a substantial challenge to manage. Gateway Earth stations must track at least two satellites at a time to ensure continuous service. Links must be handed off, just as in terrestrial cellular, but the time delays and distances involved will make this problem complicated. A MEO system will have less demands on the ground segment due to the relatively slow motion of the satellites. Still, GESs will require multiple tracking antennas to assure continuity of service as satellites pass by.

It would be extremely difficult to provide an apples-to-apples comparison of the ground segment. We can observe that LEO systems will have more GESs simply because of the greater number of satellites. In theory, we only need one GES per GEO satellite. An actual GEO MSS system will probably have multiple GESs to provide domestic access to the PSTN and thus avoid a lot of international calling. Still, the quantities ought to be less. In the absence of details concerning the ground segment, we have assumed that there is no advantage to a particular orbit. Ground segment costs have, therefore, been left out of the comparison.

9.8.5 Overall Economic Evaluation

An overall evaluation of the investment cost of the satellites is presented in Figure 9.21. This was obtained by calculating the power, weight, and cost of each satellite, which increases with altitude. The cost per satellite is multiplied by the number of satellites, according to Figure 9.20. Total investment for the space segment, including launch and insurance, is indicated in the top curve. We see that the total investment decreases monotonically as altitude increases. The reasons for this are straightforward.

First, the cost of the satellites themselves is proportional to the number of satellites. The GEO satellites are heavier and more expensive than those for the lower altitudes. However, this factor is dwarfed by the quantities. Notice the rather flat curve at the bottom of Figure 9.21, which gives the total cost of launching all of the satellites. There is almost no variation as the altitude is increased. Adding the cost of satellites to the cost of placing them in orbit, we see that the downward slope is preserved as altitude is increased. The conclusion here is that, from an investment standpoint, GEO MSS systems are less expensive than any other altitude.

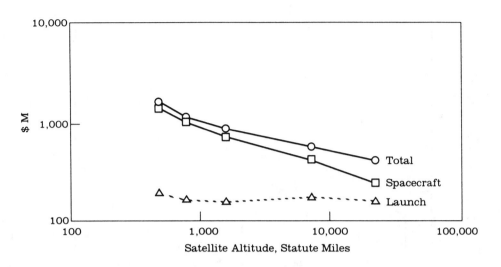

Figure 9.21 An evaluation of the relative investment cost requirement for different implementations of a global MSS system (space segment only). The three curves represent (1) the total investment needed to achieve operation, (2) the cost of the spacecraft alone (e.g., construction only), and (3) the cost of launching the constellation. The cost is driven by the number of spacecraft as the total launch cost is relatively constant across the different altitudes.

9.8.6 Revenue Requirements

The last relationship to be compared for the different orbit geometries is the revenue requirement for the service. In Figure 9.22, we plot the relative revenue requirement of the total space segment for each altitude, where GEO is set equal to 100. The disadvantage of LEO relative to GEO is by a factor of almost 7:1 when considered on an annual cost basis. This is first driven by the investment cost of initially deploying the satellites and is pushed further by the fact that the LEO satellites only last about half as long as the GEO satellites. This is because the LEOs are affected by atmospheric drag, which eventually causes them to reenter the Earth's atmosphere. MEO and GEO satellites are far beyond the atmosphere and so can sustain themselves by expending far less fuel to maintain orbit.

A MEO system suffers a modest penalty of about 40 to 50% relative to GEO. This suggests that the added cost might be traded off against the advantage of a particularly effective technical or business strategy. A global GEO system is yet to be devised, and instead the approach being taken is for a number of independent and incompatible GEO systems to be deployed in different regions of the world. I-CO Communications, on the other hand, will deploy a global MEO

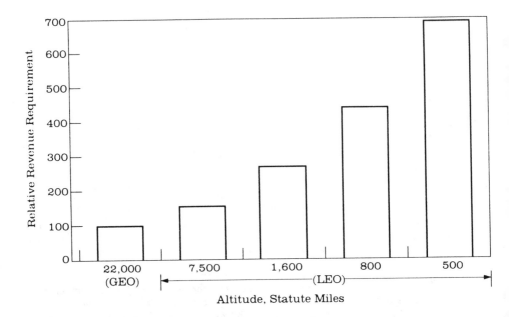

Figure 9.22 The relative revenue requirement for the global MSS system investment shown in Figure 9.21. This illustrates the high cost penalty of LEO systems, which must first achieve operation at high cost and then maintain service by continuously launching new satellites to replace those that reenter the Earth's atmosphere. Also, most of the capacity is lost to service because the satellites spend most of their time over water. MEO and GEO satellites have longer lifetimes and can direct their beams toward land users and, hence, tend to demonstrate much lower annualized costs.

infrastructure that permits ubiquitous service from anywhere on the planet. The service will be uniform and consistent, regardless of the country or region where the connection is established. If the model of the international business traveler as the predominant user is correct, then I-CO stands to do very well in the competition for market share and adequate revenues.

This basic comparison points up the need to look at the top-level system parameters before choosing a satellite network configuration. The LEO proponents seem to have made their decision based on a few technical benefits of being close to the Earth but have sacrificed simplicity and cost. We will be observing the progress that each of the proponents makes over the coming months and years. Perhaps the MEO/ICO approaches will prove viable since their fundamental costs are significantly lower than for LEO. It has been argued that the MEO architecture will win out for global networks that serve subscribers no matter where they live or travel. On the other hand, it would appear that the GEO solution continues to be the lowest in overall cost and the simplest to implement and operate as well.

9.9 PROSPECTS FOR GLOBAL MSS SERVICE

Satellite-based mobile telephone systems are still emerging, as evidenced by the relatively low penetration levels of the service at the time of this writing. Still, the promise is exciting because there are countless areas where this type of service will be valuable and probably commercially successful. The major focus now is on how to make the satellite link useful to handheld phones. This is a major challenge because of the limitations of the link to the simple UT antenna and the shortage of viable spectrum. The fact that frequency reuse is so difficult with MSS networks adds to the problem. Still, there are great benefits to the realization of this dream of ubiquitous MSS with handheld phones.

Terrestrial cellular networks will probably work better and be cheaper to use in urban environments. The places where satellite networks will shine will likely be the underdeveloped regions and in those *dead spots* that are unavoidable in land-based radio networks. The key to success of MSS will be to find those niches and to exploit them in a way that is complementary to the successful cellular networks of today.

References

[1] Whalen, D. J., and G. Churan, "The American Mobile Satellite Corporation Space Segment," *The 14th Int. Communications Satellite Systems Conf., Part 1, A Collection of Technical Papers*, March 22–24, 1992, Washington, DC, AIAA, 1992.

[2] Wagg, M., "Regional Satellite Developments—MOBILESAT," *Mobile Satellite Communications in Asia Conf.*, AIC Conferences, Hong Kong, December 6–7, 1993.

[3] Frezza, B., "CDMA: Blazing a Trail of Broken Dreams," Wireless Computing Associates, April 15, 1996.

[4] Louie, M., M. Cohen, D. Rouffet, and K. S. Gihousen, *Multiple Access Techniques and Spectrum Utilization of the GLOBALSTAR Mobile Satellite System*, Palo Alto, CA: Space Systems/Loral, 1995.

[5] *Globalstar Telecommunications Limited*, filing before the Securities and Exchange Commission (Form S-1), November 29, 1994.

[6] Kondoz, A. M., *Digital Speech—Coding for Low Bit Rate Communications Systems*, West Sussex, England: John Wiley & Sons Ltd., 1994.

[7] *Recommendations for Testing of Service Quality, Consultative Committee on Telegraph and Telephone (P-Series)*, Geneva, Switzerland: International Telecommunication Union, 1992.

[8] Macario, R. C. V., Ed., *Modern Personal Radio Systems*, London, UK: The Institution of Electrical Engineers, 1996.

Part IV:
Service and Business Development

Frequency Coordination and Regulation of Services 10

Operators of communications satellites are dependent on national governments and the rules of international frequency coordination to obtain the authority needed to deliver services. Consistent with international laws and regulations, each country controls the use of the radio spectrum within its borders. The government is the entity that obtains international recognition for its satellites and Earth stations. The international "rules of the road" for obtaining and using frequency spectrum and either GEO or non-GEO orbits provides assurance that the new system will neither cause nor receive unacceptable interference when it goes into operation. Even after the orbit position has been granted, any operator or major user must obtain the approval of the government before services can be provided. Our objective is to understand the regulatory hurdles that a new system faces and to implement an appropriate strategy to meet these challenges.

10.1 REGULATORY BACKGROUND

The international regulatory environment established under the ITU provides a broad framework for the approval process for obtaining orbit positions and authorization to transmit to and from satellites. This process is covered in detail in Sections 10.3 through 10.6. However, it is the individual governments who directly control these provisions within the domestic borders of the respective countries. The regulation of satellites and the services they provide is relatively new as compared to more established spectrum users like radio and TV broadcasting, terrestrial microwave links, and special services like radio astronomy. Three decades ago, the leading governments of the world recognized the potential value of using the radio spectrum with satellites and so the frequencies were allocated and the procedures developed even before HBO and BSkyB existed. On the domestic front, governments in the United States, Canada, and Western Europe were the first to allow privately owned satellites and Earth

412   *The Satellite Communication Applications Handbook*

stations to be brought into service. For this to happen, a wide array of technical, business, and political interests had to be balanced.

We review the regulatory environment with regard to satellites in a number of the leading countries. This provides the basis for understanding how the regulations can be adopted and applied in other countries where the process is only beginning. Later, we move from this specific experience base into the broader aspects of how the rights to the use of spectrum and the orbit arc are shared on a global basis.

10.1.1 The U.S. Regulatory Environment

The FCC has authority over the U.S. satellite industry, assigning orbit positions, granting construction permits, and licensing Earth stations. Government usage of the spectrum is handled by the National Telecommunication and Information Agency, a completely different agency, but they and the FCC cooperate in the context of the ITU. In the past, the FCC preassign slots to companies for commercial development. This has allowed satellite operators to plan, construct, and market satellite capacity with reasonable confidence. Provided that the proper regulations and law are followed, the availability of slots has rarely hampered a legitimate business from proceeding.

On the other side, a significant number of authorized satellites have never been launched. The FCC typically allows replacement satellites from the same operator. Deals to exchange orbit slots between operators have been allowed, subject to public notice and debate in an open forum. The public notice process is surprisingly effective in identifying and resolving differences.

More recently, the FCC has chosen to use auctions to decide who would be granted the license to develop a business around the use of a particular piece of the microwave spectrum. The first such auction occurred in 1996, when MCI Corp. bid and subsequently paid $682 million for 28 of the 32 Ku-band channels at the last remaining U.S. BSS orbit position (110 deg WL). Such prices confirm the high value of the spectrum that provides access to an attractive market like the United States. In comparison, Hughes and Hubbard paid nothing more than nominal filing fees for the 32 channels that they employ at 101 deg WL, which is another attractive BSS orbit position.

Auctions now appear to improve upon the use of the lottery, which in the past resulted in some recipients not having the financial resources to follow through. In those instances, the licensees simply sold off their licenses through a secondary market where values were continuously bid up. This market mechanism works to a fashion but has the disadvantage that the U.S. government and taxpayers receive none of this benefit. Having learned its lesson, the auction was introduced as a means to decide license issue. The prices received for licenses in PCS and BSS have been very high, demonstrating in clear terms the value of spectrum. Auctions, however, are only used within a country's borders and have no bearing on the international coordination process.

As a result of these measures taken over a 30-year period, the United States has seen the greatest development of satellite communication in the world. There are nearly 40 operating GEO satellites providing services throughout the United States and offshore locations. The FSS systems are foundations of the cable TV and broadcast TV industries, and several BSS systems are implemented or under development. Other services delivered through VSATs and hub Earth stations have given a strategic advantage to some of the more successful companies in a variety of fields. This environment is providing strong motivation to other nations and regions to follow suit.

10.1.2 Advances in Japan

Japanese regulatory policy has changed rapidly from a government monopoly to near-open competitive entry. They studied the U.S. model but chose something uniquely Japanese. The *Ministry of Post and Telecommunications* (MOPT) is the regulator, having a very high degree of control of spectrum use and the provision of telecommunication services within Japan. NTT, the internal public network operator, has been privatized and now has several domestic competitors. Called Type I and Type II carriers, these companies may use terrestrial or satellite systems to compete head-to-head with NTT. These competitors are backed by the major trading companies and utility companies who have been allowed to enter the telecommunications marketplace.

There are three FSS satellite operators in Japan, namely, Japan Satellite Corp. (J-Sat), Space Communications Corp. (SCC), and NTT itself. All three are currently operating Ku-band satellites and provide a variety of services in the video, data, and telephony fields. The market is still under the control of MOPT, who wishes to maintain some order of discipline. Likewise, there are international communications firms that compete with KDD, the previous monopoly international service provider for Japan. All three use satellites and fiber optic links to provide connectivity from Japan to other countries of the world. Through these initiatives, Japan has demonstrated a special ability to make a rapid conversion to introduce a competitive market in satellite and other services.

10.1.3 The European Experience in Orbit Assignments

Europe in general and the European Union (EU) lack both a common legal framework like the U.S. Communications Act and a strong central regulatory agency like the FCC. Individual countries proceed on their own and attempt to work things out on a bilateral basis either directly or through the ITU. This leads to less efficient orbit utilization and more political horse trading between governments. This hurts the business because of the time and difficulty associated with making something happen on a reasonable schedule. SES Astra relies on the very friendly Luxembourg government to back the system both financially and through its regulatory framework. Meanwhile, the monopoly PTAs are backers and users of the only regional system, EUTELSAT. The EU is trying to resolve this difficulty by creating a regulatory framework for its members. Another area of difficulty is dealing with the various protective rules and laws from country to country that are designed to protect domestic industry from foreign (and even domestic) competition. Cable systems in one country want to be protected from satellite broadcasting; satellite broadcasting in another way wants to be protected from privately owned broadcasting stations. This impedes integration on a multinational basis.

10.1.4 Expansion of Satellite Operations in Asia

The Asian environment outside of Japan is the most open and dynamic in the world, with government and commercial operators competing for orbit positions, spectrum, and ultimately the business. The international regulatory bodies are being pushed to their limits because literally every Asian country is pursuing a satellite system of one type or another. It does not matter how large or small the country may be. The first country in Asia to implement a satellite system was Indonesia. Having gone through the ITU regulatory process, this country is one of the leaders in exploiting its position as one of the largest and fastest growing countries in the region. As the largest country in the world in terms of population, China has a dominant position due to its demand for service and technical capabilities in the way of design, operation, and launch of satellites. Its regulatory environment is one of the most complicated, due to the dichotomy of central control through the Ministry of Post and Telecommunications in Beijing and decentralization afforded by the provincial telecommunication authorities in each province.

On the other hand, Hong Kong and Singapore are demonstrating their unique ability to gather businesses and produce products and services that are of value worldwide. Honk Kong, in particular, has its own regulatory setup, making it relatively easy for a local company to pursue satellite communications as a business. Singapore likewise has moved from a user of satellites to an operator, with several new ventures coming online in the next few years. As a sovereign nation, Singapore is in a unique position to exploit its right to assign radio frequencies and engage in international regulatory affairs.

The satellite systems that have sprouted in Thailand, Malaysia, and the Philippines are all backed by commercial companies. This is more like the U.S.

model, where private companies are afforded the use of the orbit for achieving both a national and a commercial purpose. Other countries are moving more slowly, in line with the ability of their economies to absorb this type of investment. Also, it continues to be attractive to rent transponders for INTELSAT and neighboring countries as long as the requirements are on the low side. Competing international satellite operators like Pan Am Sat now are providing capacity in the region, giving the smaller users even more options from which to choose.

10.1.5 Satellite Regulation in Latin America

The fourth region of primary interest in satellite regulation is Latin America, which stretches from Mexico through Central America to the full extent of the continent of South America. Mexico and Brazil were early adopters of domestic satellites and continue their leading roles as supporters of the ITU and its regulatory processes. The rapid trend toward privatization of telecommunication monopolies has influenced the development of satellite networks in this region. In particular, whereas the domestic systems were implement by the government, the next generations will more likely be owned by private interests. The government will continue to grant licenses and maintain the process of frequency coordination with neighboring countries.

This is the backdrop for considering the impact of regulation on the development of satellite networks. In the following sections, we review the basis of international regulation of orbit and spectrum usage around the world. This is an introduction and review and is not a substitute for the regulations themselves or a specific study of the situation that would exist for a given system or service. What we wish to do is to provide some broad guidance to newcomers to the field. The people who engage in the regulatory process are professionals who have the requisite experience and technical expertise. They fall into three broad categories: (1) the people who write the regulations, (2) the people who administer the regulations, and (3) the people who practice frequency coordination in order to allow a given system or service to go into operation. Many of us, particularly this author, likewise need to understand these rules so that, as practitioners, we can take the most appropriate steps in bringing the needed systems into existence.

10.2 SHARING RADIO FREQUENCIES

Satellite systems and terrestrial radio stations and wireless networks must compete with each other for valuable spectrum. Communications satellites are unique in their ability both to cause interference and to receive interference as well. Much of our attention is focused on the techniques for resolving the interference between satellite networks that operate on the same frequencies. This process is what has allowed satellites to proliferate in North America, where an orbit spacing of 2 deg is commonplace. In other regions, such close spacing is becoming a necessity because many countries wish to exploit satellite technology. What we find is that coordinating the use of radio frequencies among satellites is growing in complexity.

The mechanisms for how satellites and Earth stations can interfere with one another are covered elsewhere in this book and in [1] as well. Satellite systems are also able to cause interference to terrestrial radio stations that employ the same frequencies. The severity of the problem is reduced for the GEO because the satellites are usually far above the local horizon at a given location. The tendency has been to segregate their respective usage so that each can pursue their respective business. On the other hand, there are frequency bands that began their existence as terrestrial radio media, with satellite usage representing the intruder. This is the case at C band and to some extent at Ku and Ka bands.

Figure 10.1 illustrates the possible interference paths that can leak between the operation of a satellite link and a terrestrial microwave link. These are controlled through various regulatory mechanisms, including frequency coordination, as will be discussed later in this chapter. Normal operation of the respective links is indicated by the solid arrows, which provide full-duplex transmission. The terrestrial link in this illustration operates in both the satel-

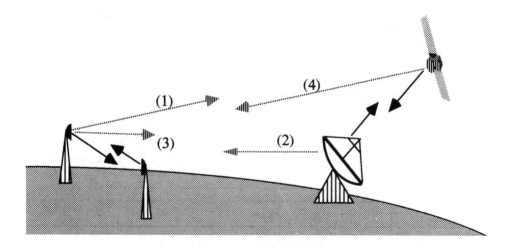

Figure 10.1 Illustration of the possible sources of interference between a satellite network and a terrestrial microwave link operating at the same frequencies.

lite's uplink band (assumed to be 6 GHz) and simultaneously in the downlink band (assumed to be 4 GHz). Other bands, such as 11 GHz and 18 GHz, are shared between terrestrial and satellite and would have similar interference paths. The interference paths, indicated by the dashed arrows, can be described as follows.

- *Path (1):* 6-GHz terrestrial radio interference into the uplink receiver of the satellite. The satellite is protected by a maximum radiated power limit from the terrestrial microwave antenna and by a stipulation that these antennas should not be directed at the GEO. As a result of the low-radiated power from terrestrial stations, interference of this type is not experienced in practice. The one possible exception is high-power tropospheric scatter links that employ billboard-sized antennas. These must not be pointed at the GEO in any instance.
- *Path (2):* 6-GHz Earth station interference into the 6-GHz terrestrial radio receiver. This the most important interference case for determining if an Earth station can be operated in a particular location. The radiation along path (2) is in the sidelobes of the Earth station antenna and propagates along a variety of paths to the terrestrial receiver. Later in this chapter, we will discuss how this radiation is assessed and the process for coordination that results.
- *Path (3):* 4-GHz terrestrial radio interference into the downlink receiver of the Earth station. A common term for this is *terrestrial interference* (TI), and it has been one of the reasons why DTH services are more popular in dedicated satellite bands like Ku band than in shared bands like C band. It might be possible to operate the downlink at a different frequency from the local source of TI. Alternatively, the Earth station antenna might be shielded from the terrestrial transmitter—a technique that has been applied to path (2) as well.
- *Path (4):* 4-GHz satellite interference into the 4-GHz terrestrial radio receiver. The radiation level from a satellite is typically too weak to be of much concern to terrestrial receivers. Even still, the ITU has placed *power flux density* (PFD) limits from the satellite on the surface of the Earth to provide a high degree of protection margin. We consider this aspect later in the chapter.

The ITU administers the worldwide radio frequency spectrum, providing the primary forum for the subdivision of the spectrum for the various radio communication services that are in use today. Their particular challenge is to allocate spectrum in such a way that the needs of all members are met and that new technology can be accommodated. Traditionally, the area of greatest interest was in medium- and high-frequency channels (below 30 MHz) used for AM radio broadcasting, particularly because these *short wave* signals can propagate well beyond the borders of a country and can interfere with one another. *Radio frequency interference* (RFI), whether harmful or just unacceptable, is something that the ITU seeks to avoid.

More recent technology like terrestrial microwave radio for line-of-sight transmission (between 1 GHz and 30 GHz) and satellite communication for domestic and international services have challenged the ITU to adapt its ways and produce workable regulations. It has been able to do this for the most part. A strong and more effective political force is the interest of developing countries, particularly those in Africa, to obtain special privileges. For the time being, such political moves are dealt with appropriately and without subverting the basic mission of the ITU.

10.3 STRUCTURE OF THE ITU

Founded in 1865 as a postal union, the ITU is the oldest and most prominent of international organizations in the field of telecommunication regulation. As a specialized agency of the United Nations and with its origins dating back before the age of radio, the ITU has proven itself in a changing and turbulent world. Today, it has over 150 member nations with its headquarters in Geneva, Switzerland. Membership is restricted to government telecommunication agencies, called Administrations; each is required to contribute to the Union's financial support more or less in proportion to its economic strength. Telecommunication carriers, equipment operators, and users are represented through their respective governments and often attend meetings along with their government counterparts.

10.3.1 Objectives of ITU Regulation

The objectives of the ITU, reflecting the shared purposes of the Administrations, are:

1. To maintain and extend international cooperation for the improvement and rational use of telecommunications of all kinds;
2. To promote development of technical facilities and their efficient operation so as to improve telecommunications services and increase their usefulness and availability;
3. To harmonize the actions of nations in attainment of these goals.

Generally speaking, the ITU does not have direct power to regulate telecommunication and radio transmission since these are considered to be the right of sovereign states. This means that the government of a country has total regulatory control of how the radio spectrum is used within its borders. Radio transmissions, however, do not honor borders and therefore can easily spill over to interfere with the legitimate radio communication activities of other nations. It is this interface with which the ITU is concerned. Because the ITU is a common body among governments, its regulations and recommendations are followed by the member nations out of the practical need for a realistic framework.

Developing countries have a special place in the ITU because of the shared objective of improving worldwide telecommunication services. A country with a poor telecommunication infrastructure cannot provide international telecommunication services of good quality. Furthermore, there is the more general U.N. mission of improving the living conditions of the world's population, for which the ITU carries out the telecommunication development program.

10.3.2 Regulatory Philosophy

There are two basic areas in which the ITU regulates international telecommunications. The area of principle interest to satellite communications is the management of the radio frequency spectrum, which is a critical role because individual countries and operators of radio transmitters need "rules of the road." Otherwise, radio stations would transmit on any frequency they choose and interference would be very common; therefore, no one would be able to count on reliable broadcasting and point-to-point radio communication. Through international conferences, technical recommendations, and its full-time spectrum management activities, the ITU oversees the use of radio frequencies and provides an effective forum for the resolution of interference difficulties. They do not actually police the airwaves, which is a function left to member nations.

The second principal area is in the interfacing of telecommunication networks with one another. In previous chapters, we discussed the interfaces between telephone, data, and other networks. Without these definitions, telephone calls could not be completed automatically across borders. With the digitizing of public networks and expansion of new data communication services, the importance of ITU activity has increased significantly in the last few years.

10.3.3 ITU Sectors and Bodies

The three areas of focus of the ITU are evident in its organizational structure, as shown in Figure 10.2. The Radiocommunication Sector carries out the mission with respect to the use of the radio spectrum by ITU members and is the

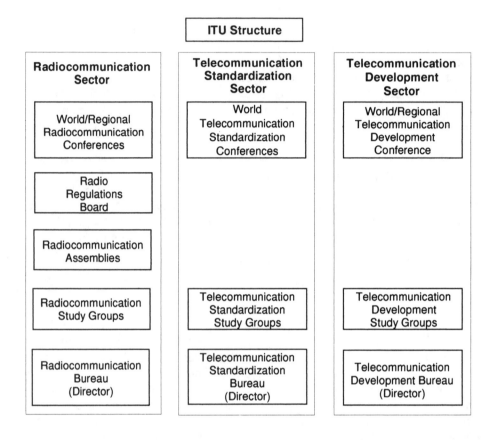

Figure 10.2 The overall structure of the ITU, indicating the main branches of the Radiocommunication Sector.

area of greatest interest in satellite communication. Next of importance is the Telecommunication Standardization Sector, where the effort concerning the interfacing of networks is centered. Many of the highly recognized standards, like SS7, ISDN, and X.25, are among the accomplishments of this sector. The third arm, Telecommunication Development, is that part of the ITU that carries out the U.N.'s mission for improvement of the telecommunications infrastructure in developing countries. The programs and processes that the three sectors carry out are set at their respective world conferences. We will speak more about the *World Radiocommunication Conference* (WRC) and the *Regional Radiocommunication Conference* (RRC), which are so important to the use and regulation of the spectrum.

Figure 10.2 does not indicate two overall administrative areas that are nevertheless important to the functioning of the ITU. These are the Plenipotentiary Conference (the Plenipot) and the General Secretariat, which is the full-time headquarters. The charter and administrative operations of the ITU are established by international treaty. Periodically, these are modified by a Plenipot, where new officers are elected and the budget and agenda for the Union are set for the future. While the international conferences have become more and more important over time, the cost of the operation of the ITU has continued to escalate. Because the Administrations ultimately pay for the operation of the Union through allocations (the United States having the biggest), there is continued pressure to reduce the number and duration of the conferences and to reduce the full-time staff. This is counterbalanced by the rapid changes in telecommunications technology that drive the need for flexible regulations. The upshot is that the ITU's budget keeps increasing as its activities are expanded.

The General Secretariat in Geneva runs the ITU on a day-to-day basis, maintaining a full-time staff and arranging future conferences. The Secretary General who heads the organization is elected at the Plenipot. All of the regulations, reports, and supporting documents are published and distributed to members. Copies of documents, which are quoted in Swiss Francs and are expensive, can be obtained from the ITU in Geneva. Another noteworthy activity of the General Secretariat is the biannual Telecom Expo held in Geneva. This is the biggest telecommunications show in the world and is attended by dignitaries from the majority of countries and telecommunications organizations, including operators and manufacturers. The total amount of money spent on just one of these shows would probably pay for a satellite and launch.

Use of the radio spectrum comes under the auspices of the Radiocommunication Sector and, in particular, the *Radiocommunication Bureau* (BR, using the French abbreviation) and the *Radio Regulations Board* (RRB). There can be confusion between the purposes of these two very important elements of the ITU. As the new name and face for the CCIR, the BR has the task of setting down the technical specifications and criteria that apply to the ITU's management of the radio spectrum. These recommendations are followed by all administrations and the full-time elements of the ITU. The way that these recommendations come into being is through the radiocommunication assemblies that the BR conducts from time for time. There will be an assembly just prior to any WRC in order to establish the technical ground rules for any proposed changes to the Radio Regulations, which are the rules of the road for frequency usage.

10.4 THE ITU RADIO REGULATIONS

The Radio Regulations of the ITU contain the rules and procedures for the planning and use of all radio frequencies by administrations. Under international law and treaty, member nations are obliged to follow the rules, procedures, and allocations contained in these regulations. The administrations have the right and responsibility, in turn, to assign individual frequencies to their respective government and private users. Experience has shown that the regulations provide a reasonably effective and flexible framework for this purpose.

10.4.1 Objectives of the Radio Regulations

The following is the preamble from the Radio Regulations, as modified at WRC-95, to provide readers with a clear understanding of their scope and intent.

Administrations shall endeavor to limit the number of frequencies and the spectrum used to the minimum essential to provide in a satisfactory manner the necessary services. To that end, they shall endeavor to apply the latest technical advances as soon as possible.

In using frequency bands for radio services, administrations shall bear in mind that radio frequencies and the geostationary-satellite orbit [the ITU term for GEO] are limited natural resources and that they must be used rationally, efficiently, and economically, in conformity with the provisions of these Regulations, so that countries or groups of countries may have equitable access to both, taking into account the special needs of the developing countries and the geographical situation of particular countries.

All stations, whatever their purpose, must be established and operated in such a manner as not to cause harmful interference to the radio services or communications of other administrations or of recognized operating agencies, or of other duly authorized operating agencies which carry on a radio service, or which operate in accordance with the provisions of these regulations.

With a view to fulfilling the purposes of the ITU set out in Article 1 of the Constitution, these regulations have the following objectives:

- To facilitate equitable access to and rational use of the natural resource of the radio-frequency spectrum and the geostationary-satellite orbit;
- To ensure the availability and protection from harmful interference of the frequencies provided for distress and safety purposes;
- To assist in the prevention or resolution of cases of harmful interference between the radio services of different administrations;
- To facilitate the efficient and effective operation of all radiocommunication services;
- To provide and, where necessary, regulate new applications of radiocommunication technology.

The application of the provisions of these regulations by the ITU does not imply the expression of any opinion whatsoever on the part of the Union concerning the sovereignty or the legal status of any country, territory or geographical area.

10.4.2 Pertinent Content of the Radio Regulations

The Radio Regulations are contained in a set three loose-leaf volumes, informally called the Red Books, organized more or less by major topics. The articles that have a direct relevance to satellite communication are listed in Table 10.1. Unlike legal codes, the Radio Regulations are not drafted by lawyers but rather by experts in frequency management and telecommunications. On first examination, the books are very imposing and it might seem that using them is an impossible chore. The language is not entirely clear and the logical flow at times is difficult to follow. Many of the paragraphs seem to read exactly the same way, differing only by a few words or numbers.

The best way, however, to approach these volumes is with a question or issue that needs resolution. For example, an issue concerning the time limit of the advance publication can be researched by looking for the applicable article and regulation. This is shown in the table of contents as being part of Article 11, Section I, RR11-1. You would then proceed to these pages and read the

Table 10.1

Key Provisions of the ITU Radio Regulations (Taken from the Table of Contents)

Article	Content
Article 1	Terms and definitions
Article 6	General rules for the assignment and use of frequencies
Article 8	Frequency allocations
Article 11	Coordination of frequency assignments to stations in the space radiocommunication service except stations in the broadcasting-satellite service and to appropriate terrestrial stations
Article 13	Notification and recording in the Master International Frequency Register of Frequency Assignments to radio astronomy and space Radiocommunication stations except stations in the broadcasting-satellite service
Article 15	Coordination, notification and recording of frequency assignments to stations in the broadcasting-satellite service in the frequency bands 11.7–12.2 GHz (in regions 2 and 3) and 11.7–12.5 GHz (in region 1) and to the other services to which these bands are allocated, so far as their relationship in the broadcasting-satellite service in these bands is concerned.
Article 27	Terrestrial radiocommunication services sharing frequency bands with space radiocommunication services above 1 GHz
Article 28	space radiocommunication services sharing frequency bands with terrestrial radiocommunication services above 1 GHz
Article 29	Special rules relating to space radiocommunication services
Article 30	Broadcasting service and broadcasting satellite service

paragraphs one by one until you discover the particular time period. As you take the time to follow the development of the particular topic, the logic and application always come through. A complicating factor, however, is that one needs not just the latest version of the Red Books but the Final Acts of the most recent WRC as well. This is because there is typically a lag of one to two years in the publication of updated Red Books.

The main areas of focus of the Radio Regulations are: (1) definitions that are used throughout; (2) general rules for using frequencies; (3) the Table of Frequency Allocations, which is central to the regulatory process; (4) special regulations for particular types of services, particularly for GEO and non-GEO satellites; (5) procedures for frequency coordination; and (6) techniques and rules for sharing bands between services (e.g., space and terrestrial). There are literally thousands of specific regulations.

There are a total of 59 articles in the Radio Regulations (the quantity will vary as a consequence of the WRCs). In addition, there are a number of appendices that lay out particular plans and analysis techniques referred to in some of the articles. For example, Article 30 contains the plan for the assignment of orbit positions and frequency channels for the Broadcasting Satellite Service in Regions 1 and 3. Other portions of the Red Books cover resolutions of past WRCs, which cover a variety of policy issues. The exact interpretation and application of the resolutions are not as clear as for the rules and procedures that proceed them.

10.4.3 Table of Frequency Allocations

A key section shows the allocations of frequency bands to particular services, such as the Broadcasting Service, the Fixed Service (for communication between radio stations which are fixed on the surface of the Earth), the Mobile Service (for ships at sea and land vehicles), and various satellite services that are for broadcasting, fixed communications, and mobile communications as well. These are summarized in Table 10.2. An allocation is made to a particular service for a particular region in a particular band. The geographical regions are region 1, Europe and Africa; region 2, North and South America; and region 3, Asia and the Pacific. The distinctions between these regions are blurring, particularly when considering the use of non-GEO satellites.

The Table of Frequency Allocations recognizes two levels of priority: primary and secondary. A primary service has the highest level of right to operate in the particular band. A secondary service may operate as long as it does not cause harmful interference to a primary service. There can be more than one primary service in the same band, in which case the two must coexist through the prescribed sharing criteria and frequency coordination procedures. Generally speaking, users seem to stay away from secondary service assign-

Table 10.2
Definitions of Radiocommunication Services by the ITU

Terrestrial Services	*Space Services*
Fixed	Fixed satellite (FSS)
Broadcasting	Broadcasting satellite (BSS)
Mobile	Mobile satellite (MSS)
Aeronautical mobile	Aeronautical mobile satellite (AMSS)
Land mobile	Land mobile satellite (LMSS)
Maritime mobile	Maritime mobile satellite (MMSS)
Radiodetermination	Radiodetermination satellite (RDSS)
Amateur	Amateur satellite
	Intersatellite (ISS)

ments, preferring instead to use bands in which the particular service is allocated as primary. If you are in the position of being a secondary service, then you must be prepared to cease transmission on first complaint from a primary service user. However, if no complaints are received, then the secondary user can continue to operate.

The Table of Allocations contains dozens of footnotes that allow Administrations to make exceptions to the rules when assigning frequencies to particular services. These are of two classifications: additional allocations, which add spectrum in particular countries or a subregion (but not an entire region); and alternative allocations, which replace an allocation in the table for particular countries or a subregion. For example, a footnote may permit a country in region 1 to assign a terrestrial microwave radio user to a frequency in a microwave band allocated exclusively to satellite communications. These result from the exceptions taken at every WRC, where countries agree to the overall principle but must protect an existing operator in their country that is doing something that is in conflict with the regional allocation.

10.4.4 The Purpose of Frequency Coordination

Through various procedures, the Radio Regulations define how an administration can employ a particular allocation of the spectrum and assign a frequency to one of its users. The procedures are often quite complex and may even lead to confusion and a lack of resolution. Theoretically, an administration checks the allocation to see if the frequency fits a predefined category. If so, it follows the procedure for informing other administrations whose radio communication services could be interfered with by the operation of the proposed radio station or stations. The affected administration may agree or disagree to the use of that frequency or band. Any disagreement must be resolved through the process of frequency coordination.

Space-based radio transmitters are more of a concern because of their greater potential to radiate into the territory of more countries. If the transmission could cause interference, then the administration must follow the convoluted process of frequency coordination, which is defined in the Radio Regulations. This gives the other administrations a chance to decide if they want to allow this particular station to go on the air. A successfully coordinated frequency assignment can be recorded in the Master International Frequency Register (the "Master Register") of the ITU and thereby gain international status.

The basis by which frequencies were assigned and coordinated for the GEO was originally the principle of first come, first served. This was adequate when the number of systems and administrations was small. During the 1970s and 1980s, it became apparent to may potential users of the GEO that this process might eventually cause new entrants to be frozen out of the orbit and spectrum. It was at WARC-77 that the principle of a priori planning was firmly established as a way to guarantee entry for countries who were late to construct their own satellite systems. This conference produced two worldwide Ku-band BSS plans that assigned at least one slot to every member of the ITU. Over the years, very few systems that exploited these plans were actually installed and put into operation. The principle maintains its attractiveness, although the original a priori plan for BSS continues to be modified at successive WRCs.

Other bands were opened up for a priori planning, but the nature of the plans were less demanding. In particular, new segments of C band and Ku band were allocated to the FSS and preassigned to administrations through the Allotment Plan. An allotment is made of a particular frequency or band segment to a particular country or area. It is contained in a plan. This gives each administration an orbit position and access to the spectrum for a variety of purposes. The need for coordination is reduced but not eliminated. The use of allotment band frequencies is increasing slowly, as the more popular segments of the FSS become very crowded. Allotment planning is considered later in this chapter.

A frequency assignment is usually made by an administration to one of its domestic users. These in turn can be registered with the RRB in order to gain international recognition and protection. Frequency assignments consider the particular power, frequency, bandwidth, and modulation type. From a domestic standpoint, a user obtains permission to use a particular frequency through a license granted by the administration. Licenses are not granted by the ITU. The way the frequencies get registered is that the appropriate administration follows the procedure for coordination, which is summarized in the following section.

10.4.5 Rules for Satellite Operations

Article 29 identifies special rules for the use of the spectrum, some of which are key to the operation of GEO and non-GEO satellites, as appropriate. The following are the abbreviated statements of these rules.

- All space stations (GEO and non-GEO) must be fitted with devices to cease emissions by telecommand.
- Transmissions to and from non-GEO satellites shall cease or be reduced to eliminate interference to FSS networks operating in the GEO (this provision was modified at WRC-95 to allow non-GEO Ka band systems to operate on a co-primary basis with GEO Ka-band systems).
- Space stations on GEO satellites shall have the capability to maintain their nominal longitudinal position to the tolerances of ±0.1 deg in bands allocated to the FSS and BSS and ±0.5 deg in other bands. These limits must be maintained unless no unacceptable interference is caused.
- Space station antenna pointing shall be maintained to be the greater of 10% of the half power beamwidth or 0.3 deg. This limit applies only if required to avoid unacceptable interference.
- Off-axis EIRP from Earth stations in the direction of the GEO should be minimized.

10.4.6 Power Flux Density Limits

There are specific provisions of Article 28 that limit satellite power in bands where FSS is co-primary with the fixed service. This is based on the PFD at the surface of the Earth, which is calculated from the formula

$$PFD = EIRP - 10 \log_{10}(4\pi D^2) - 10 \log_{10}(bw) \tag{10.1}$$

where D is the path length in meters and bw is a reference bandwidth in hertz used in the regulation.

D depends on the location of the point on the ground where the limitation is checked; while the bandwidth, bw, is specified in the regulation to take account of the modulation format of a terrestrial microwave systems that could receive interference. In most cases, bw is equal to 4 kHz (4,000 Hz in the formula) to correspond to the bandwidth of a telephone channel in an analog FDM/FM microwave system.

Table 10.3 provides the PFD limits for the satellite bands in common use. These limits apply for all conditions and all methods of modulation and relate to the PFD that would be measured on the surface of the Earth under assumed free-space propagation. The first column in Table 10.3 identifies a particular

Table 10.3
The ITU Power Flux Density (PFD) Limits, in dBW/m²/4kHz, as Measured at a Given Elevation
Angle on the Ground (WRC-92)

Frequency Range (GHz)	PFD Limit, Below 5 deg (dBW/m2/4kHz)	Escalation Rate Between 5 deg and 25 deg (dB/deg)	PFD Limit, Above 25 deg (dBW/m²/4kHz)
1.525 to 2.500	−154	0.5	−144
2.500 to 2.690	−152	0.75	−137
3.4 to 7.75	−152	0.5	−142
8.025 to 11.7	−150	0.5	−140
12.2 to 12.75	−148	0.5	−138
17.7 to 19.7	−115	0.5	−105
22.55 to 23.55	−115	0.5	−105
24.45 to 24.75	−115	0.5	−105
25.25 to 27.5	−115	0.5	−105
31.0 to 40.5	−115	0.5	−105

frequency range corresponding to an allocation to FSS, BSS, or another satellite-related service. In the second column we find the PFD limit that applies for ground receiving elevation angles between 0 deg and 5 deg. It is the lowest value permitted in the particular frequency band because terrestrial microwave antennas could be pointed along low elevation angle paths that align with a satellite. Between 5 deg and 25 deg, the PFD is allowed to increase in direct proportion to the elevation angle according to the linear escalation rate shown in the third column. The PFD limit reaches a maximum at 25 deg and remains constant all the way to the subsatellite point where the elevation angle must equal 90 deg. The reason why the PFD limit is a constant maximum over the higher elevation angles is that terrestrial microwave antennas seldom, if ever, point in such directions.

A word of caution is necessary with regard to the values in the table and their interpretation. In each case, it is presumed that the carrier is fully modulated with the intended information, whether analog or digital. This is the normal and routine operation and represents more than 99% of the time that signals are transmitted from space. However, there are situations when the satellite operator or user must transmit what amounts to an unmodulated carrier at full or reduced power, thus violating the power spectral density stated in the table. There are legitimate reasons for this apparent violation of the ITU Radio Regulations (we shall explain how this might not actually be a violation), such as the following two scenarios.

- During initial *in-orbit testing* (IOT) of the satellite, when the manufacturer and operator typically transmit unmodulated carriers at full or elevated

power. This is required to determine the operating status and performance of the individual transponders as well as the gain and coverage of the antenna system.
- Prior to start of service, a new Earth station or one switching to a different transponder or satellite must first transmit a "clean" unmodulated carrier to verify the correct polarization and power level. The impact can be reduced by ensuring that such carriers are modulated by an energy dispersal signal (typically a sweep across several megahertz).

Another type of apparent violation occurs when a satellite is tested in a temporary orbit position prior to relocation to the final position. There are a variety of good reasons for doing this, such as that the assigned position is occupied by a satellite at end of life that the new satellite will replace. All of these situations required that the specific PFD rules or the prior coordination not be followed to the letter.

The Radio Regulations, under Rule 601, require that any satellite or Earth station follow all of the rules and procedures so as not to cause harmful interference. However, any member of the ITU (i.e., the governments and their licensees) can violate a rule as long as such operation does not cause harmful interference to the services of another Administration that is following the rules. This sounds curious at first reading, but upon consideration it is quite reasonable. What it says is that you may, if necessary, operate outside of the rules (because circumstances sometimes require this), but that if you do this, you cannot interfere with other properly operating services.

10.5 INTERNATIONAL FREQUENCY COORDINATION

The process of international frequency coordination is made difficult by the fact that the spectrum and orbit space are limited resources and must be shared by all nations and users. This sharing is a complex process where cooperation is vital. However, this cooperation can break down when satellite operators enjoy or behave as if they have a monopoly position, acting as if they own the orbit positions and spectrum that they employ. Such obstructions are often overcome through the process of diplomacy that can transcend some of the self-interest that at times seems to dominate the process. For this reason, a prospective satellite operator must make a good faith effort to follow the ITU rules for frequency coordination and, at the same time, conduct its own campaign for the support of its domestic administration and possibly other administrations in the countries that could impact the outcome.

10.5.1 The First Step in the Process

The overall regulatory process is shown in Figure 10.3. Beginning at the top, the ITU develops the frequency allocations and rules for coordination; next, the administration participates in the ITU activities and oversees the assignment and user of frequencies within its domestic borders; and finally, the user applies for and obtains frequency assignments and the authority to operate from the administration. A planned user of the orbit and spectrum, such as a company proposing to launch a GEO satellite, must therefore apply to the domestic regulatory body who acts as the administration for the particular country. The user will usually be required to prepare the actual applications that are forwarded to the RRB by the administration. As reviewed in previous paragraphs, the ITU has created the international frequency coordination framework embodied in the WRC, BR, and RRB components of its structure.

There are basically two types of frequency coordination: terrestrial coordination, for land-based microwave transmitters; and space coordination, for radio transmitters and receivers on satellites. Terrestrial coordination involves any terrestrial radio transmitter that can potentially radiate signal power across a border into a neighboring country (e.g., paths 2 and 3 in Figure 10.1). In particular, the Radio Regulations and Recommendations provide technical analysis procedures to compute the *coordination contour*, which is a graphical depiction of the expected and worst case power levels from a transmitting Earth

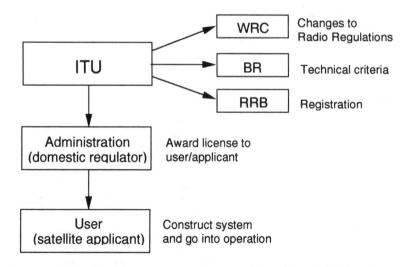

Figure 10.3 The relationship between the applicant for a satellite system; the associated domestic administration who submits the application; and the ITU, where the regulatory process is developed and managed.

station after propagating through the atmosphere. This is covered in more detail near the end of this chapter. A neighboring administration analyzes the coordination contour to determine whether or not this level of power could interfere with the operation of domestic radio receivers that employ the same frequency band. If so, then the two administrations would, on a bilateral basis, make an agreement as to which frequencies would be used or how the transmit radiation pattern of the offending Earth station should be altered. After coordination is complete, the administration can register the new frequency assignment with the RRB in Geneva.

10.5.2 Frequency and Orbit Coordination

The process for coordinating a new communication satellite is described in Article 11, "Coordination of Frequency Assignments in a Space Radiocommunication Service Except Stations in a Broadcasting-Satellite Service and to Appropriate Terrestrial Stations" (coordination of BSS assignments in the 11.7-GHz to 12.5-GHz band is covered in Article 15). Figure 10.4 summarizes the regulatory process for a typical application. This chart is intended to provide a conceptual flow and does not include all of the details of the Radio Regulations. In fact, it is highly recommended that any reader who needs to understand the specifics of the process study the text of Article 11 in its entirety.

Before an administration can launch a satellite and begin operation, it must go through the three step process: advance publication; coordination; and notification and registration. The *Advance Publication of Information* (API), provided to the ITU in the first step no later then five years nor earlier than two years before the start of operation, is a general description of the proposed satellite network to allow other administrations to assess the potential impact on their existing or planned satellite networks. After review by the RRB for compliance with the Radio Regulations, the information is published in the Weekly Circular, which is a special publication containing these applications as well as requests for coordination and notices of frequency assignment (discussed later). The required information is specified in Appendix 4 and compiled in an ITU form called the AP4.

The API step is a prerequisite for coordination, allowing any administration to submit comments on how the new satellite network could interfere with the operation of their own existing or planned satellite networks. If this is the case, they are to respond with their "comments" within four months of publication to the filing administration. The filing administration is supposed to try to resolve any difficulties raised in these comments by altering the orbit position and/or transmission characteristics of the proposed satellite. This could involve a change in the coverage pattern, power levels, or specific frequency bands. The Radio Regulations require that the filing administration

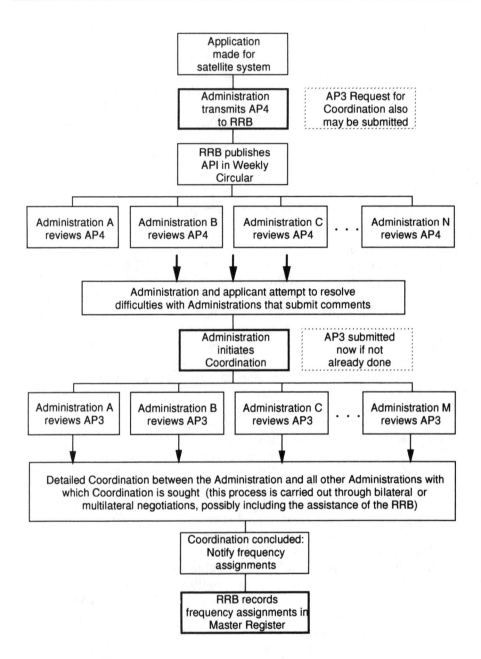

Figure 10.4 Typical regulatory process for a new FSS satellite network including the advance publication, coordination, and notification phases.

make an effort to bring this to resolution; however, if the administration making the comments does not remove its complaint, the rules do not explicitly obligate the filing Administration to do much else. At the end of the four-month period of the comments and attempts to resolve the problems, the filing administration must submit a progress report to the RRB. Additional reports may be submitted on six-month intervals. In any case, the filing administration may move on to the coordination phase when six months have passed after the publication of the API in the Weekly Circular.

After the prescribed period, the administration can initiate the coordination process for the satellite and Earth stations in the network. This is the most critical phase as it determines the priority that the new network will have over new applicants who come later. Coordination is only required with other satellite networks that fall into one of the following three categories:

1. Those that have been recorded in the Master Register of the RRB;
2. Those that have already completed coordination;
3. Those that have entered coordination before the new network in question.

The fact that any satellite network that has at least entered coordination gains priority over new comers is very significant to the entire process of launching a new satellite. Because of the importance of being early, the manager of a new satellite and Earth station project should push the respective administration to initiate the API and coordination process as soon as possible. Furthermore, modifications to the procedures made at WARC-88 allow an administration to file for coordination with the RRB *at the same time that the API is submitted.* The RRB must wait the prescribed six months from publication before taking action, but this nevertheless assures that the new network will gain status in the shortest period possible.

The information to be supplied for coordination is contained in Appendix 3 and compiled in the AP3 form. It is very much like the AP4, except that specific frequency assignments are requested. This can be accommodated by filing in the characteristics of the transponders (center frequencies and bandwidths) along with the basic specifications of some typical Earth stations and services. The Radio Regulations require that the filing administration send a copy of these forms to each administration with which it needs to coordinate. The determination of the affected administration uses the calculation of the percentage increase in equivalent link noise temperature ($\Delta T_l/T_l$), defined in Appendix 29. The threshold for coordination was initially set at 2% but was subsequently increased to 6% to reduce the quantity of coordinations that a new applicant would have to undertake. The analysis technique is very conservative in that it will indicate that unacceptable interference could occur even when it would not be the case in practice. The RRB also makes this calculation as a check on the filing administration. The idea is to assure the priority given by the Radio Regulations to any other administration that has a system that meets one of the three conditions cited above.

The requested administration must evaluate the data in the AP3 filing to see if the calculated interference levels would be acceptable. It must then: (1) acknowledge receipt of the data; (2) examine the data to determine if interference would be caused to its lawful frequency assignments; and (3) within four months of receipt, inform the requesting administration of its agreement that the interference is acceptable or its disagreement, giving also the technical details upon which its disagreement is based, including relevant characteristics about its system not previously notified and its suggestions that it can offer with a view to a satisfactory solution to the problem.

An affected administration will evaluate the technical characteristics of the proposed satellite network and inform the filing administration if it agrees to its operation. This is not a unilateral process, and the techniques for making the assessment have become fairly standard over the years. The basic criteria for acceptable interference levels are set by the ITU or BR at relevant WRCs or through recommendations. Also, if the two administrations have come to agreement on criteria in the past, then the same criteria should be used unless the two sides agree to something else. Any special agreements between the two administrations will not "prejudice" the negotiations with and among other administrations.

Coordination of satellite networks is a bilateral activity where the newcomer must approach the incumbent and obtain their agreement regarding the potential for interference between systems. Such discussions and negotiations will take a year or more, particularly in difficult or acrimonious situations. This difficult process could be helped by a new arrangement under the WARC 88 Final Acts, wherein a *multilateral planning meeting* (MPM) can be called by an administration. The concept is that the parties involved would get together more or less outside of the Radio Regulations and work to achieve a settlement.

The RRB is the full-time body of the ITU that, sitting in Geneva, analyzes the proposed frequency assignments submitted for review by administrations. The RRB "examines" the assignment to make sure that it fits the Table of Frequency Allocations and that the coordination procedures have been followed. In particular, the new assignment should not cause unacceptable interference to an existing assignment that has already entered or completed the review and coordination. While not conveyed policing authority, the RRB does have power over administrations because of the status given to frequencies that have been recorded in the Master Register. When a frequency assignment is in coordination between two administrations, the RRB can assist by performing calculations of the expected level of interference and can make recommendations to the parties on how the interference could be prevented.

Most of the time, coordination and registration are accomplished in a straightforward manner, taking anywhere from six months to three years, depending on the number of administrations involved and the complexity of the technical analysis of potential interference. Occasionally, an administration locks it heels and refuses to allow coordination of a new frequency assignment by another administration. The question arises as to what motivates agreement on coordination in an environment of newcomers who have to obtain agreement from entrenched providers. Experience has shown that the following principles, while not necessarily written down, apply in practice.

1. The force of the rules and their interpretation by the RRB is such that all administrations shall make all possible mutual efforts to overcome the difficulties, in a manner acceptable to the parties concerned.
2. Various types of interactions and meetings can be employed, particularly:
   a. Bilateral meetings between the administration requesting coordination and the Administration that has raised the objection;
   b. Multilateral meetings as needed and where appropriate (not necessarily an MPM introduced by WRC-88).
3. All parties recognize that:
   a. Every administration has a legitimate right to put up satellites;
   b. Any administration must be willing to make adjustments in its planned or existing operation to allow a new applicant to enter into service (these adjustments are typically only on paper and have little real impact);
   c. A new entrant must approach the existing operators with the proper attitude, recognizing that the principle of first come, first served is firmly embodied in the Radio Regulations and international law;
   d. An established operator must allow a new entrant to move forward with a plan for satellites;
   e. That today's new entrant becomes tomorrow's established operator; also, any established operator will eventually want to introduce a new satellite network and hence will find itself in the position of being a new entrant.

The completion of the coordination process will take time and effort. There are often difficult problems with one or more other administrations, and some of these problems may involve a serious incompatibility between the new and the existing systems. The important point is that the new applicant and the associated administration must continue to work through problems and to creatively find solutions with which everyone can live. The following concepts and techniques are offered as loose guidelines for reaching agreement.

- Hold the meetings with the other administrations and be sure to come prepared to discuss all of the technical issues and options.
- Request that the other party do their part.
- Involve the ITU and the RRB, and seek their counsel and help by telephone and in person.
- Be assertive, but not overly demanding.
- Prove to the other administrations that your side is actually going ahead and therefore must be taken into account.
- Make proposals on how to allow safe operations without unacceptable interference.
- Be willing to make reasonable adjustments, but don't sacrifice the economic viability of your system.
- Be prepared to use as much influence as possible to aid the process.
- Be patient; coordination can be a lengthy process.

Notification of frequency assignments, the last step in the regulatory process, should be accomplished before the service is initiated. This is done by compiling the AP3 forms with the characteristics of the satellite network after the completion of coordination. New forms are needed because it is likely that coordination has resulted in changes to the original values. The RRB will verify that the entire coordination process is complete with all of the appropriate administrations. If so, they will record the assignments in the Master Register and the new operator is free to begin using them to provide service. But what if this cannot be completed before the satellite is launched and the Earth stations put on the air? There have been instances where this was the case and the satellite network was activated nonetheless. The key point, however, is that the applicant has entered into coordination on a timely basis and all of the potentially affected administrations are fully aware of the new system and its operating characteristics. In no instance may the new system cause harmful interference to an existing system that is properly registered. If this were to happen, the new operator would have to immediately remove the interference when requested. Having followed the procedures properly, this situation cannot result.

10.5.3 Terrestrial Coordination of Earth Stations

Terrestrial coordination is the process intended to protect terrestrial microwave stations in other countries from transmissions from new Earth stations that operate in a shared frequency band. The bands where this is required are those where the FSS and the fixed service are coprimary. Referring to Figure 10.1, we are trying to control interference along path (2) where the Earth station interferes with the terrestrial station, and path (3) where the inverse situation can exist. An applicant for a new Earth station starts with data that describes the radiation pattern of the antenna 360 deg of azimuth along the horizon. This is used to compute the amount the RF energy that could propagate from the Earth station location to locations in a neighboring country where terrestrial receivers could be located. The ability of the Earth station to receive interference is also needed to see how the distant terrestrial transmitter could create difficulty for its operation. Additional shielding from and to the potential interference is provided by the existing terrain around the Earth station site.

The mechanism used to perform this assessment is the coordination distance, which is the calculated distance over which interference could potentially result. This distance is calculated with formulas in Appendix 28 to the Regulations and considers attenuation produced by propagation over the surface of the Earth along the great circle path. Without line of sight, there is actually substantially more path loss for great circle paths than for free space. The techniques are complicated and are best carried out with appropriate computer software.

The technique computes the two different coordination distances; that is, d_1 corresponds to the great circle, and d_2 is that produced from rain scatter propagation. Formulas and graphs for computing d_1 and d_2 are found in Appendix 28. These consider the radiation pattern of the Earth station antenna, where the longest distance occurs along the axis of the main beam. The best way to view the coordination distance is in the form of a coordination contour, which is a polar plot of d_1 and d_2 as a function of azimuth around the Earth station. A simplified example is shown in Figure 10.5. The minimum distance provided for in the rules is 100 km, while 200 km is more typical (as shown in the figure). The elongation of the contour is produced by the main beam of the antenna.

An administration who receives the request for coordination, including the coordination contour along with the other information in Appendix 3, must determine if interference could result from the operation of the new Earth station. Receipt of this data must be acknowledged within 45 days of receipt. The examination would consider:

- Interference to terrestrial stations existing or to be operating before the Earth station enters service or within three years, whichever is longer;
- Interference to the Earth station by such terrestrial stations.

Each affected administration notifies the requesting administration within four months of one of the following:

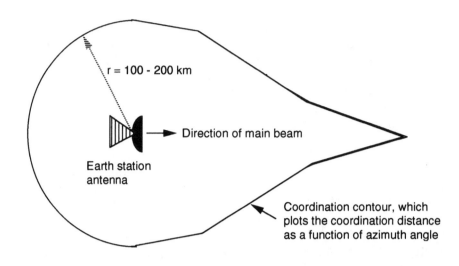

r = 100 - 200 km

Direction of main beam

Earth station
antenna

Coordination contour, which
plots the coordination distance
as a function of azimuth angle

Figure 10.5 Conceptual model of the coordination contour for an Earth station in the FSS. Terrestrial coordination is required if the coordination contour, which contains the coordination area, extends into the territory of another country.

- Its agreement to the proposed Earth station with a copy to the RRB;
- Its desire to include specified terrestrial stations in the coordination;
- Its disagreement.

In the last two cases, the notification should include a diagram showing the location of existing or planned terrestrial stations within the coordination area (which is the area inside the coordination contour) and suggestions for solving the interference problem. A copy to the RRB should also be provided.

Terrestrial coordination is considered to be very routine by all administrations due to the popularity of both satellite and terrestrial microwave communication. The requests for terrestrial coordination are generally handled on a routine basis and should produce agreement in a reasonably short period of time. This also considers the fact that all countries have some previously recorded terrestrial links and Earth stations and so no one is in a position of control of the spectrum based on first come, first served. In cases where it is difficult to get agreement, the RRB can be counted on to provide assistance. They have even been in the position of acting as a disinterested party and honest broker in finding the needed solution to the problem.

10.6 WORLD RADIOCOMMUNICATION CONFERENCE

A WRC is held every two years or as required to deal with the changing needs of Administrations for radio communications services. These conferences are where the Radio Regulations are established and rewritten periodically. Even less frequently a RRC is held by one of the three ITU regions. The last RRC met in 1985 to develop a plan in Region 2 for the Broadcasting Satellite Service. Regions 1 and 3 already had developed their plans at the WARC-77, which was convened to create the planning process for this new service.

The policy of holding a WRC was instituted by the ITU at WRC-95 in recognition of the changing nature of technology coupled with the accelerating demand for spectrum. Also, instead of focusing each conference on a particular service or set of regulatory issues, the WRC has an open agenda to allow literally any topic to be addressed. A typical WRC lasts two months and is held at the conference center immediately adjacent to ITU headquarters in Geneva. Preparation for the next conference begins almost immediately after the last one is over. Preparatory committees meet at the national and global levels. Since many of the issues have to do with sharing frequency bands and coordinating systems, there is a need to provide a solid technical foundation for the conference. This includes the development of technical criteria and possibly special computer software to analyze the impact of a particular approach or plan.

After the conference begins, it takes many weeks to refine the two complex aspects until a general consensus can be worked out. Each country can express its position at general "plenary" meetings and can participate in smaller working groups that delve into critical issues. The Final Acts of the conference must be written, translated into the various languages (where French is the governing language of the ITU, or the UIT as it is known in the French language), and approved by the entire body. Recommendations and studies of the BR are used as a basis for technical positions that the WRC takes and incorporates into future provisions of the Radio Regulations. Some sections of the Final Acts become provisions of the Radio Regulations, and others become instructions to the RRB and the BR. Following the WRC, the respective governments have the opportunity to confirm the Final Acts, incorporating them into domestic law. There are instances where a particular government was not in agreement with the overall conference and so took exception to some provision or perhaps the Final Acts in their entirety.

The WRC that concluded in October 1988, took three months to effectively rewrite the procedures for initiating new satellite communications systems. It was a difficult conference with a wide agenda that addressed [2]:

1. Changes to the current frequency coordination rules for the *fixed satellite service* (FSS), the allocation used by all domestic and international telecommunications satellites. The modified coordination procedures simplify matters for administrations but do not alter the basic approach, this being the principle of "first come, first served."

2. Allotting new frequencies for the "expansion bands" to provide at least one dedicated orbit reservation for each administration. This satisfies the political concerns of developing countries.

3. Allowing "additional uses" in the expansion bands that can be introduced on a temporary basis before the more permanent allotments are applied in a particular region of the orbital arc. This opens up the expansion bands for experimentation with new services such as mobile communications and private international networks.

4. Adding frequencies to the *Broadcasting Satellite Service* (BSS) plan, which was first established by the WRC in 1977. The new frequencies are assigned to each country for the *feeder link*; from the uplink Earth station to the broadcast satellite.

Under the revised rules, a separate allocation of spectrum has been set aside for all Administrations, particularly those who have no current launch plans. The complete provisions of the plan are contained in Appendix 30B, "Provisions and Associated Plan for the Fixed-Satellite Service in the Frequency Bands 4500–4800 MHz, 6725 to 7025 MHz, 10.70 to 10.95 GHz, 11.20 to 11.45 GHz, and 12.75 to 13.25 GHz." Each administration is allotted resources as follows:

- A nominal orbital position;
- A bandwidth of 800 MHz (uplink and downlink) divided into the following bands:
  - 4500 to 4800 MHz (space to Earth),
  - 6725 MHz to 7025 MHz (Earth to space),
  - 10.70 GHz to 10.95 GHz (space to Earth),
  - 11.20 GHz to 11.45 GHz (space to Earth),
  - 12.75 GHz to 13.25 GHz (Earth to space);
- A service area for national coverage, which is based on an elliptical antenna beam shape;
- Generalized parameters as defined in Annex 1 to Appendix 30B, which is a rather complicated collection of standard characteristics that allow the ITU to develop and maintain a plan of this magnitude; it is independent of modulation method but makes a series of assumptions about satellite and Earth station antenna characteristics and noise temperatures;
- A *predetermined arc* (PDA), which is a segment of the GEO centered about a nominal position intended to provide flexibility in the plan; thus, the final position of the satellite can be adjusted to improve the compatibility with nearby satellites in the plan. For a system in the predesign stage, the PDA is equal to ±10 deg, subsequently being reduced in stages to ±5 deg in the design stage and 0 deg for a system in the operational stage.

The allotment planning approach is designed to give the developing countries and other new ITU members the confidence that a position in orbit will always be available for them. What these countries had been concerned about is that the existing rules for coordination tend to favor the industrialized countries that already have satellites in orbit and who can afford to launch additional ones in the near future. (Ironically, Indonesia, India, Mexico, China, and Brazil were among the existing GEO operators at the time of the conference.)

Some satellite systems can operate in the 800 MHz of spectrum covered in the allotment plan even though the are not directly included. Referred to as existing systems, they are satellite networks that satisfy one of the following:

- Which are recorded in the Master Register;
- For which the coordination procedure has been initiated;
- For which the information relating to advance publication was received by the RRB before August 8, 1985.

Some existing systems were in operation at the time of this writing, including satellites operated by SES and AMSC. They are listed in Part B of the allotment plan.

The first of the conferences to actually carry the title of World Radiocommunication Conference was WRC-95, which concluded in Geneva on November 17, 1995. The primary thrust of the conference was to open up some of the spectrum demanded by the non-GEO systems under development. This involved primarily feeder link frequencies in the following bands:

- C-band:
  - Uplink: 5091 MHz to 5150 MHz (must protect aeronautical radionavigation) and 5150 MHz to 5250 MHz,
  - Downlink: 5150 MHz to 5216 MHz (reverse of uplink) and 6700 MHz to 7075 MHz (protection to GEO satellites under Article 29 no longer provided);
- Ku-band: No spectrum allocated to non-GEO satellites;
- Ka-band: 19.3 GHz to 19.6 GHz and 29.1 to 29.4 GHz (protection to GEO satellites under Article 29 no longer provided), 19.6 GHz to 19.7 GHz and 29.4 GHz to 29.5 GHz, and 15.4 GHz to 15.7 GHz and 19.3 GHz to 19.6 GHz (reverse band).

WRC-95 began a process of review of the early BSS Ku-band plans, which will be an important agenda item for WRC 97. This resulted from a general feeling among Administrations that the plans are not entirely practical for introduction of services widely around the world. In fact, quite a few of the systems that use the BSS channels have been implemented by making modifications to the plans. This points up the shortcomings of attempting to establish assignments well in advance of the construction of the system.

The process of preparation and execution of the WRCs is vital to the regulation of the spectrum by the ITU. As a result, the Radio Regulations will continue to be a living document that changes on a two-year cycle. Practitioners will need to maintain their understanding of the procedures and allocations in a current state through review of the documents and contact with people who participate at the WRCs.

10.7 ADDITIONAL REGULATORY APPROVALS

The discussion up to this point has concerned the international regulatory process as it relates to the radio spectrum and orbit resource. There will be other regulatory approvals that will need to be obtained in each and every country where satellite communication services will be provided. Unlike the ITU with its forced consistency, the domestic approval process varies greatly from country to country. As discussed in [3], the degree of complexity and potential for success depend on the nature of telecommunication policy and degree of openness to competition. These are summarized in Table 10.4.

The information provided in this section is for general guidance and should not be considered to be recommendations on how to enter a market in a country. Obtaining the necessary approvals is a very complex and involved process, requiring a commitment of the necessary resources. Readers who wish to approach this particular topic for a particular project should seek the aid and assistance of people and organizations who have experience in the particular country and type of service.

It is vital that a new entrant identify the particular environment in question and gain a thorough understanding of the rules that apply to the type of services to be provided. In the following paragraphs, we briefly review some of the generic types of approvals that should be experienced in a majority of cases. Just how the particular government chooses to address these areas will depend on the factors listed in Table 10.4. In protected environments where the government exercises a high degree of control, it will be much more difficult to find a workable solution. This could involve the use of forms of influence which are outside of the regulatory process. Some examples of this are suggested in our previous work.

10.7.1 Operation of Uplink Earth Stations

The terrestrial coordination procedures discussed previously are carried out by the administration for the benefit of an Earth station operator. Many of the

Table 10.4
Six Telecommunication Market and Regulatory Structures that can be Encountered During
the Process of Obtaining Approvals in Different Countries

Level	Classification	Characteristics
1	Government monopoly	The classic case of a monopoly vested in a government ministry or department; this monopoly may be based in law or by administrative fiat or convention.
2	Public corporation government monopoly	A government monopoly that is instituted through a government-controlled corporation rather than an administrative department.
3	Government competition	A structure wherein both the government and private sector entities compete with each other in the market.
4	Regulated monopoly	The classic case of private ownership of facilities (one entity) with regulation by a government department.
5	Regulated competition	The extent of competition, the number of competitors, the products or services that can be offered by each supplier are subject to authorization, licensing, type approval, and the like, by a government department or agent of the government.
6	Liberalized entry	A market situation characterized by the absence of all official government rules and regulations that serve to regulate market entry, structure or conduct coupled with complete reliance on private sector entities for the provision of goods and services.

satellite applications covered in this book require some form of large transmitting Earth station. In all cases, the applicant will have to satisfy the domestic regulator that the operation of the Earth station will be safe and in accordance with local standards.

Many countries require that the government or government monopoly, as appropriate, actually own and operate the station. This can be arranged through the *build, operate, and transfer* (BOT) or *build, transfer, and operate* (BTO) schemes that are common around the world. With BOT, the service provider or user constructs the Earth station and completes its acceptance test. It may then operate the facility for a specific period of time (say, five years), after which ownership is transferred to the government entity. Service may or may not be provided thereafter depending on the terms of the original deal. The BTO scheme reverses the sequence of the last two steps. Another situation is where the government or monopoly operator must actually provide the uplink or hub capability as a service to the user. This would be paid for by a combination of initial construction charges followed by monthly or annual operating fees. The equipment would always be owned and operated by the organization providing the capability. Combinations of direct ownership, BOT, BTO, and service provision are possible.

10.7.2 Type Acceptance of Terminals

The terminals required in applications like VSAT networks and MSS services are too numerous to expect to obtain individual operating permits as one would have to do for a major Earth station. The type acceptance approach is an expedient approach for clearing user-oriented Earth stations through the regulatory approval process. By type acceptance we mean that the regulatory body is given a sample (or samples) of the type of terminal so that they may conduct whatever tests they think are appropriate. This would consider electromagnetic compatibility with other devices and services, safety, and orbital interference potential.

Once type acceptance is obtained, the network provider can begin to introduce the terminals into the market. Follow up inspections and tests may be required to ensure that terminals continue to satisfy the requirements of the original type acceptance. The tests may be conducted in a government laboratory by a third party who engages in this activity for a business, by the manufacturer, or by the network operator.

10.7.3 Importation of Equipment

User terminal or other equipment that is produced within the country where service is to be provided can usually be introduced into the network once type acceptance is obtained. It could be a completely different story if the equipment must be imported. One reason for this concern is that the inflow of possibly advanced equipment may harm domestic manufacturers. Also, the government may not want the particular type of device or service to proliferate without adequate control on distribution and use. These concerns are heightened with satellite communication because of the ease with which the terminals can transmit data out of the country and receive information of a type that the government may need to restrict. Reasons for this involve business, economic, social, and economic issues, which are beyond the scope of this book.

To clear this question, the service provider or distributor must obtain an import license. This could be nothing more that an addendum to a type acceptance certificate, or it may involve a totally different process and part of the government. The import license would identify the rules and restrictions as well as the particular import duties that would be levied on the overall operation and each user device as well.

There is generally one area where terminal equipment may be brought in under less restrictions. This is where the equipment would be for temporary use, after which it will be removed from the country. Activities where this is applied would be in emergency situations where communications need to be restored in a hurry as well as for demonstrations of new technologies or services. The latter is important when the operator is in the process of obtaining the approvals and needs to show the government what is involved in providing service in the first place. Such temporary importation may or may not be subject to duties and taxes.

10.7.4 Usage Restrictions

Having obtained all of the approvals needed to introduce the equipment and network operation, there remains the possibility of usage restrictions on the service. An example would be for a DTH service in a country that does not allow selling movies over pay TV. In this case, the business must be based on advertiser-supported programming and possibly flat subscription movie channels. But, a PPV service would clearly be out of the question.

Another example taken from MSS would be a restriction of international calls over the satellite network. In this instance, all calls would have to be connected to the domestic GES or, if technically possible, to another UT within the borders of the same country. A restriction of this type might be required by law in order to protect a monopoly international telephone service provider. The network might be technically capable of making international connections, but the government might have to be satisfied that such calls are inhibited through hardware or software blocks that can be audited from the outside.

10.7.5 Competitive Entry

Over and above the operational issues cited in the previous paragraphs, there is still the overriding question about whether a particular government will allow a new and potentially competing service to be introduced. This goes back to the levels of competitive entry cited in Table 10.4. If the service is particularly advanced and of value to the new entrant, then there is more willingness to take on the entrenched monopoly. This will involve the political processes of the particular country, possibly including the introduction of new laws and regulatory procedures. This was the case in Japan prior to the creation of their domestic satellite communications service industry in the late 1980s.

10.7.6 Licensing

The licensing of individual terminals may be required since, as a transmitting device, it is a form of radio station. Type acceptance could be used in lieu of licensing. However, it is up to the country in question about how they would handle the matter. A license from the government might be obtained by the user when registering the unit as an operating terminal. This would apply whether the unit was imported or domestically produced and provides a vehicle for the government to obtain additional revenue from the license fee. Licensing is different from a one-time activation charge by the network operator, although such charges provide the vehicle for the government to collect from every user (e.g., a license).

10.7.7 Other Roadblocks

The final category of additional regulatory requirements is referred to as road-blocks. Governments and monopoly service providers can be very innovative and persistent in the way that they interact and control the new operator. The potential list could be very long, being different from country to country. Actually, some of the advanced and most industrialized countries of the world have the most difficult and convoluted roadblocks. An area that continues to get much attention is the control of cross-border data flows to protect the privacy of domestic citizens. As mentioned previously, satellite services do not obey national borders.

Just how one identifies and resolves these other roadblocks will depend on the particular situation. There is no substitute for a thorough examination on the ground in the country in question. This might be accomplished with a series of field trips. In many other situations, one will actually have to establish a permanent activity to learn about and subsequently address the concerns of the domestic authorities.

References

[1] Elbert, B. R., *Introduction to Satellite Communication*, Norwood, MA: Artech House, 1987.
[2] Taylor, L. A., "Depoliticizing Space WARC," *Satellite Communications*, January 1989, p. 28.
[3] Elbert, B. R., *International Telecommunication Management*, Norwood, MA: Artech House, 1990.

The Business of Satellite Communication 11

After being in business for over three decades, we know satellite communications systems and services to be a mature industry. Literally any organization with the financial means can invest in a space segment capacity as well as the associated Earth stations for communications services. As we shall see, this can be done as an integrated offering, such as in the MSS and BSS fields, or segmented to focus on a particular aspect, which is common for FSS systems. The financial rewards in this business vary widely, just like in real estate and oil exploration. We discuss many of the factors necessary to establish a satisfactory market for the systems and the services satellites deliver.

11.1 THE SATELLITE MARKETING CHALLENGE

Satellite communications is a capital-intensive industry, like gas pipeline operation and long-haul fiber links, where the investment cost for entering the business is high and operating cost for delivering service is relatively low. Furthermore, the investment represents a sunk cost that is very difficult to liquidate if the revenues are not satisfactory. This aspect is like constructing a large commercial building that could end up being unfilled by tenants. However, unlike a fixed building that is located in one particular part of town, a satellite system is unique in its ability to serve an entire nation or region. Its capacity can be subdivided and sold in small increments like lumber or airline seats. Another issue that impacts the business performance of a satellite is its fixed lifetime. This means that capacity that is not earning revenue due to a lack of customers represents a permanent loss for the time that it is not used.

Figure 11.1 indicates the major elements in a satellite network that is used for an integrated service such as DTH or data communication. The pair of satellites along with the TT&C stations and satellite control center are owned and managed by a satellite operator. Most of the expense is tied up in the initial capital needed to construct and launch the satellites (including launch

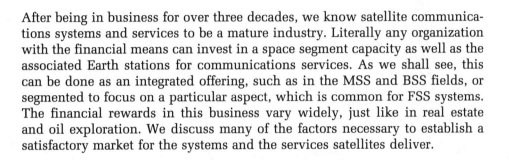

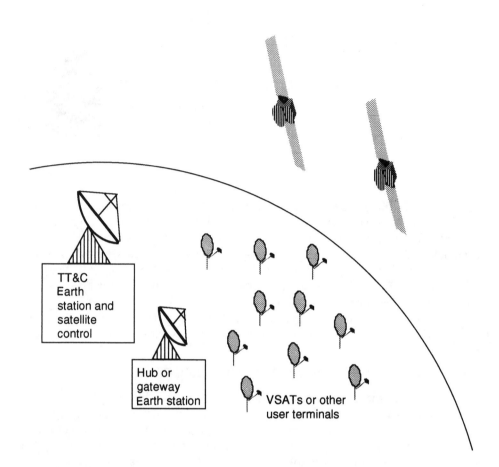

Figure 11.1 Key elements of a typical satellite system from the standpoint of marketing and business. The satellites and TT&C/satellite control facilities make up the space segment, owned and maintained by the satellite operator. The ground segment is often provided in pieces by different organizations who either use the capabilities themselves or offer their services to outside customers.

insurance) and to install the ground control elements. The application portion, shown here as a hub or GES and a collection of VSAT-type user terminals, has a cost that is more or less in proportion to the number of customers served. Whether the application elements are owned by the satellite operator or not depends on many factors.

For example, the most successful DTH service business in the world is BSkyB in the United Kingdom. The satellites for this service are owned and operated by SES in Luxembourg, who in turn has no interest in BSkyB other than to supply transponders to distribute programming. In this case, the DTH service provider concentrates on that particular business and leaves the investment and operating demands of the satellites to SES. In the United States, the case of the integrated owner/operator seems to be more common, with DIRECTV being the best example. Here, Hughes Electronics Corp. has two separate subsidiaries that provide the service: DIRECTV, Inc., is the DTH operator, with ownership and control of the application segment, while Hughes Communications, Inc., controls the satellites.

Perhaps the most difficult challenge to achieving satisfactory financial rewards from a satellite investment is the likely small size of the potential customer base. In transponder marketing, the users must posses their own transmit and receive Earth stations, which can be a significant investment for many organizations. In the case of cable and broadcast TV, the customer could already have the necessary antennas but may be using another satellite. The only way to acquire the business is to undercut the existing operator in price or to offer additional features at the same price. The buying opportunity comes when the existing satellite is nearing its end of life, which is a situation that presents itself to every incumbent transponder provider. This is why an operator should have an effective plan for replacing satellites without loss of continuity.

A DTH system, including sufficient satellite capacity and program content, is a large commitment by anybody's measure. An organization wishing to explore satellite communications as a value-added service but on a much smaller scale can enter the VSAT network service business. New entrants in this market still have the hurdle of building up an adequate customer base. A new network provider must make it as easy as possible for subscribers to install the necessary user equipment and access the service. Break-even points for a VSAT shared hub operation as compared to a DTH system might be 2,000 data communication users as opposed to one million or more pay TV subscribers.

11.1.1 Selling Hardware

The range of possible elements of hardware covers many manufacturing sectors in aerospace, telecommunications, electronics, construction, and heavy industry. This affords manufacturers many possible opportunities to contribute to any number of equipment or service businesses. Due to the large investment involved in the launch and introduction of a major satellite system, manufacturers of spacecraft, launch vehicles, or ground systems largely deal with organizations that can make the necessary capital investment. They in turn must have market power in their proposed area of service. This business is like the aerospace industry, where the customer buys through the request for proposal or request for tender (RFP or RFT) process and usually chooses the lowest bidder. There is also a certain degree of development risk that the system could fail to meet specifications, be delivered late, or cost substantially more than the original estimate. Technologies that are internal to the satellite as well as those needed to develop and manufacture the satellite make entry into this business very difficult. There are few spacecraft manufacturers that have been financially successful at it. It takes a lot of highly qualified people to produce spacecraft, and these must be flexible due to the fact that the requirements and resulting designs differ to some degree for each customer.

Manufacturing and selling Earth stations has evolved from a similar structure into something more like a "mom and pop" industry. Originally dominated by large electronics and defense firms like Northrop, Hughes Aircraft Company, and NEC Corp., most large Earth stations are provided by small specialists like Satellite Transmission Systems (STS) of Hauppauge, NY, and ETS of Melbourne, FL. This is possible because these stations are put together like custom houses from off-the-shelf electronic units produced by still other specialists like Miteq, Scientific Atlanta, Hughes Network Systems, and NTL. The situation with regard to user terminals is different again because we are talking about a device that is more akin to a piece of consumer electronics like a VCR. This is exemplified by the first generation of digital DTH set-top boxes by Thomson Consumer Electronics, designed for the DIRECTV and USSB networks in the United States. However, unlike a VCR, a DTH box is usually tied to a particular DTH service, rendering it useless if the service is discontinued. A standard voice-band modem used to access the Internet or World Wide Web with a PC is not so constrained, making this type of communication device almost as versatile as a fax machine or, for that matter, a telephone.

Success in the sale of hardware is dependent on the key factors of reputation, technology base, and economy of scale. As we indicated, large Earth stations are usually produced by small companies who specialize in the type of business that is called "rack and stack." Satellite manufacturing is restricted to the most established producers, and few try to enter this business at this late day. Regarding user terminals, the field tends to be wide open to any company that either has a technical edge or can flexibly produce quality consumer electronics at affordable prices. Different players from the two ends of this particular spectrum can integrate their strengths through licensing agreements or other types of joint ventures.

11.1.2 Selling Services

Marketing services from satellites is in some ways much easier because there are usually many more potential customers with which to deal. The corollary to this is that it takes many customers to make a successful business out of it. Most potential customers are only looking for communications and have no concept of how the satellite fits in. The real challenge is to make the service very user friendly. That a satellite is used is not sufficient to entice a customer onboard.

A satellite operator can sell space segment alone or offer partial or full network services. The marketing of satellite capacity in the FSS has been called *plain old transponders*, a parallel to the *plain old telephone service* (POTS) term used to describe the conventional PSTN. The principle advantage of the plain, old transponder business is that it greatly reduces the demands on the satellite operator and tends to maximize the range of services that the satellite might deliver. The cost of operating the system in this manner is relatively small in relation to the cost of capital. Customers must acquire and operate their own ground segments in order to make the satellite useful for business purposes.

The other approach is called *value-added services*, where the operator adds the ground segment elements needed to provide end-to-end capabilities for customers. However, including ground facilities in the offering increases business risk because the delivery of services is tied to a number of fixed points on the ground. This in itself can represent a good business for the operator of one or more Earth stations, providing such services as DTH networks, shared hub VSAT services, and fixed telephony communication. Many companies have found this to be a good business when the local economy is strong. On the other hand, the construction of too many such facilities within a limited area can result in a kind of glut. Price competition forces down revenues until one or more of the providers must sell out or close down the operation. Converting the installed network to enter a different service is extremely difficult.

The satellite operator would have taken development, launch, and initial operations risk. Our focus is on selling the space segment after the satellite is placed into service. With such facilities as delivery-in-orbit satellite contracts from the leading spacecraft manufacturers and adequate launch insurance, these risks are of reasonable magnitude. Once the satellite has been successfully placed into service, it is a rare occurrence for there to be a catastrophic failure that would bring the entire system down. Rather, the operator must deal with the normal problems of building an effective marketing and operations organization along with planning for the replacement of the satellites as they near end of life.

There are some other services that would allow the satellite operator to add value for customers and potentially increase revenue and, hopefully, profits. Customers who require satellite capacity also need to install and manage Earth stations. This poses a number of problems for the newcomer to the business. A satellite operator can assist the customer by providing engineering and installation services, along with the associated maintenance after the sale. Some, in fact, offer to "outsource" the operation and maintenance of the facilities. Actually, this is not a new concept, just a new name for the provision of *operations and maintenance* (O&M) service, something that old-line contractors like RCA (now GE), Lockheed-Martin, and Hughes have done in the government sector for decades. O&M service can be profitable if the customer is willing to pay the true cost along with a reasonable profit on top. The opportunity (or problem) for the satellite operator lies in the fact that many equipment vendors, in their wish to increase hardware sales, throw in warranty and maintenance services at very little additional cost to the customer.

11.2 SELLING THE SPACE SEGMENT

The space segment market has provided the greatest net returns to operators of any particular segment of the overall business. This, of course, is not guaranteed, as the key determining factor is the location of the satellite tied to the country market that it serves. This is referred to in the "neighborhood" in the trade. A really good neighborhood is one where several services aggregate, attracting more subscribers and, as a consequence, more services. If your market is based on such an excellent neighborhood and you have a leading position, then the financial rewards are likely to be outstanding. A prime U.S. example is Galaxy I, a cable TV satellite that achieved 100% fill with top line services like HBO, CNN, ESPN, MTV, and the Discovery Channel. In Europe, the Astra 1 series proved that the neighborhood principle also applied in a region with several country markets. The converse too is true; that is, without a good neighborhood, financial returns are far less satisfactory, as witnessed by several defunct satellite operators in the United States and Europe.

11.2.1 Transponder Segmentation

For a company that has selected the space segment market for its area of business, there are several ways that its salable capacity can be offered to prospective buyers. As indicated in Figure 11.2, capacity can be divided up in terms of time or frequency. This then prescribes the types of applications that can employ the capacity and therefore the market segments that can be addressed. The approaches to these particular markets are provided in the following paragraphs.

Before considering how to sell this capacity, we first need to examine how rewarding, in financial terms, each of these options could be. This last, but vital, factor is indicated by the shading of the particular segment of the market shown in the Figure 11.2. The most attractive segment shown in the upper right-hand corner consists of offering full transponders to the market on a full-time basis. Since the customer is committed to all of the capacity, the problems associated with maintaining the "fill" are reduced to a minimum. Customer commitments could be for a period as short as six months to as

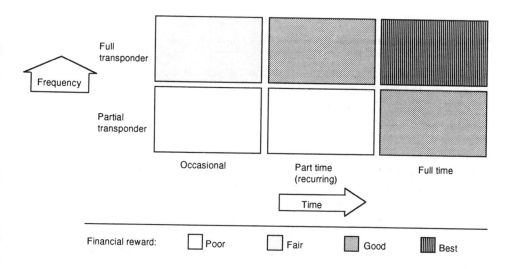

Figure 11.2 Division of plain old transponder services, indicating their relative value and importance to the satellite operator or service provider. The most attractive services are
represented by full-time commitments to customers in the TV and data communications fields.

long as the life of the satellite. In addition, the revenue generated from full
transponder sales or rentals is typically greater on an aggregate basis than from
the other segments.

Moving along the top toward the left, we encounter the part-time (recurring) and then the occasional segments. When we approach the market from
this perspective, we accept the possibility of lower maximum revenue in
exchange for the ability to serve more customers whose needs are not full time.
Part-time users commit to specific time periods on a daily or weekly basis. If
you can line up enough part-time users, you may come close to filling the
transponder from a financial standpoint (periods of time after, say, midnight
and before 6 AM are difficult if not impossible to sell, so the idea is to charge
a bit of a premium for the more desirable hours). Occasional users make no
specific commitment for recurring services, instead reserving the desired block
of time as if it were a seat on a commercial airliner. Transponder resources are
"booked" through a reservation desk, and the service is subsequently delivered
at the prescribed time by the network operations staff. The aggregate revenue
could approach that of the previous two full transponder services, but only if
there is a large enough pool of users and transponders.

The other way to "slice up" the transponder is to divide the bandwidth
and power among several users. We are then supporting one or more FDMA
applications in the same transponder, where each user is assigned a separate

carrier frequency. The maximum power that is available is reduced by the required output backoff (typically 5 dB). A small customer could take a very narrow slice of perhaps 30 kHz and only 1% of the total power (after the backoff). On the other extreme, one customer might commit to a significant fraction of the transponder, say 45%, and then subdivide it further for their own internal use. The rest is available for other customers. The financial reward for this is as good as part-time/full-transponder service or, in exceptional cases, as good or better than a full-time, full-transponder commitment by one customer. As will be discussed later, the difficulty comes in pulling in enough partial transponder users within a reasonable amount of time (because before adequate fill is achieved, a lot of potential revenue is lost).

The white blocks in the lower left of Figure 11.2 indicate part-time and occasional services in the partial transponder segment. Experience has shown that it is much too complicated to make assignments and track usage in this type of arrangement. There are users who may wish to employ an occasional slice of bandwidth for applications such as video teleconferencing and emergency communications. However, the business they generate is usually much less than the difficulty they cause. The better approach is to offer a dedicated FDMA channel to the user; they in turn employ it on a part-time or occasional basis. The frequency is therefore dedicated to them and they will not have a problem gaining access to it. As a satellite operate, you would be able to depend on a continuous and consistent flow of revenue from this customer, even though their usage is erratic. Whether it makes sense or not depends on the customer's particular situation and resources.

The typical satellite operator will market space segment services in search of the most revenues with the least effort at selling. The degree of penetration in the occasional and partial transponder service markets will depend on the amount of disposable capacity, the available market, and the technical and operational resources of the operator. Some of the fine points of space segment marketing in each of these segments is covered in the following paragraphs. The particular types of services and the applications they address are listed in Table 11.1.

The market for transponders has evolved over the years and nearly represents a commodity business because of the standardization of key parameters like frequency, bandwidth, and power. There will be instances where a customer will not be able to locate two or more potential sources of comparable transponders, in which case the satellite operator will have a decided upper hand in the negotiation. On the other hand, if multiple sources exist, then there tends to be a significant degree of price competition. Other important aspects have to do with the term of the agreement and the nature of backup service that the satellite operator can put on the table.

Table 11.1
Space Segment Services and Applications

Overall Market Segment	Sales Approach	Application
Transponder marketing	Outright sales	Cable and TV networks, DTH
	Leases (long and short term)	Cable and TV networks, video backhaul
Partial transponder services	SCPC	Radio networks, data broadcasting, point-to-point trunks
	Fractional bandwidth	VSAT networks
Part-time and occasional video services		Video backhaul, syndication, business TV, distance education

Another factor is the neighborhood principle that was introduced previously. If this factor is involved, then the pricing and availability of the transponder for the particular use could be severely impacted. For all of these reasons, there is no such thing as a standard price for a transponder, and prices can vary not by percentages but by integer factors. During the mid-1980s, for example, a standard C-band transponder on U.S. domestic satellite such as Westar 4 rented for $60,000 per month, while at the same time a cable TV network might be paying $250,000 for a transponder with the same technical performance on a cable satellite like Galaxy I. The difference in neighborhood value is the result of the number of cable households that the Galaxy 1 transponder can reach.

11.2.2 Space Segment Provision

Having decided on the satellite with its associated coverage footprint, frequency band, bandwidth, and EIRP, a satellite operator still has a variety of ways to offer the full-time capacity of a transponder. Users are concerned about the type of transaction, the term of the contract, and the degree of backup (protection) that the operator is willing and able to provide. The alternatives that are commonly applied in the industry are identified in Table 11.2 and reviewed in the following subsections. Each has technical, operational, financial, and legal implications. Because of the wide variety of user needs and the competitive nature of the market, there are other combinations and variations that are applied from time to time.

Table 11.2
Typical Arrangements for the Provision of Space Segment Service

Arrangement	Term	Features
Transponder sales	Life of the satellite	Warranty payback provision can be applied
Long-term leases	Life of the satellite	Comparable to transponder sale, without warranty
	3 to 8 years	Not considered a sale, since the transponder reverts to operator
Short-term leases	Fixed term (non-cancelable)	Used for business startup and limited duration applications
	Month-to-month	Itinerant usage for non-critical application; extension may be by mutual agreement
	Ability to cancel after notice	Offers flexibility, but provides less security to the operator
Part-time and occasional usage	Operator reserves capacity for specific events	Most flexible service (discussed later in chapter)

11.2.2.1 Transponder Access Guarantees and Backup

Full-time or part-time services are tied to a particular satellite and frequency. Customers assume that the capacity is provided as specified in their contract. However, there are situations where a particular transponder does not perform as required and action must be taken to restore service as soon as possible. We consider below the standard provisions for guaranteeing backup by the satellite operator. In any of these situations, it is assumed that the operator would have already attempted to employ a spare amplifier switched into the same transponder channel (such redundancy would be employed to maintain service and therefore revenue).

- *Protected service*: The satellite operator will provide backup in the event of transponder failure, using capacity drawn from reserve transponders or by preempting service to a lower priority (preemptible) customer. The capacity can come from the same satellite (intrasatellite protection), another satellite in the operator's fleet (intersatellite protection), an in-orbit fleet spare satellite that would be drifted into place (total replacement).

- Unprotected service where no backup is guaranteed, but this customer cannot be preempted to restore a protected service.
- Preemptible service where no backup is provided and service can be pre-empted to restore a protected customer. A predetermined order of preemption would normally be assured in each customer's contract.

Customers who need the guarantee of protected service should expect to pay a substantial premium, whereas others who accept preemptable transponder would normally get a discount. In the first transponder service tariffs published by AT&T for their Telstar 3 satellites, they identified three levels of service with three corresponding prices. The Gold, Silver, and Bronze transponder services corresponded generally to the protected, unprotected, and preemptible levels of guarantee. Likewise, Hughes Communications employed the classifications of Primary and Standard to reflect their protected and preemptible transponder sales contracts (the nonpreemptable classification was not made available until the company began to rent transponders to customers). For most customers, the nonpreembtable class has the best balance of price and security since access to the transponder cannot be interrupted unless the operator cannot bring the service back to life.

Whenever service is stopped, the corresponding action will depend on the nature of the contract or tariff that the customer is employing. A sale of a transponder will come to an end if there is a total failure. In some contracts, the customer is entitled to a rebate of a portion of the purchase price in proportion to the unused remaining life of the satellite. This would be based on an assumed minimum lifetime, perhaps 10 years of the expected lifetime of 12 years. A more harsh arrangement would be that the customer simply loses out and no refund is forthcoming. The case of a transponder lease or rental is more straightforward: if the transponder cannot be restored, then the customer simply stops paying.

11.2.2.2 Transponders Sales and Ownership

Users of satellite capacity have a range of options for acquiring transponder capacity. Subject to local regulatory control and availability, they may be able to purchase transponders on a condominium. This was pioneered in the United States by Hughes Communications with the sale of all 24 transponders on Galaxy I to cable TV networks. Some of the advantages of such ownership are:

- Access for the life of the satellite, which is subject to preemption conditions, if any; also, backup may be provided (see Section 11.2.2.1);
- Control of all transmissions through the transponder and greater influence over the operation of the satellite, which will depend on the provisions of the purchase agreement;
- Ability to finance with leveraged leasing or other innovative financing means;
- Knowledge of who your neighbors are—this may be stipulated in the purchase agreement.

Purchase agreements are usually private transactions that allow the buyer and seller to negotiate the terms of service, guarantees, and financing (if applicable). Contracts of this type seldom become public, except if the buyer is a publicly traded company in the United States, in which case it must reveal any major transactions and commitments. Satellite operators prefer long-term commitments and generally would like to sell out the satellite prior to launch. From that point, there is no further marketing for the particular satellite and the responsibilities relate to the proper operation and control of harmful interference. Early failure of a transponder will cause termination of the agreement unless alternate transponders are included in the deal.

11.2.3 Selling Occasional Video Service

Many video users need only a few hours of satellite transmission at one time. The events could be of a one-time nature or could be repeating on a weekly or monthly schedule. Owners of unused transponders find that *occasional video* is a way to get revenue without tying up the capacity long term. Satellite operators typically have the most capacity to offer to the market and are taking over this business. Generally speaking, the revenue from *occasional video*, while important, is less than can be had from a full-time user. Some users only require occasional services and hence can save considerable money by not committing to full-time transponders. The carrier may wish to restrict *occasional video* capacity since it can cannibalize long-term deals with captive TV networks.

The market of occasional video service in the United States is the most developed in the world and covers many applications and types of customers. Among the most established are:

- Entertainment program syndication;
- News backhaul, such as bureaus and on location;
- Sports backhaul, such as stadiums, race tracks, and Olympic venues;

- Private broadcasts (business TV);
- Distance education.

Many of these applications are not large enough to justify a full-time transponder. This means that without the availability of the service, many uses would not exist to a significant degree. Of course, there could be an organization that is large enough and has sufficient resources to make the commitment. The move to technologies like digital compression and demand assignment will tend to extend the versatility of satellite services even in the absence of occasional full transponders.

A satellite operator or other service provider that wishes to enter this market will need to put into place an adequate service infrastructure. This consists of, as a minimum, an order desk to receive requests from customers, a network operation center to coordinate access to the transponders and record usage, and a billing organization to invoice customers and verify payment. As the pool of transponders grows along with the customer base, the operator will find it necessary to automate some of these activities through a computerized reservation and billing system. Virtually all of these systems that have been created are customized jobs for the specific operator. This is the best way to ensure that the computer software and the business work effectively hand in hand. A system of this type can be extended to include reserving outside facilities like terrestrial links and uplink Earth stations, along with studio facilities that might be required to complete the end-to-end service.

11.2.4 Partial Transponder and SCPC Services

The satellite operator can extend its market to smaller customers by offering fractional transponder power and bandwidth on a full-time basis. However, there are several issues that should be addressed.

- A transponder will be committed to this service even if there is one carrier operating. Other customers will have to be added as rapidly as possible to achieve an adequate fill factor.
- Each user will pay a relatively small fraction of the required revenue; revenue will increase over time according to the fill and churn (e.g., customers who terminate their service and are subsequently replaced by new customers).
- Many customers must be marketed to (each one perhaps infrequently).
- Partial transponder services are susceptible to interference from video signals in cross polarized transponders and on adjacent satellites. Assignments should be made with this in mind, and actions must be taken rapidly when interference occurs.

- Users can easily interfere with each other since they share power and bandwidth.
- Carrier power levels and frequencies must be monitored by the satellite operator. Any significant deviations by users must be addressed quickly to deal with financial and performance issues.

Buyers of partial transponder capacity usually do not care about the neighborhood that the satellite provides. The FDMA technique that is part and parcel to partial transponder service allows individual customers to operate almost independently of each other. VSAT networks, data broadcasters, and private audio networks that do not connect with one another can share the same transponder. Therefore, there is no synergy among these users and price is a major differentiating factor in satellite selection.

An exception to this rule would apply to radio networks that deliver radio programming to local radio stations. Users prefer either full-time channels or, at a minimum, fixed time blocks for the duration of daily programs. As we indicated before, noncontinuous usage of SCPC transmissions are very difficult to administer. Radio networks can also be supported through a value-added network that is operated by a third party. The manner in which radio, data, video, and telephony networks can be delivered on a value-added basis is covered in the next section.

11.3 VALUE-ADDED SERVICE OFFERINGS

Satellite communication was originally seen as a domestic telecommunications business that could compete with conventional terrestrial carriers such as AT&T and MCI. What has evolved, however, is that satellite communication is a mutual effort of the satellite operator, the user, and perhaps a third party specialized service provider. Oddly, the terrestrial carriers have an important role since they provide PSTN access, backhaul connections, the last mile between satellite link and user, and backup services over the public network.

Value-added services cover a wide range of capabilities and offerings, making it difficult to generalize. Table 11.3 provides a way of viewing the scope of the industry, extending from the simple case of maintaining a customer's equipment, through more complex arrangements like sharing an uplinking Earth station, to the most extensive offerings of complete integrated networks that are transferred to the customer on a turn-key basis.

It is relatively easy in the United States to piece together a complete service from the available elements. In other parts of the world, users will find this more of a challenge and therefore may subcontract the necessary engineering and operations. Satellite operators and other companies in the industry attempt

Table 11.3
Satellite Communication Industry Scope, Showing the Range From Space Segment Only
Through Full Range of Value-added Services

| Application Segment | Space Segment Only | Degree of Value Added | | |
		Customer Assistance	Systems Integration	End-to-End Service
Video	Full transponder	Event arrangement, maintenance, shared uplink	Uplink Earth station, equipment manufacture, network control	Video distribution network, DTH systems
Data	Full and partial transponder	Shared hub, maintenance	Hub implementation, equipment manufacture, network control	Data distribution, Internet access, backhaul
Telephony	Full and partial transponder	Shared hub, maintenance	Hub implementation, equipment manufacture, network control	Rural telephony, MSS systems, telephone trunks

to satisfy this need by establishing a version of "one-stop shopping" for satellite communication. Companies that offer these value-added services may form joint ventures to minimize the risk of the provider. On the other hand, the only way to be totally responsive to users is to control all aspects of the service. Furthermore, marketing expenses can be very high since a working network capability is usually a prerequisite.

There are many examples of successful and unsuccessful service offerings and companies built upon them. Successful ones run the risk of declining into unprofitability as the technology is diffused and low-cost providers enter the market. Also, new technology can make the service concept obsolete. Some examples of vertically integrated service offerings, representing the right-hand column of Table 11.3, are:

- Packaged video transmission services, from end-to-end including the uplinks and downlinks;
- VSAT networks via shared hubs and terminals provided under contract (with on-call maintenance service);
- Application systems that deliver a certain end-to-end capability for a specific industry could be bundled in with the network;

- Video teleconferencing networks and meeting room services;
- DTH systems, including hardware, software, and possibly programming;
- Mobile satellite communications through GEO or non-GEO constellations.

11.3.1 Selling Value-Added Services as a Systems Integrator

A company wishing to be involved with value-added services may enter the market as a systems integrator. This eliminates much of the risk associated with building your own network and offering services to the market. Traditionally, systems integration was the domain of aerospace companies that performed under contract. Philco Ford built a tropospheric scatter backbone throughout South Vietnam during the 1960s. Hughes Aircraft introduced satellite communication to Indonesia in the 1970s; Aerospatiale did likewise for the members of the Arab League in the 1980s. EDS extended the concept to corporations wishing to contain data processing and internal telecommunications expenses.

Satellite communication offers systems integration opportunities on a smaller scale because it takes less in the way of equipment and operational capability to create a working network. The satellite is usually in service and able to provide the critical linkages. Earth stations are installed at the prescribed locations using substantially off-the-shelf components. User terminals are distributed to end users in much the same way as cellular phones and DTH dishes. The systems integrator manages the design and implementation according to the contracted cost and schedule. If appropriate, the relationship could extend to operating and maintaining portions or all of the system. The latter is discussed in the next section.

11.3.2 Maintenance Services

Once the ground segment is placed into service, the operator or user is faced with the costly and difficult question of providing adequate maintenance of the systems for their useful lifetime. Anyone who has worked this side of the business knows the following.

- Systems and facilities, like an automobile, must receive proper maintenance if quality service is to last the useful lifetime of the equipment.
- Engineers who design and install the systems seldom think about the O&M phase. As a consequence, maintenance people have to *reverse engineer* many aspects of the system in order to figure out (1) how it works and (2) how to fix it when a failure occurs.
- If everything is working properly, no one says anything good (or bad).

- If there is a failure that affects service, everyone is upset (customers, managers, and marketing people).
- The last lesson learned is that there is never enough money available for (1) adequate staff or services to maintain the system, (2) adequate test equipment, (3) enough spare parts and units, (4) training of maintenance engineers and technicians, and (5) additional items (a long list).

Maintenance support may be obtained from the original vendors and third-party maintenance organizations. Otherwise, adequately skilled staff will take on the responsibility for trouble shooting and repair down to the circuit board level. This is where the service provider can step in to help the customer and keep the network or facility in good operating condition. It turns out that the number of people who perform this task is much less than it might appear. The key to success as a maintenance service provider is to have a motivated team of professionals with a thorough understanding of every element of the system. This sounds much tougher than it is. The principle of specialization can be used to deal with a complex system such as a DTH broadcast center or VSAT shared hub. Also, if the Earth stations are dispersed over a wide area, then the maintenance staff will need to be dispersed as well. Under this type of situation, there is a vital need for good communication among the maintenance team members. This can use simple devices like pagers and cellular phones. Communication in remote areas is now possible with MSS networks that are coming online.

This brings our discussion to another area of maintenance where the traditional concepts often fail. Software is now a central element in every modern telecommunications and video network. Introduction and maintenance of sophisticated software is a major concern, particularly in satellite control systems and data networks. Internal software development is often undertaken because of the relative ease of maintenance and modification after delivery. However, there are substantial risks whenever a major software development project is undertaken [1]. Mark Norris has suggested the following simple process to encourage success.

- Articulate what you want.
- Get the right supplier for the job.
- Ensure that you keep control of the design.
- Keep tabs on the supplier.

Computer systems can provide an effective system for administering the network and user access. This includes the classic network management functions such as:

- *Configuration management*: inventory control, installation, configuration, version control of network hardware and software;
- *Fault isolation and repair management*: identification, resolution, and tracking of failures; contingency fall-back plans and disaster recovery;
- *Security management*: encryption, password requirements, physical device security, and security policies;
- *Performance management*: real-time and historical statistical information about traffic volume, bandwidth occupied, resource usage, and congestion;
- *Accounting management*: tracking costs of usage, such as programs watched, connect time, network data storage used, and pages printed.

There is a tendency to build up staff so that services are maintained at their optimum level. Such an approach is greatly appreciated by users but can cause costs to escalate out of control. Therefore, it is important to minimize the required number of positions. If there are too few people, then there is no time to rest and errors will occur more frequently. Too many people is bad because of the cost and the boredom that result.

The proper operation of a satellite application system depends on the quality of the O&M. If this is the responsibility of the service provider, then there should be a process for measuring the effectiveness of the O&M function and how well it fulfills its obligations. The following are some suggested metrics that this author has found useful on an ongoing basis.

- *Availability*: The most important measure of operational performance, considering all factors and measuring only the percent of time that the service is available.
- *Mean-time-to-repair*: This is the classic maintenance metric, used also in availability calculations.
- *Number of trouble calls and average length of an open "trouble ticket"*: A trouble ticket is simply the record of a complaint from a user, indicating the nature of the problem, who reported it, and the actions taken to resolve it. The ticket is closed when the problem is resolved or the user declares the issue moot (because the user closed down the facility, left, or canceled the request for some other reason).
- *Customer satisfaction, measured through a survey*: The customer is asked to rate the service quality, the people, and the company on a scale of, say, 1 to 5, five being the best. These numbers are collected monthly and provided to all participants. The most effective survey is taken personally either over the telephone or face-to-face. Simply mailing the survey usually is not enough. Or, if the survey is mailed, it should be followed up with a telephone call or visit.

11.4 THE MARKETING ORGANIZATION

The fundamental role of marketing is to establish the business and technical capabilities that attract good customers. From this, a solid business and sustainable strategy are developed. This area is usually ignored by prospective satellite operators, who view marketing as order taking. Building a new market is difficult and expensive. This is discussed in the next section. If you are fortunate to have a solid business, then you must protect that position through continued marketing activity.

A successful marketing strategy is built on people who have the necessary familiarity with the important elements of the business. Marketing teams should consist of professionals from different functional areas, such as engineering, finance, sales, operations, and promotion/public relations. For a satellite operator, the cost of creating and supporting such a team is small compared to the investment in space segment assets. Therefore, a little spent on excellence in this area will usually pay for itself in increased revenues. Moving toward value-added services, the cost of marketing and adding customers can exceed the value. For this reason, careful study of the total expense of marketing should be undertaken before commitments are made.

The following is a suggested list of the more significant costs of a strong marketing organization.

- *Marketing staff*: the multidisciplinary team that analyzes the markets and customer needs, identifies the right type of offering, and creates the infrastructure and systems needed to enter the particular market or segment.
- *Sales staff* (may be part of marketing): the high-energy people who take what the marketers have developed and expose it to live customers. In satellite communication, sales people may take on the principle customer-relations role but leave the technical and business dealing to other professionals with greater content knowledge.
- *Service demonstration capabilities*: consists of temporary and permanent facilities that allow customers to experience the service capabilities without making a major commitment.
- *Travel and related expenses*: direct contact with customers, strategic partners, regulatory bodies demands that members of the marketing team must travel a significant percentage of time.
- *Advertising and supporting materials*: the promotion campaign to inform prospective customers of the availability of the service capability and to generate a positive impression of the organization.
- *Subcontractor arrangements*: any service or offering will involve components supplied by outside organizations.
- *Legal expenses*: to prepare and negotiate contracts and service agreements.

- *Tracking of revenue and costs*: success in the business is not guaranteed. The business activity must be know where it is from a financial standpoint at any time.
- Corrective action is justified when performance in a particular segment is not satisfactory.

The particular makeup of the marketing organization will depend on the special needs of the business and the structure of the organization. Since we are involving satellite communications in the marketing mix, it will be essential that some members of the team be conversant in the technology and economics. As time goes by, the optimum organizational structure will change. By monitoring progress and making the appropriate adjustments, the best match to the market will be obtained. Having versatile people on the team will tend to facilitate this adjustment and prevent the hardship of staff reductions. What we hope for is a service opportunity that allows people to grow in professional and financial terms.

11.5 FINANCING A SATELLITE SYSTEM

Any satellite communication system requires capital to get started and so service providers must approach many of the traditional financial markets for funds. A satellite system is a major capital commitment, representing a fixed asset with a given lifetime. Once the satellite is launched, the life is determined. Fortunately, a satellite with wide area coverage and conventional transponders is very versatile as it is not tied to a particular network configuration. The operating cost of the space segment of a satellite system is low relative to the cost of capital. Almost any organization can acquire satellite technology because of the defusion of expertise that has occurred over the decades since Early Bird was launched.

The cost of the ground network can equal or exceed that of the space segment. Capital and operations expenses are more in balance; that is, the total capital of a ground station project is approximately equal to the annual cost of operation, including return on investment. Fortunately, the ground segment operator can repair and upgrade ground facilities; while the satellite, once launched, cannot effectively be modified. An important trend is that users are purchasing and operating their own ground stations, greatly reducing the operating and capital costs (and risk) or satellite operators.

11.5.1 Elements of Capital Budgeting Analysis

The first step in financing a system is knowing what it costs. The cost of implementing a satellite system is relatively easy to estimate because there are only a few items to consider, each of which is quite expensive. From the investment, it is also relatively straightforward to estimate the annual financing and operating expenses. In some satellite systems, some revenue comes in lump sums from transponder purchases or prepaid leases, while other revenue comes over time from service charges and lease payments. The profit projection of the system is obtained by taking the difference between the revenue and the cost at each point in time. These can be converted back into a present value of the profit by discounting the profit with a cost of money factor (typically in the range of 15% to 25% when considering U.S. dollars).

The following are some typical examples of each type of cost.

Sum of capital assets is the sunk cost of the system, determined by adding the purchase price of equipment and facilities along with the cost of initiating the business. They include:

- First cost of system hardware;
- Cost of facility construction or improvement;
- Management and other expenses, tied to initiation of the service or business.

Annual cost factors are costs that will be expended each year in the running of the business. They include:

- Recovery of investment—that is, converting the capital cost into an annual cost, using depreciation factors, which provides for the recovery of the original investment to the sources of capital;
- Return on investment—that is, the cost of money and profit on the operation;
- Internal operating costs—which are routine and nonroutine expenses incurred each year in the running of the service business;
- External operating costs—in other words, money paid to outside entities, including other service providers for PSTN connectivity, maintenance of hardware and software, rent and utilities;
- Marketing expense—that is, the money expended to acquire customers (discussed in previous paragraphs);
- Overhead, general, and administrative expense—meaning the sum total of indirectly related expenses for running the overall business, which includes office rentals, secretarial support, human resources, procurement support, and corporate offices.

This background can be made clearer through an example. Say that we are considering investing in a Ku-band satellite to serve cable TV systems in a developing market. The cost of the satellite and launch are assumed to be

$150 million, and the premium to insure it until on station, at 20%, is $30 million. The investment in TT&C ground equipment might be another $10 million. For the sake of simplicity, we assume that this total investment of $190 million is all spent in the year prior to launch. Say that we can operate the satellite for one year at a cost of $5 million (this could either be for our own staff or under an outsourcing agreement with another operator). The rest of our organization that gets involved with marketing the services and running the administrative part of the company might cost another $3 million per year. To make these annual expenses more realistic, we can assume that there is inflation of 5% per year.

On the revenue side of this example, let us assume that our satellite has 32 transponders and that we have been able to sell six of them for a total of $50 million at the time of launch. From that point, our annual revenue from transponder leasing and other services (occasional and SCPC) add up to $15 million in the first year. Our owners have informed us that we must give them a return of 15% before tax on their initial investment (assuming that we have no debt, which is a big assumption).

Putting all of this together, we obtain the spreadsheet in Figure 11.3. To satisfy the 15% return requirement, the revenue growth rate from year to year must be at least 25% for there to be a positive *net present value* (NPV). This means that by the tenth year of operation, the annual revenue has grown from $15 million (not counting the initial transponder sales) to $112 million, increasing by a factor of 7.5. Such a *ramp-up* of revenue is possible because the operator should be able develop a market position that can be exploited more and more over time. On the other hand, if this cannot be done, then obvious negative results will ensue.

This simple example is not far from the truth. The cost of ownership and operation of a satellite is easy to determine. The harder part is finding a satisfactory market and assuring adequate growth of revenues. It would be nice if our little satellite company could sell out the transponders in the first year, in which case we could keep our management expenses to a minimum. A real company might incur greater costs to find customers and to attract them over time. This is highly dependent on the degree of competition and the obstacles that customers face in using our services.

11.5.2 Sources of Capital for New Satellite Systems

The money for financing a satellite communications business can come from a wide variety of sources, both internal and external. Internal sources are within the corporate or governmental structure of the overall enterprise. If funding from internal sources is adequate, then there is no real need to approach capital markets. Obtaining these types of funds requires an understanding of the inter-

	Year:	1	2	3	4	5	6	7	8	9	10
Investment											
Satellite	80										
Launch	70										
Insurance	30										
TT&C ground	10										
TOTAL INVESTMENT	190										
Annual expenses											
Inflation rate	5%										
Satellite operation		5	5	6	6	6	6	7	7	7	8
Marketing		1	1	1	1	1	1	1	1	1	2
Administrative		2	2	2	2	2	3	3	3	3	3
TOTAL ANNUAL COST		8	8	9	9	10	10	11	11	12	12
Revenue											
Growth rate	25%										
Transponder sales		50									
Leases and services		15	19	23	29	37	46	57	72	89	112
TOTAL REVENUE		65	19	23	29	37	46	57	72	89	112
NET CASH		57	10	15	20	27	36	46	60	78	99
Discount rate	15%										
Discounted cash flow		50	8	10	11	13	15	17	20	22	25
Present value	191										
Net present value	1										

Figure 11.3 Example of a financial spreadsheet analysis for a hypothetical space segment consisting of one operating satellite.

nal capital budgeting process and little else. Therefore, this section concentrates on external sources.

Considering now the two possible sources, debt and equity, we first define the terms. Debt is nothing more than a direct loan from a bank or other institution that must be repaid with interest. Typical forms of debt include direct loans and bonds. The holder of the debt is usually not an active participant in the business. Equity, on the other hand, represents a purchase of some percentage of ownership in the business, usually without a payment guarantee like the interest that applies in the case of debt. Equity can be defined by privately held shares of stock in the corporation (or simply a percentage of the company's ownership) or publicly traded shares that can be resold on a public stock exchange somewhere in the world. Public shares can be bought and sold more or less freely, while private shares usually cannot.

Between the basic categories of debt and equity, there is a wide variety of external sources. Listed below are some of the more popular ones that have been used or attempted in the satellite communication industry from time to time. For any particular service business, the sources will be used or combined in innovative ways to optimize the cost of funds. Another important factor is the desire of the initial owners of the business to maintain adequate control, which is an issue with both categories of funds. With equity, there is a direct loss of control (called dilution) as new investors are granted their respective share of control in exchange for capital. The control that remains for the founders could be driven to zero. In the case of debt, the lender will almost always demand "covenants" in the debt instrument, the purpose of which is to preclude the borrower from weakening the financial position of the lender. These covenants would reduce the flexibility of future sales of debt or equity and could actually force the business to be run in a certain manner.

Forms of debt include:

- Bank loans;
- Government loans or loan support;
- Commercial bonds, low-yield, high-quality;
- Commercial bonds, high-yield, "junk" bonds;
- Convertible bonds and warrants;
- Project finance;
- Leases, sale/leaseback;
- Vendor financing, debt.

Forms of equity include:

- Stock, private placement;
- Stock, initial public offering;
- Limited partnership;

- Joint ventures, "strategic" partnership;
- Vendor financing, equity.

The most popular sources of funding for satellite ventures at the time of this writing are: (1) bank loans with government guarantees; (2) initial public offerings; (3) vendor financing, both debt and equity based; (4) junk bonds; and (5) private placements from "strategic" partners. The three largest private financings of the 1990s include AMSC, a publicly traded company; Globalstar, both a limited partnership and publicly traded company; and Orion and APT Satellite, two privately held satellite operators. Each arrived at their respective positions through a multistep process of using more than one of the available financing means. In some cases, it was intentional, and in others, it was a case of what to do for the next round.

One way to approach the capital markets is to locate an experienced financial advisor. Many of the firms in this field are subsidiaries of major international banks, like Bank of America, Chase Manhattan, JP Morgan, and ING Bank. Others are smaller companies without banking interests. Examples include Argent Group, Space Machine Advisors, and Good Target. These companies may carry out a portion of the financing itself in return for a percentage fee, or they may charge directly for their services. It is a good idea to investigate several of these firms before making a selection.

11.5.3 Evaluating Venture Viability

A satellite communication venture must stand up to the same scrutiny applied to other business opportunities. Business people like to focus on the bottom line (the profit). However, the more critical part is the top line (the revenue). This can be very difficult to estimate, particularly for a new venture such as DTH or MSS. To help focus on the top line, we offer the following basic questions to be asked whenever considering the viability of a new service venture.

- Where will the revenue come from (who are the customers)?
- How will *they* get *their* money?
- What services are you offering that they will find attractive?
- What are the blockages that will prevent your customers from using your service?
- Who will provide any parts of the service that you cannot?
- How will you sell your service?
- If your market does not support the system financially, what is (are) your backup marketing plan(s)?

The technical and operational side of the system must also be investigated before giving approval. The following aspects should be well thought out ahead of time:

- A solid technical concept for the network with contained risk;
- Presale commitments for transponders or a credible marketing concept;
- Orbital positions or coordination of them;
- Startup money to carry through the first year;
- A contract or commitment from a spacecraft manufacturer or other integrated supplier (for a turn-key delivery);
- A launch commitment or reservation;
- An experienced organization to manage the construction and operation of the system.

Our last bullet point can be the most critical success factor, particularly in a competitive market (most satellite markets are moving in this direction). Earlier in this chapter, we discussed the issues involved in the operation and maintenance of a network or system. These apply to the satellite operation itself. Fortunately, the quantity of personnel involved in satellite operation is relatively modest. It is more important that they have adequate backgrounds for the tasks at hand. In this business, you cannot beat the experience.

In the 1970s, entering satellite communications meant buying an entire satellite system, including spacecraft, TT&C, and all of the Earth stations. Over the years, smaller network opportunities opened up through cable TV programming, video networks, syndication, data broadcasting, radio networks and music distribution, teleports, and private networks (business video and VSATs). These networks each require space segment capacity to function.

Satellite system operation is not typically a start-up business for entrepreneurs (although the late Rene Anselmo, founder of PanAmSat, was a noted exception). You need to have a solid business strategy in order to survive on a long-term basis. In addition, there must be an adequate source of funds over this term. The technology resources must be provided and maintained as well over the years and will continue to do so. Gaining access has become less of a challenge, and new entrants appear from literally any quarter of the field. The future portends exciting times with new applications and new businesses. Perhaps you, the reader, will start one.

Reference

[1] Norris, M., *Survival in the Software Jungle*, Norwood, MA: Artech House, Inc., 1995.

About the Author

Bruce Elbert is a leading authority on satellites and telecommunications, having thirty years of working experience at Hughes, COMSAT, and the U.S. Army Signal Corps. In addition to an international industrial background, he is a prolific writer of books and articles and has been featured as a speaker at numerous conferences and seminars. He is also the editor of Artech's Technology Management and Professional Development Series. At UCLA, Mr. Elbert teaches satellite communications, information networking, and high-tech marketing to practicing engineers. He holds BEE and MSEE degrees from City College of New York and the University of Maryland, respectively, and an MBA degree from Pepperdine University.

Index

The Artech House Telecommunications Library

Vinton G. Cerf, Series Editor

UNIX Internetworking, Second edition, Uday O. Pabrai

Videoconferencing and Videotelephony: Technology and Standards, Richard Schaphorst

Wireless Access and the Local Telephone Network, George Calhoun

Wireless Communications in Developing Countries: Cellular and Satellite Systems, Rachael E. Schwartz

Wireless Communications for Intelligent Transportation Systems, Scott D. Elliot and Daniel J. Dailey

Wireless Data Networking, Nathan J. Muller

Wireless LAN Systems, A. Santamaría and F. J. López-Hernández

Wireless: The Revolution in Personal Telecommunications, Ira Brodsky

Writing Disaster Recovery Plans for Telecommunications Networks and LANs, Leo A. Wrobel

X Window System User's Guide, Uday O. Pabrai

For further information on these and other Artech House titles, contact:

Artech House
685 Canton Street
Norwood, MA 02062
617-769-9750
Fax: 617-769-6334
Telex: 951-659
email: artech@artech-house.com

Artech House
Portland House, Stag Place
London SW1E 5XA England
+44 (0) 171-973-8077
Fax: +44 (0) 171-630-0166
Telex: 951-659
email: artech-uk@artech-house.com

WWW: http://www.artech-house.com